Electrodeposited Nanowires and Their Applications

Edited by

Nicoleta Lupu

Electrodeposited Nanowires and Their Applications
Edited by Nicoleta Lupu

Republished by InTech
Janeza Trdine 9, 51000 Rijeka, Croatia

Edition 2014

Technical Editor Teodora Smiljanic

Cover Designer Dino Smrekar

Additional hard copies can be obtained from orders@intechopen.com

Electrodeposited Nanowires and Their Applications
Edited by Nicoleta Lupu
p. cm.
ISBN 978-953-7619-89-3

Preface

Within the last years, the nanomaterials science and technology represented one of the most attractive subjects for physicists, chemists, biologists, medical doctors and engineers. These materials present a special interest from the point of view of basic scientific understanding, but also their potential applications are very attractive. Additionally, nanomaterials and nanostructures represent the basis for the development of new technologies, systems and equipments. The continuous development of miniaturized devices for different applications is demanding novel multifunctional materials which can perform different functions simultaneously.

Natural connections between physics, chemistry and life sciences are becoming much closer by means of nanotechnology, leading to complex and very useful applications. Biomedical sciences, which by their special social status are occupying the first place in the hierarchy of the priorities, benefit in unitary and almost completely way by the other sciences support, especially physics and experimental chemistry. The detection and quantification of the biological and chemical species represents the main objective of medicine, starting with certain diseases diagnosis and ending with the discovery of new drugs. Nanostructures, such as nanowires, carbon nanotubes or nanoparticles, exhibit new and sometimes unique opportunities for medical purposes. Some metallic nanowires are presenting electrical, optic and magnetic unique properties that can be exploited for (bio)detection and imagistics.

Nanowires are among the nanomaterials playing an important role in nanoscience and nanotechnologies, due to their specific behaviour as well as the number of their potential applications: optical and electronic components, connectors in electronics, high density magnetic recording media, biotechnology, etc.

Nowadays, nanowires are of interest for both fundamental research and potential applications. Nanowires arrays can be produced using different templates and methods, one of them being the electrochemical deposition into nanoporous templates. The fabrication of such complex simple and multilayered nanowires arrays (specific stacks of metallic nanowires, both crystalline and amorphous, as well as wafer structures of nanowire arrays having different compositions and physical properties) by electrodeposition, mainly in polycarbonate and anodized aluminium templates, opened up new directions in what concerns the applications of such complex nanostructures in spintronics, engineering and bioengineering.

This book describes some nanowires fabrication by electrodeposition technique and their potential applications. Understanding the design, size dependent properties and working principles of electrodeposited nanowires described here, requires a multidisciplinary background of physics, chemistry, materials science, electrical engineering, or bioengineering.

The book is organized in nine chapters. In the first two chapters, some theoretical considerations concerning the computational tools to study and predict the stability, as well as to model the properties and potential applications of nanowires are presented. Then, the fabrication of functional metallic nanowires using electrodeposition technique is described in detail pointing out the most important factors to be considered during the preparation process. The annealing effect on the metal and semiconductor electrodeposited nanowires in alumina template as well as the influence of the template on the electrodeposition process and the macroscopic behaviour of the nanowires are presented in the next chapter. After that, the electrochemical alumina template synthesis and the fabrication of both single-component and multilayered/superlattice metallic nanowires and nanotubes and their major applications are described. The importance of understanding both the individual and global properties of magnetic nanowire arrays, in order to implement them adequately in devices, is described in detail in the next chapter by means of the very promising first-order reversal curve (FORC) method. Finally, some applications of electrodeposited nanowires in magnetorheological fluids, elastomer composites and biomedical applications are described.

The book offers a new and complex perspective on the fabrication and use of electrodeposited nanowires for the design of efficient and competitive applications. While not pretending to be comprehensive, the book is addressing not only to researchers specialized in this field, but also to Ph.D. students, postdocs and experienced technical professionals.

Editor

Nicoleta LUPU

National Institute of Research and Development for Technical Physics, Iaşi
Romania

Contents

1

Computational Tools to Study and Predict the Long-Term Stability of Nanowires.

Martin E. Zoloff Michoff[1], Patricio Vélez[1],
Sergio A. Dassie[2] and Ezequiel P. M. Leiva[1]
INFIQC,
[1]Departamento de Matemática y Física
[2]Departamento de Fisicoquímica,
Facultad de Ciencias Químicas, Universidad Nacional de Córdoba
Argentina

1. Introduction

"The era in which the number of transistors on a computer chip doubles at a constant rate is drawing to a close". This is not the prophecy of an obscure mind, but is more or less the conclusion drawn by none other than the man who coined Moores´ law[1]. In an interview held in 2007, Gordon Moore recognized that by about 2020, his law would come up against the laws of physics. Furthermore, he recognized a change in a paradigm: the replacement of the top-down approach currently used for building circuits by a bottom-up procedure, where chips would be assembled using individual atoms or molecules. This is nothing but the realm of nanotechnology, while there is some consensus that the elementary switches of these circuits should be molecules with some feature allowing for the on/off status required for the components of logical devices, many questions remain concerning their stability. In the case of micrometric circuit components temperature may be an issue, but in the case of single molecules thermal effects may be overwhelming, since current flow occurs across a single bond. The lifetime of this bond, will determine the lifetime of the circuit component. Under these conditions, circuit engineering will be coming unexpectedly close to chemical kinetics.

It still is far from clear which will be the technological procedure for the massive production of these molecular circuits. However, there are a number of experimental techniques for the study of their properties that are well established. These are shown schematically in Fig. 1. Fig. 1d shows a method devised to study the structure of monatomic nanowires (NWs). It has been developed by Kondo and Takayanagi (Kondo & Takayanagi, 1997) using *High Resolution Transmission Electronic Microscopy* (HRTEM) and allows the generation of suspended NWs. In this approach nanowires are generated *in situ* by focusing an electron beam on adjacent sites of a self-supported metal thin film (*ca.* 3 nm), making holes and allowing them to grow until a nanometric bridge is formed inside or between grains. The

---

[1] Times Online September 19, 2007,
http://technology.timesonline.co.uk/tol/news/tech_and_web/article2489053.ece

relatively long lifetimes of these metallic nanowires (of the order of seconds and even minutes) allows for a detailed characterization of their geometries. This method has even been used to study the structure of alloyed metal monatomic nanowires (Bettini et al., 2006). However, because of the preparation procedure, these nanowires are susceptible to contamination by light impurity atoms (Galvão et al., 2004; Legoas et al., 2002; Legoas et al., 2004), not observed directly in the HRTEM, that affect their structure upon insertion in the monatomic chain.

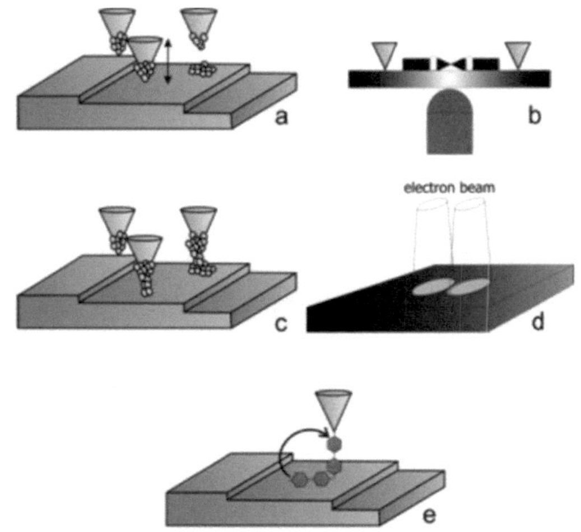

Fig. 1. Different methods employed to generate suspended nanowires: a) metallic jump-to-contact, b) mechanically controllable break-junction, c) electrochemical growth, d) electron beam punching, and e) molecular jump-to-contact.

Another procedure that has been used to study the properties of monatomic metal contacts is the so-called *Mechanically Controllable Break-Junctions* method (MCBJ, Fig. 1 b). In this technique, a metal wire is mechanically broken by mounting its two ends on a bendable substrate. In this way, the length of the contact can be adjusted by controlling the bending of the substrate. Lithographic designed MCBJ provides highly stable electrodes, with standard deviations of less than 1 pm over 24 h at low temperatures. A detailed discussion on this technique has been given by van Ruitenbeek et al. (van Ruitenbeek et al., 2005). The metal neck at the MCBJ can be also tuned by electrochemical methods as long as the metal involved can be deposited electrochemically. The main drawbacks of this technique are the lack of information on the atomic configuration, and the fact that both electrodes must be made of the same metal. The former situation was challenged recently by Yoshida et al. (Yoshida et al., 2007), who developed MCBJ inside a HRTEM chamber that allows for simultaneous performance of conductance measurements and electronic imaging of the atomic chain.

In the case of metallic *Scanning Tunnelling Microscopy* (STM) and conducting *Atomic Force Microscopy* (AFM) break junctions (Ohnishi et al., 1998; Rubio-Bollinger et al., 2001; Xu et al., 2003a; Xu & Tao, 2003), the sharp tip of the scanning probe microscope is approached to the surface and brought into contact with it (Fig. 1a). The surface may be made either of the same material as the tip or of a different one containing islands of the tip material. Upon retraction, an atomic chain of atoms occurs, whose properties are analyzed. When employed in an *Ultra High Vacuum* (UHV). The main drawbacks of this technique are the need for an

intensive cleaning procedure and the implementation of thermal and mechanical stability conditions, as it is the case of STM. On the other hand, this methodology can be used *in situ* for electrochemical experiments. In fact, it has been employed by Tao and co-workers to perform conductance measurements with Au nanowires under a variety of experimental electrochemical conditions (He et al., 2002; Xu et al., 2003b), including the properties of the nanojunction when molecules are inserted into it (Xu et al., 2003a; Xu & Tao, 2003). Efficient electrochemical methods for the generation of nanowires have also been developed. Li and Tao (Li & Tao, 1998) managed to bridge the gap between an STM tip and a substrate by a suitable potential control of both the tip and the surface(Fig. 1c). In a similar procedure, the gap between two supported metal pieces was filled by a nanometric wire using an adequate feedback electronic setup (Li et al, 1999). The latter procedure has the advantage of removing thermal drift problems.
Finally, the method developed Haiss et al. moves more into the spirit of bottom up nanostructuring (Fig. 1e). In this procedure, a molecule bridges spontaneously the gap between an STM tip and a surface (Haiss et al., 2004; Haiss et al., 2006). The substrate-tip connection is verified by jumps in the tunnelling current measured.

2. Structure and stability of pure and contaminated metallic monatomic nanowires

2.1 Experimental measurements

The conductance measured through metallic monatomic nanowires is quantized in units of $G_0 = 2e^2/h$ (where e is the charge of an electron and h stands for Planck's constant). The force during the fabrication and breaking of a gold nanowire was measured using a specific STM supplemented with a force sensor at room temperature (Rubio-Bollinger et al., 2001). Force and conductace curves were thus obtained simultaneously. The later displayed a steplike behavior down to a value close to one conductance quantum (G_0), which corresponds to a one-atom contact, while the force curve showed a sawtooth like signal decreasing in amplitude in a sequence of elastic stages separated by sudden force relaxations. In this experiments the one-atom contact of gold was further stretched a distance of about 1 nm while the conductance remained close to G_0, which signals the formation of a chain of about four atoms long that finally breaks. This corresponds to a monatomic neck of ~4 gold atoms. Similar observations were made in STM experiments at 4 K (Yanson et al., 1998), and from direct observations by means of HRTEM measurements (Ohnishi et al., 1998; Rodrigues & Ugarte, 2001a; Rodrigues & Ugarte, 2002; Rodrigues et al., 2000).
Due to the inherent irreproducibility of the contacts formed, and therefore of the measured conductance curves, it is useful to construct histograms with a few hundred measured curves. A force histogram for the Au nanowire showed a narrow distribution centered at 1.5 ± 0.3 nN for the force needed to break one single bond in the chain (Rubio-Bollinger et al., 2001).
Although gold has been the most prominent metal studied so far, mechanical elongation has shown to be useful to fabricate nanowires of Ag (Rodrigues et al., 2002), Co, Pt (Rodrigues et al., 2003), Pd (Kiguchi & Murakoshi, 2006; Rodrigues et al., 2003) and Cu (Bettini et al., 2006), as well as for Au and Ag alloys of varying compositions (Sato et al., 2006b).
Two striking experimental features have drawn the attention of the researches in the last few years: first, the existence of unusually large Au – Au distances in the order of 3.6 – 4.0 Å (Kizuka, 2008; Koizumi et al., 2001; Ohnishi et al., 1998; Sato et al., 2006b), longer than the bulk distance of 2.88 Å. These experimental observations along with the simulation of

HRTEM measurements suggest that the atomic-sized wires are complexed with light elements (such as H, C, S for example) (Kizuka, 2008; Koizumi et al., 2001).
Second, the long-term stability of gold nanowires at room temperature, in many cases in the order of seconds (Legoas et al., 2002; Ohnishi et al., 1998; Rodrigues & Ugarte, 2001a; Rubio-Bollinger et al., 2001; Takai et al., 2001; Yanson et al., 1998), which is an extremely long time in relation to those characteristic of molecular motion. For instance, Fig. 2 shows the time evolution of gold chain under elongation, which is in the order of ~1 second.

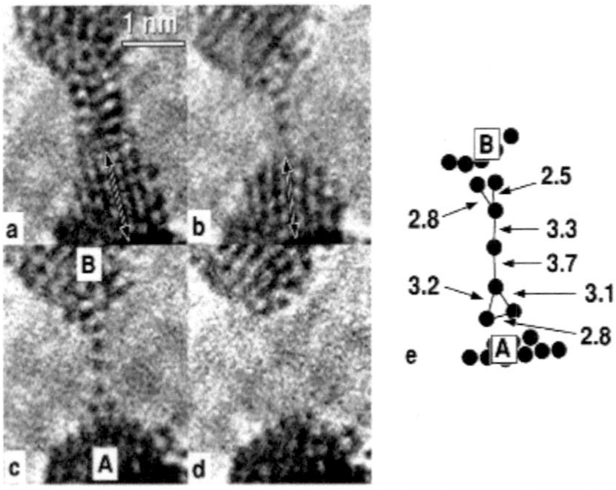

Fig. 2. Time sequence of atomic resolution images of the formation, elongation and fracture of a suspended chain of gold atoms: (a) 0 s; (b) 0.64 s; (c) 1.12 s; and (d) 3.72 s. Atomic positions appear as dark lines or dots. A schematic representation of the chain structure is shown in (e) (distances are marked in Å and the error bar is 0.1 Å); the letters A and B indicate the apex position in (c). Note that the chain is attached to the tips through a two atom structure. The double arrows in (a) and (b) have been drawn to indicate the movement (rotation) of the lower apex. Reprinted figure with permission from Rodrigues, V.; Ugarte, D. *Phys. Rev. B*, Vol. 63, No.7, 073405 (pp 1-4), 2001. Copyright 2001 by the American Physical Society.

The stability and breakdown of gold nanojunctions at different stretching rates has recently been measured by Huang et al. (Huang et al., 2007a). Information about the lifetime of the Au - Au bond is extracted from the length of the last plateau in the conductance traces of Au - Au point contacts. The most probable stretching distance for this last plateau (corresponding to an atom-sized contact between the electrodes as verified by its conductance at a G_0 value) maintains a constant value of ~0.1 nm at low stretching rates (0.8 - 8.3 nm/s). At high stretching rates (45.9 - 344 nm/s) a maximum plateau of ~0.17 nm is reached. Between these two plateaus, a linearly proportional regime is observed between the most probable distance and the logarithm of the stretching rate. This three-phase regime (illustrated in Fig. 3) has also been observed in the breakdown of biological molecules measured by AFM (Auletta et al., 2004; Merkel et al., 1999; Schönherr et al., 2000; Zou et al., 2005). The linear increase in stretching distance is related to the stretching rate according to Eq. 1 (Evans & Ritchie, 1997; Evans, 1999; Evans, 2001).

$$L^* = \frac{k_B T}{k_s x_\beta} \ln\left(\frac{t_{off} k_s x_\beta}{k_B T}\right) + \frac{k_B T}{k_s x_\beta} \ln \upsilon \quad (1)$$

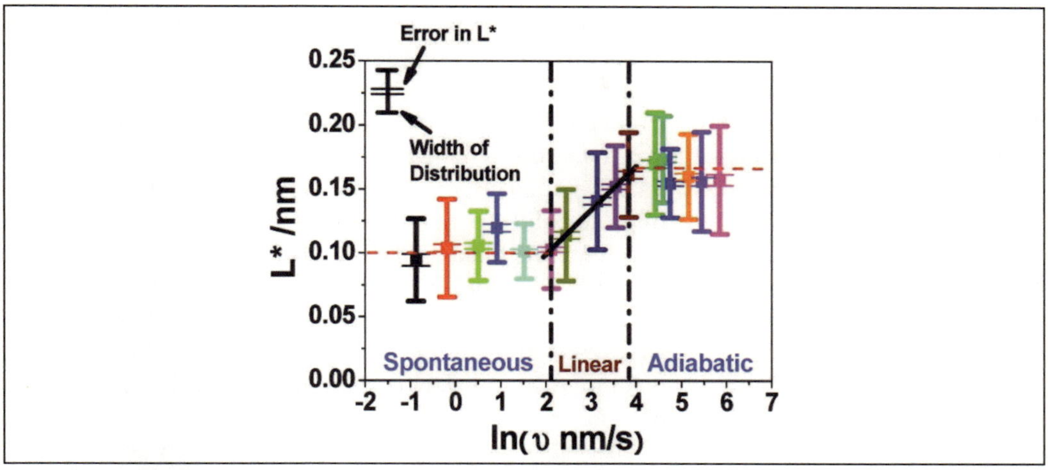

Fig. 3. Most probable stretching distance, L^*, vs logarithm of stretching rate for a Au - Au point contact. The black line is the linear fit of L^* with the logarithm of the stretching rate, according to Eq 1. Reprinted with permission from Huang, Z. F.; Chen, F.; Bennett, P. A.; Tao, N. J. *J. Am. Chem. Soc.*, Vol. 129, No.43, 13225-13231, 2007. Copyright 2007 American Chemical Society.

In Eq. 1 L^* is the most probable stretching distance, k_s is the effective spring constant of the bond, x_β is the average thermal bond length along the pulling direction until dissociation, t_{off} is the natural lifetime, and ν is the stretching rate. By fitting the linear regime observed for the Au - Au point contacts to the above equation, a value of t_{off} = 81 ms was obtained.

2.2 Computational simulations

Experimental observations such as those described above, have triggered numerous theoretical approaches to gain further insight into the structure and transmission properties of these metallic nanowires. The mechanical structure and evolution of a tip-surface contact has been modelled by means of different computational techniques. In early works, *Molecular Dynamics* (MD) and effective medium theory potentials have been used to simulate the mechanical deformation of atomic-sized metallic contacts under tensile strain for Au, Ag, Pt, Ni (Bahn & Jacobsen, 2001; Dreher et al., 2005; Rubio-Bollinger et al., 2001; Sørensen et al., 1998). The relative utility of different semiempirical potentials for MD simulations of stretched gold nanowires has been recently reviewed (Pu et al., 2007a). In this work the authors find that the second-moment approximation of the tight-binding potential reproduces well the energetics of finite Au clusters. Also, the calculated tensile force just before the nanowire breaks is around 1.5 nN, consistent with the experimental result The stretching of gold nanowires has also been simulated in the presence of solvent (Pu et al., 2007b).

Tight binding MD simulations using the Naval Research Laboratory potentials, along with *ab-initio* quantum calculations within the *Density Functional Theory* (DFT) framework, provided evidence that a one-atom thick 5-atom long necklace is formed for gold nanowires under a stretching force (da Silva et al., 2001; da Silva et al., 2004). Before breaking, relatively long Au - Au distances, of the order of 3.0 - 3.1 Å are obtained, in agreement with experiments.

The main drawback of the MD simulations mentioned above is that the stretching rate used (typically in the order of 1 -2 m/s) is around 9 - 10 orders of magnitude larger than the experimental values (in the order of a tenths to a few nm/s).

The use of *ab-initio* quantum methods to study the stability of monatomic metallic nanowires is usually performed using two different methodologies. One of these procedures involves stretching the nanowire in steps of a certain elongation, minimizing the energy of the system at each step, until the breakage of the nanowire is achieved (Bahn et al., 2002; da Silva et al., 2004; De Maria & Springborg, 2000; Häkkinen et al., 2000; Nakamura et al., 1999; Nakamura et al., 2001; Novaes et al., 2003; Okamoto & Takayanagi, 1999; Rubio-Bollinger et al., 2001; Sánchez-Portal et al., 1999; Skorodumova & Simak, 2003). This method has been successfully used to, for example, predict the pulling force necessary for breaking the nanowire (da Silva et al., 2004; Nakamura et al., 1999; Novaes et al., 2003; Rubio-Bollinger et al., 2001). An alternative procedure, and more costly from a computational point of view, consists in performing *Ab-initio Molecular Dynamics* (AIMD), which has been employed to obtain a detailed description of the elongation process of pure gold nanowires (Hobi et al., 2008; Torres et al., 1999), and gold NWs in the presence of organic molecules (Krüger et al., 2002), or light weight elements as contaminants (Hobi et al., 2008; Legoas et al., 2004). AIMD takes into account the thermal motion of the system, for which in a first approach seems a more appropriate approximation to obtain reliable information about the mechanical proprieties and the stability of metallic atom-sized wires. Nonetheless, the experimental elongation times are far larger than those accessible by an AIMD simulation, typically in the order of nanoseconds.

As it was mentioned above, one of the possible explanations for the rather long Au - Au distances observed in a monatomic neck is the presence of light weight elements, such as C, H, O or S, intercalated between gold atoms. This possibility has only been investigated so far by the means of computational simulations, since these light weight elements would have a low contrast against the much heavier Au atoms, and thus rendering their direct visualization by today's electronic microscopes very difficult. Carbon is a frequent contaminant in bulk gold (Legoas et al., 2002), while hydrogen (O'Hanlon, 2001) and oxygen (Bahn et al., 2002) are impurities very difficult to extract even in the best UHV conditions. Although the possibility of hydrogen acting as a contaminant has been ruled out by an AIMD study (Legoas et al., 2004), this result has been subsequently challenged (Hobi et al., 2005).

Legoas et al. (Legoas et al., 2002) modelled monatomic gold chains contaminated with carbon by means of geometry optimization at DFT-LDA level. The authors used an isolated linear chain and no tension was exerted on the system. Their results showed that long Au - Au distances of around 4 - 4.5 Å could be explained by the presence of two consecutive carbon atoms (C_2) inserted into the gold chain. Whereas, another set of anomalously long bond in the order of 3 - 3.7 Å could be a consequence of a mixture of pure Au - Au bonds with contamination of such a bond by a single carbon atom.

Skorodumova and Simak (Skorodumova & Simak, 2003) showed, using DFT-GGA calculations, that the unusual structural stability of monatomic gold wires could be explained in terms of hydrogen contamination. Stretching the nanowire, the authors observed that the chain takes a linear structure with hydrogen atoms intercalated and a Au - Au distance of 8.8 Å before the nanowire breaks. The cohesive energy of the contaminated gold wire was found to be 2-fold higher than a pure chain of gold atoms. This last result was attributed to a partial charge transfer from gold to hydrogen. Subsequently, the influence of carbon was explored using the same computational methodology (Skorodumova et al., 2007; Skorodumova & Simak, 2004), also finding that carbon can enhance the stability of linear gold chains yielding large interatomic distances.

Novaes et al. (Novaes et al., 2003) studied through *ab- initio* calculations the effect of H, B, C, N, O and S impurities on a gold nanowire electronic and structural properties. The authors find that the most likely candidates to explain the distances in the range of 3.6 Å and 4.8 Å are H and S impurity atoms, respectively.

The main drawback of the procedures presented so far is that the presence of the impurity is simply assumed, and no description is obtained of how, when, or with what probability it migrates to the position it was assumed to have. To overcome this limitation, an AIMD study of the formation and growth of gold chains with a variety of impurities (H, C, O, S), without any assumption of their initial positions was performed (Anglada et al., 2007). One or two impurity atoms were introduced randomly in an amorphous column of 50 - 150 gold atoms. These amorphous solid columns were stretched during 4 - 18 ns until they broke. Hydrogen was always found to evaporate before formation of the monatomic chain took place. Carbon and oxygen were found in the final chains with low probability (~ 10 %), while sulphur was found participating in it with a high probability (~ 90 %). The mean distances between gold atoms bridged by C, O and S were 3.3, 3.4 and 5.0 Å, respectively, in good agreement with experiments.

Inasmuch as this last study provides a level of accuracy and reliability superior to those mentioned before, the stretching rate is still much higher than those typically used in experiments, and the simulations last only a few nanoseconds, while experiments take place in the order of 0.001 - 1 second.

The study of the effect of impurities on the structure and stability of gold nanowires is an ongoing investigation topic. Some of the most recent theoretical work can be found in (Jelínek et al., 2008; Novaes et al., 2006; Zhang et al., 2008). Although gold is by far the most prominent element of interest in the formation of nanowires, some recent computational studies have also involved copper nanowires (Amorim et al., 2007; Amorim et al., 2008; Sato et al., 2006a).

3. Single molecule nanowires

3.1 Experimental measurements

Building a device in which a single molecule bridges two metallic electrodes is of major interest, since one could easily tailor the nanojunction electronic properties by changing the molecule or even only a substituent in the molecule. This opens an enormous range of possibilities in the field of molecular electronics.

Most of the experimental (Cui et al., 2001; Haiss et al., 2004; Haiss et al., 2006; Huang et al., 2006; Huang et al., 2007a; Huang et al., 2007b; Li et al., 2007a; Li et al., 2006a; Xu et al., 2003a; Xu et al., 2005; Xu & Tao, 2003) and computational work (Batista et al., 2007; Hou et al., 2005; Hou et al., 2006; Kim et al., 2006b; Li et al., 2005; Li et al., 2006b; Li et al., 2006c; Paulsson et al., 2008; Perez-Jimenez, 2005; Stadler et al., 2005; Wu et al., 2005) performed with this systems deal with the measurement and/or theoretical determination of the conductance of the molecular nanojunction, particularly those in which the linker atom to the metallic electrodes is either S or N. For a recent review on some of these aspects readers can refer to (Vélez et al., 2007).

The chemical identity of the linker moiety plays a fundamental role in determining both the electrical and mechanical properties of the molecular nanojunction. Without discussion the most widely used anchoring group is thiol (Chen et al., 2006; Haiss et al., 2008; Haiss et al., 2009; Huang et al., 2006; Huang et al., 2007a; Huang et al., 2007b; Li et al., 2007b; Li et al., 2006a; Ulrich et al., 2006; Xu et al., 2003a; Xu et al., 2005; Xu & Tao, 2003), although pyridine (Xu et al., 2003a; Xu & Tao, 2003), isocyanide (Beebe et al., 2002; Kiguchi et al., 2006; Kiguchi et al., 2007; Kim et al., 2006a), selenium (Patrone et al., 2003a; Patrone et al., 2003b; Yasuda et al., 2006), amine (Chen et al., 2006; Hybertsen et al., 2008; Kamenetska et al., 2009; Kiguchi et al., 2008; Park et al., 2007; Park et al., 2009; Quek et al., 2007; Quek et al., 2009; Quinn et al., 2007; Venkataraman et al., 2006a; Venkataraman et al., 2007; Venkataraman et al., 2006b), phosphines (Kamenetska et al., 2009; Park et al., 2007) and carboxylate (Chen et al., 2006; Martín et al., 2008) have proved to provide enough binding strength to yield a stable contact. In most of these works, the metallic electrodes that the molecule bridges are made of gold, but some recent reports showed the utility of platinum electrodes for those purposes (Kiguchi et al., 2007; Kiguchi et al., 2008).

Xu et. al. reported the first electromechanical measurement of a molecular junction (Xu et al., 2003a). The authors determined simultaneously the conductance and the force under mechanical stretching for the octanodithiol (ODT) and 4,4′-bipyridine (BYP) nanojunctions. The quantum conductance for BYP resulted 40 times larger than that of ODT, while the force quantum was 0.8 ± 0.2 nN, considerably smaller than the 1.5 ± 0.2 nN determined for ODT. This last value is the same as that required to break a Au - Au bond (Rubio-Bollinger et al., 2001). Thus, the authors concluded that the breakdown of the ODT nanojunction involves a Au - Au bond rupture, whereas in the case of BYP, the lower breaking force would indicate that a Au - N bond is breaking. These results are in agreement with the notion of the Au- S bond being stronger than the Au-N bond (Stolberg et al., 1990). In a subsequent study, Huang et. al. found that the behavior of the ODT nanojunction as a function of the stretching rate is essentially identical to that of a pure gold point contacts (Huang et al., 2007a).

One useful experimental parameter to determine how strong is the molecule bonded to the metallic electrodes is the length that the junction formed by a single molecule can be stretched before it breaks. This allowed Kiguchi et. al. to establish the following order in binding energies for 1,4-disubstituted benzenes with Au and Pt electrodes: Au-NH_2 < Pt-NH_2 ~ Au-S < Au-isoCN < Pt-isoCN ~ Pt-S (Kiguchi et al., 2006; Kiguchi et al., 2007; Kiguchi et al., 2008).

The statistical analysis of the stretching length was also used to establish the following order in the sense of increasing binding strength: Au-COOH < Au-NH_2 < Au-S (Chen et al., 2006).

3.2 Computational simulations

As mentioned above, a large proportion of the theoretical work on molecular nanowires involves the calculation of the conductance. Only a few of these have addressed some aspects of the thermodynamic stability of such nanocontacts.

As it respects to the mechanical properties of a molecular nanojunction, one or the first *ab-initio* studies was performed by Krüger et. al. (Krüger et al., 2003). A thiomethyl radical bonded to a 5-atom planar cluster was used as a model for a typical Au - S contact. The junction was elongated, with geometry optimization at each step, obtaining an isomerization of the cluster into a linear chain which finally breaks at a Au - Au bond with a force of ~ 1.5 nN. In a Car-Parrinello molecular dynamics the same authors showed that when ethylthiol attached to a gold surface is pulled, this leads to the formation of a monoatomic gold nanowire, followed by breaking a Au - Au bond with a rupture force of about 1.2 nN (Krüger et al., 2002).

In a different study, the stretching and breaking behavior of a benzene dithiol molecule sandwiched between two Au(111) slabs was studied using DFT calculations (Lorenz et al., 2006). It was found that breakage occurs through a dissociation of one of the Au - S bonds with a maximum force of 1.25 nN in the case when the molecule is directly attached to the surface, and of 1.9 nN when an adatom is placed between the sulphur and the gold slab.

Similar studies were carried out for nanojunctions involving 4,4'-bipyridine (Stadler et al., 2005; Vélez et al., 2005), pyrazines (Vélez et al., 2005; Zoloff Michoff et al., 2009), amines (Hybertsen et al., 2008; Kamenetska et al., 2009), and alkylphosphines (Kamenetska et al., 2009). Binding energies and rupture forces are the parameters that can be obtained from these types of computational simulations that can be related to the mechanical stability of the molecular nanojunction. It should be noticed that temperature activated processes are not considered in these calculations.

4. Long term stability of metallic monatomic nanowires

As can be gathered from the summary of the computational work performed so far with the aim of obtaining valuable information about the stability of NWs, the main challenge remains to develop models that would allow to use this valuable computational information to extrapolate to the experimental time scale, and taking thermal motion into account. In the following sections a simple kinetic model based on the *Transition State Theory* (TST) is presented. The utility of this model along with DFT calculations and an exploration of the energy landscape with a suitable algorithm will be illustrated for pure and contaminated gold monatomic NWs, as well as for single molecule contacts.

4.1 The minimum energy path and the transition state theory

A common and important problem in theoretical chemistry and solid state physics is to identify the path with the lowest energy for the reorganization of a group of atoms from a stable configuration to another. This path is referred to as the *Minimum Energy Path* (MEP) and is often used to define a *Reaction Coordinate* (RC) for transitions of the type of chemical reactions, conformational changes in molecules or diffusion processes in solids. The maximum potential energy along the MEP is referred to as the *saddle point*, and provides the activation energy for the occurrence of the process. To calculate the transition rate constant is of central importance in the TST (Eyring, 1935; Vineyard, 1957), as will be discussed in section 4.2.

Different methods have been developed to find the reaction path and saddle points (Michael & Michael, 2007). We focus our attention on methods that make use of two boundary conditions: the initial and final configurations for the transition. These settings should normally correspond to two local minima in the multidimensional potential energy surface. These minima may be obtained from different energy minimization techniques such as the simulated annealing, conjugate gradient, etc.

These methods require only the calculation of first derivatives of the potential energy. They generate a chain of images or replicas of the system between the initial and final configurations. All the intermediate images are simultaneously optimized in some concerted way of the potential energy surface that should be as close as possible to guarantee the convergence to the MEP. The method called *Nudged Elastic Band* (NEB) (Henkelman et al.,

2000; Henkelman & Jonsson, 2000; Mills & Jónsson, 1994) works in the scheme of these methods and its implementation is particularly simple. The NEB method has been successfully applied to a variety of problems, such as studies of diffusion on metal surfaces (Villarba & Jónsson, 1994), the dissociative adsorption of molecules on a surface (Mills & Jónsson, 1994), and the formation of a contact between a STM tip and a surface (Sørensen et al., 1996).

4.2 Kinetic model

As it was shown by Krüger *et. al.* in the computational simulation studies described above (Krüger et al., 2002; Krüger et al., 2003), the creation of a Au nanowire takes place in a number of steps, involving an isomerization process. Our approach will only deal with the final stage, where the NW breaks but the model could be extended to a multi-step process. We will assume that, for a given elongation of the wire, it may exist either in a *broken* (*b*) or *unbroken* (*u*) state (see Fig. 4). The related system energies are denoted with E_b and E_u, respectively. When shifting from state *u* to *b*, the system will find an energy barrier (activation energy) that we will denote with $\Delta E^{\ddagger}$. We will neglect the reverse process in all the treatment we give below. In principle, wire reformation could be easily introduced in the model. However, we must take into account that in order to surmount the activation energies involved, the system must gain a considerable amount of energy, which will be released in the downhill stage after crossing through the maximum. This excess energy will rapidly take the systems to other more compact configurations of the final state. To be more illustrative, immediately after the rupture the wire (state *b*) will find itself in a situation where the Au atoms are in a very low coordination, so that they will stabilize by merging to

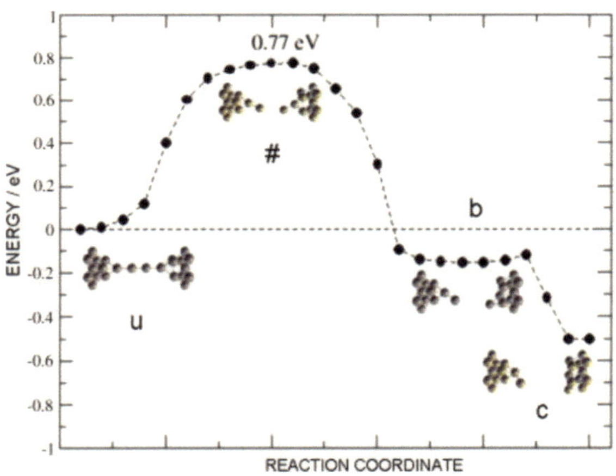

Fig. 4. Scheme of the different stages of the rupture of a Au NW. The states labelled as *u*, *b* and # correspond to the *unbroken*, *broken* and *activated* states, respectively. The state labelled as *c* indicates a situation where the broken wire has evolved towards a more compact state, in which the pieces of the broken NW are attached to the bulky pieces of the junction. Although the present figure is meant with illustrative purposes, the configurations and energies were obtained from calculations using the modified embedded atom method potentials (Baskes et al., 1989) and the nudged elastic method (Henkelman et al., 2000; Henkelman & Jonsson, 2000). Reprinted from Vélez, P.; Dassie, S. A.; Leiva, E. P. M. *Chem. Phys. Lett.*, Vol. 460, No.1-3, 261-265, 2008 with permission from Elsevier.

a bulky piece of metal of one of the tips making the junction (state *c*). These qualitative assessments are based on semiempirical calculations[2].

Following the TST, we have for the forward process (wire breaking) the frequency ν_f given by Eq. 2.

$$\nu_f = \nu_u \exp\left(-\frac{\Delta E^{\ddagger}}{k_B T}\right) \tag{2}$$

where k_B is Boltzmann's constant, T the absolute temperature, ν_u is a factor showing a weak dependence on temperature that represents an average oscillation frequency and $\Delta E^{\ddagger}$ is calculated as $E^{\#} - E_u$. Since $\Delta E^{\ddagger}$ and ν_u are in principle functions of the wire elongation, this will also be the case of ν_f. Typical values of ν_u are 3.5 – 7.0 x 10^{12} Hz (Bürki et al., 2005; Todorov et al., 2001). We have found that the assumption of constant (elongation independent) values in this range leads essentially to the same qualitative and quantitative predictions we show below. However, in order to get a parameter free description of the problem, it would be desirable to get not only $\Delta E^{\ddagger}$ but also ν_u from first-principles calculations. Sánchez-Portal et al. (Sánchez-Portal et al., 1999) have calculated the transversal and longitudinal phonon frequencies of a Au NW by means of the frozen phonon method at different wire lengths. In order to get an estimation of ν_u for our problem, we have parameterized the results of these authors for the transversal mode as a function of the wire elongation and introduced it in our equations for ν_u (Δz).

Let us now consider a differential elongation of the wire $d\Delta z$ performed in a period of time dt. The number of crossings from state u to state b over the barrier in dt will then be given by Eq. 3.

$$dn = \left[\nu_u \exp\left(-\frac{\Delta E^{\ddagger}}{k_B T}\right)\right] dt \tag{3}$$

dn being the average number of times that the system moves from the *unbroken* to the *broken* state in the time dt. If we want to calculate the number of possible crossings in a finite period of time we integrate between 0 and t, which is given Eq. 4.

$$\Delta n = \int_{t=0}^{t=\tau^*} \left[\nu_u \exp\left(-\frac{\Delta E^{\ddagger}}{k_B T}\right)\right] dt \tag{4}$$

We can estimate the lifetime of the wire τ^* by setting in $\Delta n = 1$ (Eq. 5).

$$\int_{t=0}^{t=\tau^*} \left[\nu_u \exp\left(-\frac{\Delta E^{\ddagger}}{k_B T}\right)\right] dt = 1 \tag{5}$$

Note that in the present formulation no assumption has been made on how the wire is elongated. We turn now to consider two different possibilities:

[2] P. Vélez, S. A. Dassie, E. P. M. Leiva unpublished results.

Static rupture of the nanowire: In this case, rupture of the NW is studied at a constant elongation Δz. Under these conditions, the argument of the integral in Eq. 5 becomes independent of time and the breaking time can be straightforwardly obtained by solving Eq. 6, given below.

$$\left[\nu_u \exp\left(-\frac{\Delta E^{\ddagger}}{k_B T} \right) \right] \tau^* = 1 \tag{6}$$

Rupture of the nanowire at a constant elongation rate: If we assume that the NW is stretched at a constant rate, $d\Delta z / dt = \nu_e$, then we have Eq. 7, where Δz_0 is the elongation at t=0, Δz^* is the breaking elongation for the NW and we have written the activation energy as $\Delta E^{\ddagger}(\Delta z)$ to emphasize the dependence of the quantity on the elongation. ν_u was also considered to be elongation dependent and calculated as pointed out above. However, the dependence of the predictions on this parameter is rather weak. Once the dependence of $\Delta E^{\ddagger}(\Delta z)$ on Δz is given, Eq. 7 can be solved numerically to get the dependence of Δz^* on the elongation rate ν_e.

$$\int_{\Delta z=\Delta z_0}^{\Delta z=\Delta z^*} \left[\nu_u(\Delta z) \exp\left(-\frac{\Delta E^{\ddagger}(\Delta z)}{k_B T} \right) \right] d\Delta z = \nu_e \tag{7}$$

4.2 Results for pure metallic nanowires

In order to illustrate the method, we consider a system consisting of a Au nanowire made of a supercell containing 4 atoms which are periodically repeated in space (Au_4 NW). Fig. 5 shows a scheme of the unit cell employed to simulate the Au_4 NW considered here. We used 4 atoms because the elongation distances at which a NW of this size breaks are in the range between 0.11 to 0.14 nm, which is the value that Huang et al. (Huang et al., 2007a) have found experimentally (see Fig. 9a below). Although this is a rather small system, the rupture of the wire has been found to be the displacement of one of atoms perpendicular to the wire axis (Ke et al., 2007), so that the interaction of the breaking bond with the rest of the system should be minimal. Atom 1 is fixed and the length of the supercell is stretched. For a given stretching of the NW, the energy of the system can be minimized with respect to all the atomic coordinates. Let us denote with E_{min} (Δz) the minimum energy for a given $\Delta z = L - L_0$, where L_0 is the chain length at equilibrium for each system and L is the total chain length; at some Δz , say Δz_{min}, E_{min} will present a minimum, that we denote with E_{min} (Δz). In the following discussion, we refer all the stretching lengths to Δz_{min} and all the energies to E_{min} (Δz).

While the unbroken state is clearly defined, some uncertainty remains concerning the broken one. With this purpose, *ab initio molecular dynamics* simulations were performed near the point where the NW breaks, finding structures which are very similar to that proposed below in the broken state of Fig. 7, which was obtained by minimizing the energy of the systems, as were the remaining configurations. It is interesting to point out that the broken configuration is the same as one of the most stable Au_4 clusters reported in (Bonacic-Koutecky et al., 2002).

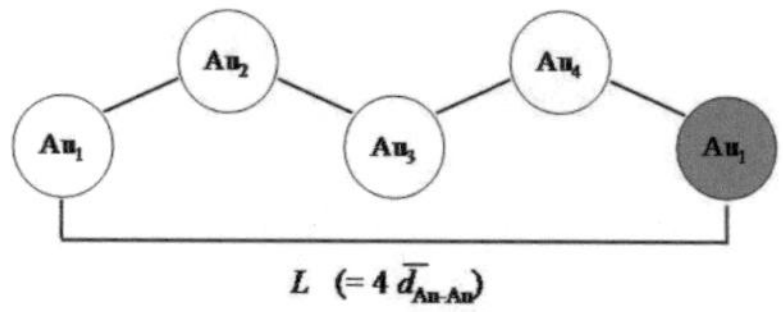

Fig. 5. Schematic representation of the unit cell employed in the present calculations. L is the total chain length and $4\overline{d}_{Au\text{-}Au}$ is the average Au-Au separation distance. Reprinted from Vélez, P.; Dassie, S. A.; Leiva, E. P. M. *Chem. Phys. Lett.*, Vol. 460, No.1-3, 261-265, 2008 with permission from Elsevier.

This configuration was then adopted to obtain the broken state for different elongations. This was achieved by compressing the system to a cell size corresponding to the desired length *L*, and then performing a conjugate gradient minimization to obtain the *broken* state for each system. The *minimum energy path* between the *unbroken* and *broken* states at each Δz, was then obtained by means of the NEB algorithm.

Curves for the energy of the *unbroken* state E_u, the *broken* state E_b, and the activated state $E^{\#}$ as a function of the wire elongation are shown in Fig. 6a. The energy curves of the *broken* and *unbroken* state meet for an elongation of $\Delta z_{u\text{-}b}$ = 0.1565 nm. At this point the activation energies for the backward and forward reactions are the same and they are equal to 0.33 eV.

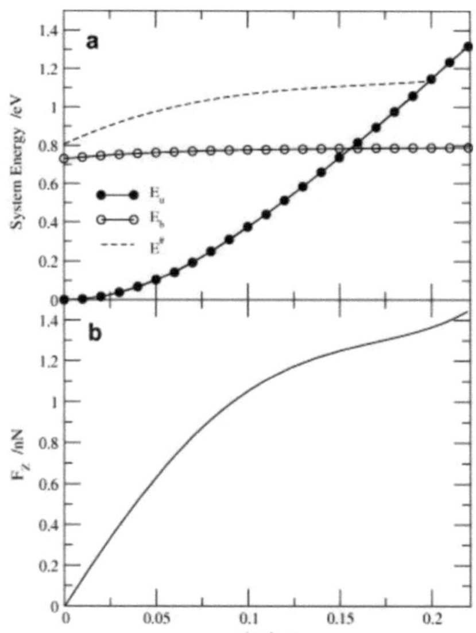

Fig. 6. System energy and force curves upon system elongation for the Au_4 nanowire. a) Energy of the *unbroken* state, E_u, energy of the *broken* state, E_b, and energy of the *activated* state, $E^{\#}$, as a function of the stretching of the NW, Δz. In all cases the energy of the system is referred to the minimum value that the system takes for the set of Δz considered. The activation energy for NW rupture is $\Delta E^{\ddagger} = E^{\#} - E_u$. b) Force calculated according to Eq. 8 as a function of Δz. Reprinted from Vélez, P.; Dassie, S. A.; Leiva, E. P. M. *Chem. Phys. Lett.*, Vol. 460, No.1-3, 261-265, 2008 with permission from Elsevier.

This would yield a switching frequency between the *broken* and *unbroken* state of the order of 10^6 Hz at room temperature, going down to the order of 1 Hz at 135 K. However, as we stated previously, after the bond is broken the system will immediately rearrange to other more stable configuration so that the forward reaction can be considered as irreversible under the usual stretching conditions. On the other hand, the energies of the *unbroken* and activated state meet at an elongation $\Delta z_{u\text{-}\#}$ = 0.1995 nm. This represents a non-activated rupture process for the NW, and thus the Au - Au average distance at this point (*ca.* 2.88 Å) represents an absolute limit for wire stability.

Fig. 6b shows the force of the system, calculated as the component of the force applied on the NW at each step from Eq. 8.

$$F_z = -\frac{\partial}{\partial \Delta z} E_u(\Delta z) \quad (8)$$

In Fig. 7 we can observe curves for the energies of the system between *broken* and *unbroken* states at some sample elongations. From these curves it can be appreciated that the transition state is shifted towards the *broken* state. The configurations of the system at the *broken*, *unbroken*, and activated state are also shown. It is remarkable that the structure of the *broken* state is similar to the Au fragment attached to an ethylthiolate molecule, as it was found in an AIMD simulation (Krüger et al., 2002).

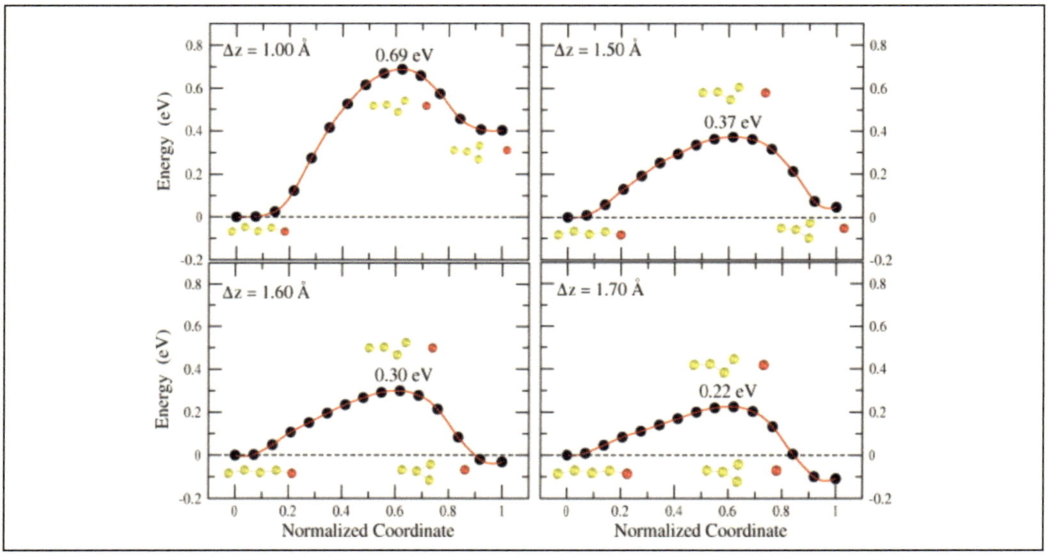

Fig. 7. Energy of the system at the reaction path between *broken* and *unbroken* states at some sample elongations. The energy is in all cases referred to the *unbroken* state. The configurations of the system at the *broken*, *unbroken*, and *activated* states are depicted within each frame. Reprinted from Vélez, P.; Dassie, S. A.; Leiva, E. P. M. *Chem. Phys. Lett.*, Vol. 460, No.1-3, 261-265, 2008 with permission from Elsevier.

Static rupture of the nanowire: A logarithmic plot for the static rupture of the NW calculated according to Eq. 6 is shown in Fig. 8. The lowest stretching rates employed in the experiments allow the rupture of NWs in times of the order of 0.1 s. Fig. 8 shows that in this order of times the wires should break at distances close to 0.105 nm. This is very close to the experimental value of 0.1 nm (Huang et al., 2007a) obtained at the lowest stretching rates. Thus, the present results also support the general idea that long Au - Au distances such as those found in the experiments of (Legoas et al., 2002; Ohnishi et al., 1998; Rodrigues & Ugarte, 2001a; Rodrigues & Ugarte, 2001b) at room temperature cannot occur for pure Au NWs. On the other hand, at 150 K and below, pure Au NWs can be considerably stretched beyond that point.

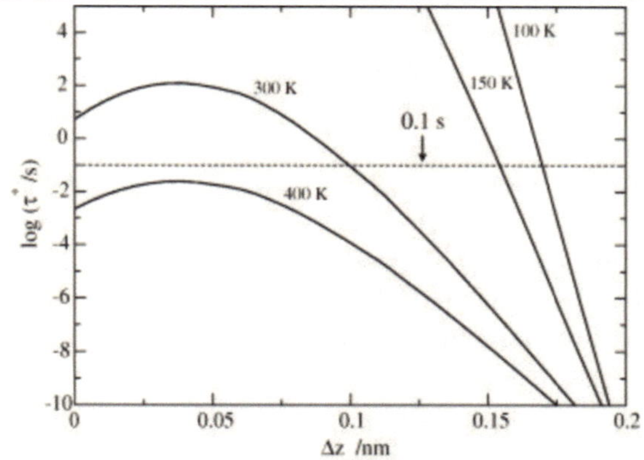

Fig. 8. Decimal logarithm of the lifetime τ^* of a Au_4 nanowire at different elongations calculated according to Eq. 6 for different temperatures. The dotted line denotes a constant lifetime of 0.1 s. Reprinted from Vélez, P.; Dassie, S. A.; Leiva, E. P. M. *Chem. Phys. Lett.*, Vol. 460, No.1-3, 261-265, 2008 with permission from Elsevier.

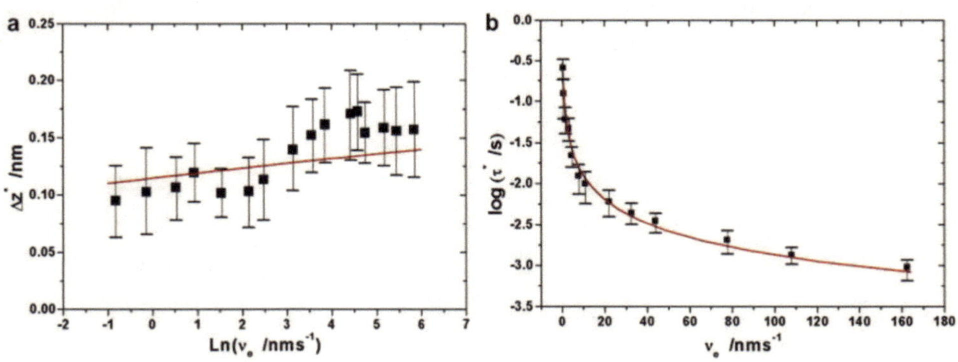

Fig. 9. Dynamic rupture of a Au_4 nanowire: a) Breaking distance as a function of the logarithm of the elongation rate v_e. The full line shows the calculations according to the present model and the squares are data for gold point contacts taken from (Huang et al., 2007a). The segments show the width of the distribution reported there. No fitting parameters were considered. b) Logarithm of the lifetime of the wire, τ^*, as a function of the decimal logarithm of the elongation rate, v_e. The full line shows the calculations according to the present model and the squares are experimental data for the rupture of the Au - ODT taken from (Huang et al., 2007a). The segments show the width of the distribution reported there. No fitting parameters were considered. Reprinted from Vélez, P.; Dassie, S. A.; Leiva, E. P. M. *Chem. Phys. Lett.*, Vol. 460, No.1-3, 261-265, 2008 with permission from Elsevier.

Rupture of the nanowire at a constant elongation rate: The dynamic rupture of a Au_4 nanowire was studied according to Eq. 7. Fig. 9a shows the breaking distance as a function of ln (ν_e) for elongation rates between e^{-1} and e^6 in comparison with the experimental date taken from Huang et al. (Huang et al., 2007a). It can be observed that the calculated results resemble the experimental trend in the general features.

A further point that can be analyzed through the present calculations is the lifetime of the wires as a function of the elongation rates. These results are given in Fig. 9b. There is no experimental data available for a straightforward comparison with our calculations. However, AIMD simulations (Krüger et al., 2002), and considerations based on experiments (Huang et al., 2006; Huang et al., 2007a; Li et al., 2006a) indicate that the rupture of a nanocontact made of an alkanethiol and Au contacts should break at a Au - Au bond, so that comparison between the present results may be made with the experiment of Huang et al. (Huang et al., 2007a), who have studied the rupture of single molecule junctions involving Au contacts and ODT. The experimental data of Huang et al. is included in Fig. 9b, where it is found that the calculated lifetimes closer resemble those from the experiment; specially taking into account that no fitting attempt was made seeking for agreement.

4.2 Results for contaminated metallic nanowires

Fig. 10 shows a scheme of the unit cell employed to simulate the pure (**Au-p**) and contaminated Au NWs considered here. The light weight elements taken into account as contaminants are a H atom (**Au-H**) and a C atom (**Au-C**). The grey circles represent Au atoms, which remain fixed at their positions during the relaxation processes. The latter consists in a local energy minimization procedure by means of the conjugate gradient method or a search of a MEP by means of the NEB method.

In the case of contaminated NWs the circle marked with an X represents the location the contaminant atom. This figure also shows the definition of the α_1, α_2 and α_3 bond angles, determined by the atoms relevant for the analysis of the rupture of the NW.

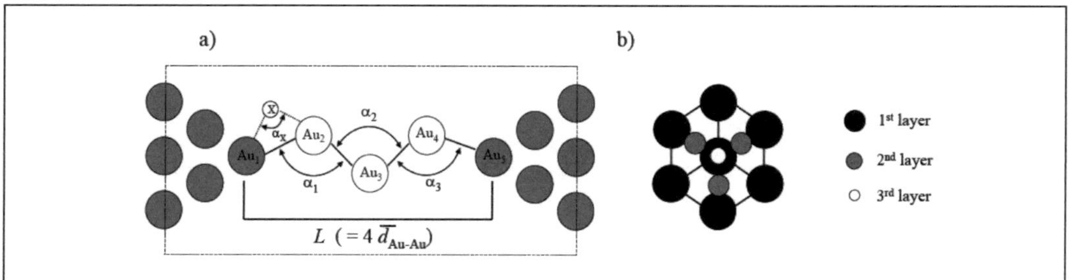

Fig. 10. Schematic representation of the unit cell employed in the present calculations: a) the rectangle indicates the extension of the unit cell. The grey circles represent the Au atoms fixed during the simulation. α_1, α_2 and α_3 are the bond angles between the Au atoms, and α_X is the angle defined by the atoms Au_1 - X - Au_2. L is the total chain length and $\bar{d}_{Au\text{-}Au}$ is the average Au - Au separation distance. b) Front view of the pyramid of Au atoms shown in a).

The length of the NW, L, is defined here as the distance between the atoms Au_1 and Au_5. We also define an average Au - Au separation, as $\bar{d}_{Au\text{-}Au} = L/4$. Fig. 10b shows the front view of one of the pyramid of Au atoms, represented as grey circles in Fig. 10a. The present systems contains as a whole 18 atoms. It must be emphasized that in the case of calculations for pure Au NWs, the present systems deliver results that are close to those presented in the previous section for only four atoms in the unit cell. This indicates that the mechanical properties of the NW appear to be quite local, with a rather slight dependence from the bulky atomic environment.

The atomic impurity was located between the atoms Au_1 and Au_2. This choice was made because in the literature we found first principles calculations, similar to those performed here, where the H and C were positioned at a similar place, as well as between the Au_2 and Au_3 atoms, with similar results (Novaes et al., 2003; Skorodumova et al., 2007). As it will be found below, the present results agree with those where the impurity was located at another sites of the chain.

Structure and energetics of pure and contaminated Au NWs. For all the systems we shall refer to the *equilibrium* state as that where the derivative of the energy with respect to the elongation is equal to zero (ie. the external force, or stress, acting on the system is null). The rupture force will be considered to be the value of the force F_z where it presents a maximum at long elongations, being the force defined in Eq. 8 (da Silva et al., 2004; Jelínek et al., 2008; Novaes et al., 2003; Novaes et al., 2006; Rubio-Bollinger et al., 2001; Vélez et al., 2005; Vélez et al., 2008). Accordingly, we refer in the following to the *"at rupture"* state as that where the coordinates of the atoms are such that $\partial F_z / \partial \Delta z = 0$ (maximum force). With this definition, we are trying to address the status of the system just at the point where the NW is breaking by further force application. The structural information at the *equilibrium* and *"at rupture"* states of the NWs is summarized in Table 1 for the three systems types considered.

Considering the length difference between the *"at rupture"* and *equilibrium* states of the different systems, it is found that the stretching lengths of the **Au-p** and **Au-C** systems are 1.9 Å and 1.6 Å respectively, while the elongation of the **Au-H** system is considerably larger (2.6 Å). This fact bears direct consequences for the force constant k_z, as will be seen later on. Inspecting Table 1 it becomes clear that the equilibrium geometries of the contaminated

	Equilibrium geometries			*„At rupture"* geometries		
	Au-p	**Au-H**	**Au-C**	**Au-p**	**Au-H**	**Au-C**
L_e or L^* / Å	9.80	9.60	10.8	11.7	12.2	12.4
d (1,2) / Å	2.63	**2.76**	**3.78**	2.85	**3.58**	**3.87**
d (2,3) / Å	2.64	2.62	2.59	2.95	2.75	2.67
d (3,4) / Å	2.63	2.67	2.66	3.05	3.16	3.15
d (4,5) / Å	2.63	2.61	2.62	2.85	2.71	2.71
α_1 / °	139.5	149.6	161.8	179.3	178.6	178.5
α_2 / °	125.2	108.6	116.5	179.7	177.0	176.5
α_3 / °	139.7	157.4	136.0	179.9	178.5	178.2
α_X / °	---	**102.5**	**168.5**	---	**179.2**	**179.5**

Table 1. Structural information of the pure Au (**Au-p**) and hydrogen (**Au-H**) or carbon contaminated (**Au-C**) NWs at the equilibrium and close to rupture situations. In the first row, L_e is the equilibrium length of the NW, whereas L^* corresponds to length just before breakage occurs. Refer to Fig. 11 for an illustration of each of the geometries analyzed. The d (1,2) bond distance and the α_X bond angle of the impurity are in boldface.

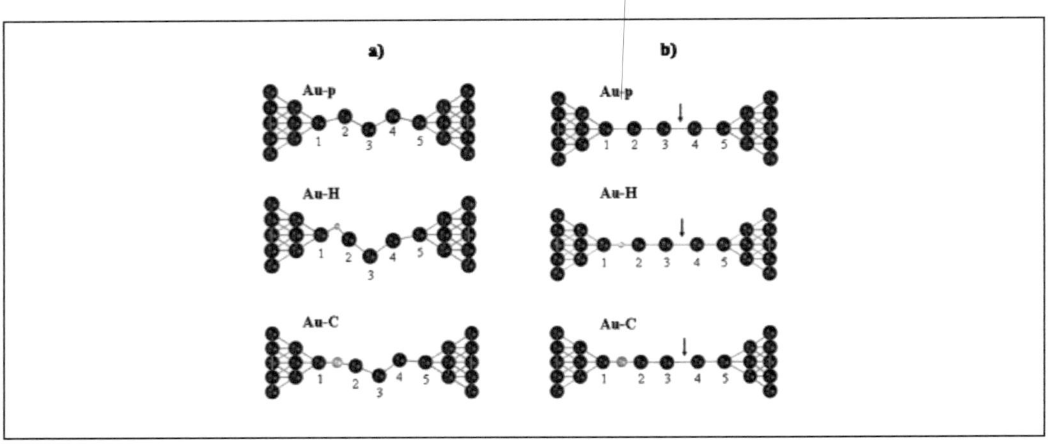

Fig. 11. a) Equilibrium geometries for the pure gold NW (**Au-p**) and contaminated with H (**Au-H**) and C (**Au-C**). b) Geometries just before rupture for stretched NWs. The arrows indicate the bond at which the NW breaks.

systems are very different from each other. Close to equilibrium, the presence of the H atom changes only slightly the d (1,2) bond distance, while the carbon atom inserts itself almost completely in the middle of this bond. This fact can be appreciated in the characteristics of the equilibrium configuration (see α_X bond angle and d (1,2) bond distance).

At the situation of the rupture, all three systems present a linear configuration and the distance of the Au_3 - Au_4 bond where it occurs is very similar (~3.1 Å). The contaminated systems arrive to the point of rupture with large Au - Au distances at the Au_1 - Au_2 bond; 3.58 Å for the **Au-H** system and 3.87 Å for the **Au-C** system. These figures indicate that our results are in a very good agreement with the Au - Au separation distances found experimentally (Kizuka, 2008; Kondo & Takayanagi, 2000; Legoas et al., 2002; Ohnishi et al., 1998; Rodrigues et al., 2000; Yanson et al., 1998) and with the first-principles calculations from other research groups (Novaes et al., 2003; Skorodumova et al., 2007).

Up to now, we have only explored the possibility of the incorporation of a single atomic impurity. In relation to this, it must be recognized that the extremely large Au - Au distances of 4 - 5 Å (Legoas et al., 2002), as well as consecutive distances of 3.5 - 4 Å (Kondo & Takayanagi, 2000; Ohnishi et al., 1998; Rodrigues & Ugarte, 2001a; Rodrigues et al., 2000) cannot be reproduced by our calculations. To tackle this point, more impurities should be considered, and probably other molecular species as proposed by other groups that performed first-principles calculations (Bahn et al., 2002; Galvão et al., 2004; Legoas et al., 2002; Novaes et al., 2003; Skorodumova & Simak, 2003; Skorodumova & Simak, 2004). However, the present approach is not devoted to predict all the Au - Au distances observed, but to understand the effect that an atomic impurity produces on a monatomic Au NW from energetic, geometrical, electronic and kinetic viewpoints.

Analysis of the time-stability of pure and contaminated NWs. In the following the kinetic aspects of the rupture process for the **Au-p**, **Au-H** and **Au-C** systems are considered. Fig. 12a shows the energy curves for the *unbroken* (u), *broken* (b) and *activated* (#) states as a function of the elongation (Δz) for **Au-H**. For **Au-p** the respective curves are similar to those shown in Fig. 6 for the 4-atom gold nanowire. The results obtained for **Au-C** have qualitatively the same features as those obtained for **Au-p**.

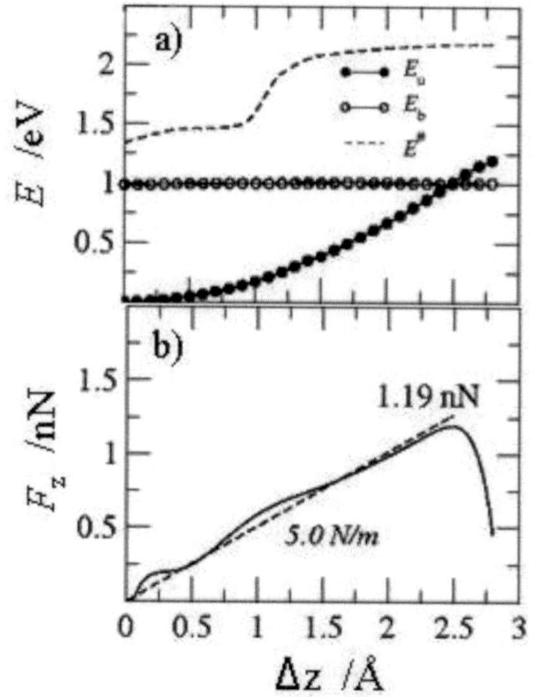

Fig. 12. a) Energy values corresponding to the *unbroken* state E_u, *broken* state E_b and activated state $E^{\#}$ of the system **Au-H** as a function of the elongation. b) Continuous curve: z component of the force acting on the system, F_z, calculated according to Eq. 8, as a function of Δz for **Au-H**. Broken line: Linear fit to F_z between zero and the maximum force. The maximum force F_z^{max} is indicated in normal typeset, while the force constant, k_z (slope of the broken lines), is denoted in italics.

In order to find candidate configurations for the *broken* (final in TST jargon) state at each elongation, molecular dynamics runs were performed at an elongation corresponding to the "*at-rupture*" configuration. This led to structures that looked like those illustrated for **Au-H** for the *broken* state configurations in Fig. 14 below.

We will perform a more detailed discussion of the reaction path for the rupture of the NWs below. We first consider the behaviour of the forces along the stretching procedure, since they are closely related to the energy curves of the *unbroken* state. In fact, the energy curves for the *unbroken* states in Fig. 12a may be used to calculate the longitudinal force F_z acting on this system according to Eq. 8. In Fig. 6, it was found that the stability limit of a monoatomic Au NW is reached at a force of 1.45 nN, which is close to the experimental value of 1.5 ± 0.3 nN (Kizuka, 2008; Rubio et al., 1996; Rubio-Bollinger et al., 2001). Fig. 12b shows F_z for the **Au-H** as a function of the elongation Δz. The maximum force F_z^{max} is also reported there. Table 2 compiles relevant information for this and the other systems under consideration in this study (**Au-p**, **Au-C**). For the **Au-p** system, the maximum force is 1.57 nN, also in agreement with the experiments and other theoretical values (da Silva et al., 2004; Rubio-Bollinger et al., 2001; Vélez et al., 2008). In the case of the contaminated systems, this value is somewhat lower (1.19 nN for **Au-H** and 1.14 nN for **Au-C**). First principles results from literature show the same trend (Novaes et al., 2003; Skorodumova et al., 2007).

	F_z^{max}, nN	k_z, N/m	$\Delta E^\ddagger(\Delta z)$, eV(Å)	
Au-p	1.57	9.1	0.99 (0.0)	0.22 (1.8)
Au-H	1.19	5.0	1.33 (0.0)	1.02 (2.7)
Au-C	1.14	8.2	1.12 (0.0)	1.10 (1.8)

Table 2. Maximum tensile force sustained before breakage (F_z^{max}, nN), force constant (k_z, N/m) and activation energy ($\Delta E^\ddagger$, eV) for the minimum energy path at the *equilibrium* and "*at-rupture*" geometries, for the systems **Au-p**, **Au-H** and **Au-C**. The activation energy for the rupture of the NW at each Δz is calculated as $\Delta E^\ddagger = E^\# - E_u$.

In their studies on the mechanic properties of monatomic Au NWs, (Rubio-Bollinger et al., 2001) found that these chains are five times harder than the massive electrodes. They evaluated from the experimental results the slopes of the force curves in the last stage of the elastic deformation, before the rupture of the NW, from a set of 200 experiments. The average value of the force constant for an average chain length was 8 N/m. The present calculations show that the behaviour of the force curves of the **Au-p**, **Au-H** and **Au-C** systems is quite elastic, as can be inferred from the linear fit of F_z between zero and the maximum force, shown in Fig. 12b for **Au-H** as a broken line. In the elastic deformation region, the force F_z may be written as:

$$F_z = k_z \, \Delta z \quad (9)$$

where k_z is the force constant of the system. The value of k_z found from the present calculations for **Au-p**, 9.1 N/m, is in perfect agreement with the experimental value reported in references (Rubio-Bollinger et al., 2001; Xu et al., 2003a). The contaminated systems present a remarkably different behaviour when compared between each other. The **Au-C** system has k_z =8.1 N/m, a very similar value to that of **Au-p**, but 1.6 larger than the value of this property for **Au-H** (k_z=5.0 N/m). These are interesting predictions, since up to date no measurements have been performed comparing pure and contaminated systems. A suitable fitting of the experimental data of this property could help to shed light on the type of impurity present when a NW exhibits large Au - Au separation distances.

Fig. 13 shows the energy of the system, E, as a function of the normalized reaction coordinate (NRC) for the system **Au-H** for some sample elongations. The corresponding configurations are presented in Fig. 14.

Fig. 13 shows that the **Au-H** system presents an energy curve along the minimum energy path that is more complex than those for **Au-p** and **Au-C** (not shown here). The distinctive feature for of **Au-H** is that at short elongations the energy curves present a minimum. This behavior is related to the incorporation of the hydrogen atom into the wire that takes place as the elongation proceeds. However, for the **Au-C** system, the C atom is incorporated into the wire since the beginning of the elongation. The minimum in the minimum energy path profile of the **Au-H** system disappears gradually with increasing elongation of the NW, remaining a shoulder at long elongations (see Fig. 13b).

The images of monatomic Au chains showing long separation distances (3.6 - 4 Å) were obtained by TEM and HRTEM using electron beam lithography for the fabrication of the NWs (Kizuka, 2008; Kondo & Takayanagi, 2000; Legoas et al., 2002; Ohnishi et al., 1998; Rodrigues & Ugarte, 2001a; Rodrigues et al., 2000). Under these experimental conditions, the elongation rate is not controlled and in principle not known. However, an estimation can be made looking at the pictures provided in some of these publications, as the one illustrated in

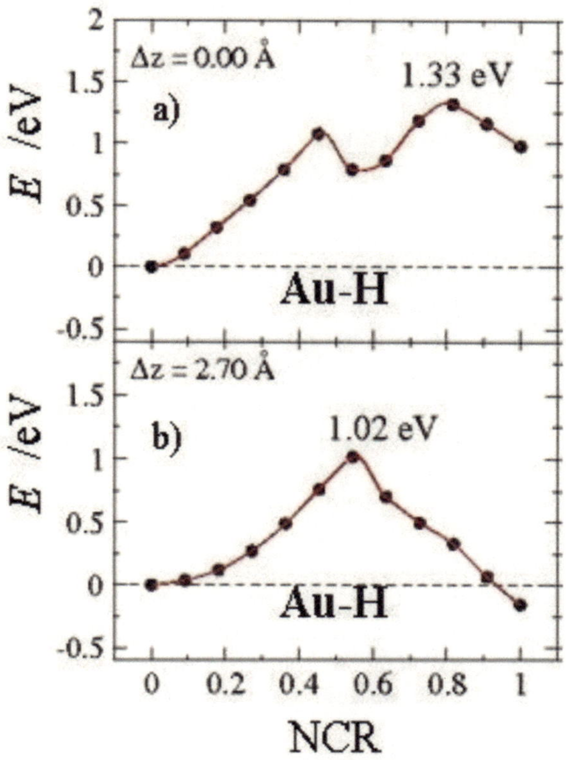

Fig. 13. System energy, *E*, as a function of the normalized reaction coordinate (NRC) along the reaction path between *unbroken* (u) and *broken* (b) states for different elongations Δ*z* of the NW. a), b) and c) correspond to the **Au-p** system, d), e) and f) to the **Au-H** system and g), h) and i) to the **Au-C** system. In all cases the energy of the system is referred to the energy of the *unbroken* state (u). The activation energies $\Delta E^{\ddagger}$ and the corresponding elongations are given for each case.

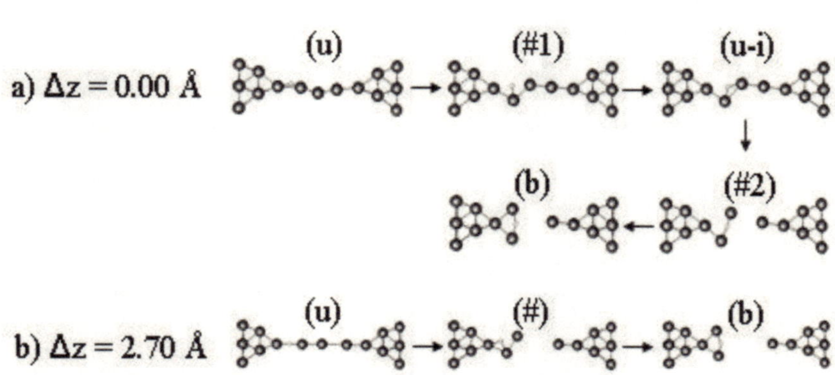

Fig. 14. Configurations of the *unbroken* (u), *activated* (#) and *broken* (b) states of the NW for each of the elongations Δ*z* whose energy curves along the reaction path are shown in Fig. 13 for **Au-H**.

Fig. 2 (Rodrigues & Ugarte, 2001a), and is found to be very slow, of the order of 0.1 nm/s in the final stage, allowing for a complete equilibration of the system at the atomic scale all along the elongation process. Thus, the rupture of the NWs in these experiments takes place in the timescale of a second, so that it can be inferred that the activation barriers must be quite high. In the static limit, the lifetime τ^* of a NW can be estimated from Eq 6. In this equation, we have used $\nu_u(\Delta z) = \nu_u = 3 \times 10^{12}$ Hz and T = 300 K. Similar values were used in references (Bürki et al., 2005; Rubio et al., 1996; Todorov et al., 2001; Vélez et al., 2008). Fig. 15 shows a plot of the decimal logarithm of the lifetime τ^* of the NW, as a function of the elongation force F_z for the **Au-p**, **Au-H** and **Au-C** systems.

It becomes evident from this figure that the contaminated NWs live considerably longer than Au-p NWs for all elongations. It is found that pure Au NW become unstable in the experimental time scale from an elongation which corresponds to $d_{Au\text{-}Au}$ = 2.74 Å / atom on. In fact, for larger F_z ($d_{Au\text{-}Au}$), pure NWs should live less than 0.1 s. The remarkable behaviour of the **Au-H** system is due to the shape of the activation energy, discussed above in Fig. 13. Fig. 15 also supports the idea that impurities are responsible for the large Au - Au distances observed experimentally. These impurities, as stated above, modify geometrically the structure of the NWs, giving place to the occurrence of the anomalous large Au - Au distances. Furthermore, they modify the NW chemically by charge transfer and rearrangement, which in turn changes the potential energy surfaces so that for each elongation the contaminated NWs present higher activation barriers for the rupture than the pure NWs. Fig. 15 makes also plausible that stretched (contaminated) NWs may have lifetimes of the order of the second, sometimes even of the order of minutes (Ohnishi et al., 1998).

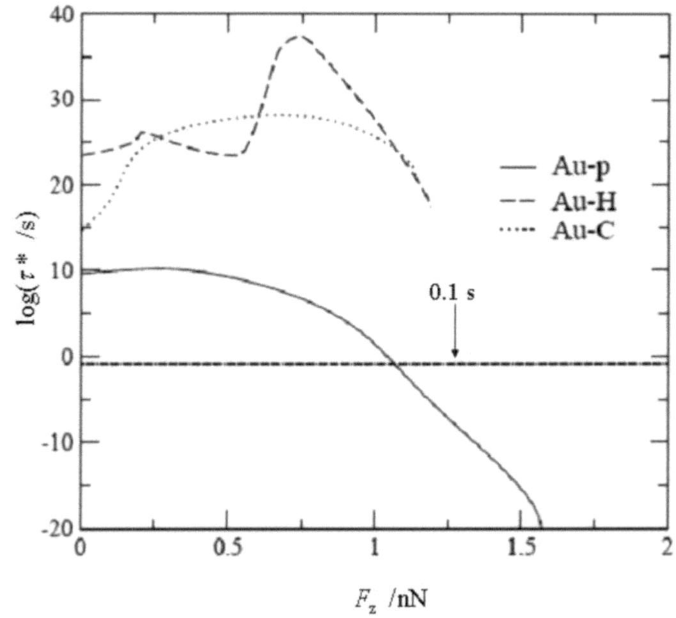

Fig. 15. Decimal logarithm of the lifetimes of the NWs corresponding to the systems **Au-p** (—), **Au-H** (---) and **Au-C** (• • • •) as a function of the elongation force F_z, calculated according to Eq. 6.

5. Long term stability of molecular nanowires. Effect of substituents on the Au – N nanojunction.

The formalism proposed in the previous sections could, in principle, be applied to study the mechanical and kinetic stability of single molecule nanowires. To the best of our knowledge, no such study has been yet attempted. One possible approximation to the long term stability problem is to estimate the kinetic barrier for the rupture of the nanowire by means of the *binding energy* of the system (E_b).

The mechanical properties of *ortho* substituted pyrazines bonded to planar gold clusters of 6 and 7 atoms have been studied using the same methodology described for the monatomic metallic nanowires. In Table 3, we summarize the values obtained for the binding energies (E_b) and rupture forces (F_{rup}) for the substituted pyrazines bonded to the 6-atom (Au_6) and the 7-atom (Au_{6+1}) gold clusters.

The systems studied are illustrated in Fig. 16. These systems are taken as a model for the last stage in the stretching of a Au - bipyridine nanojunction, and allow an assessment of the effect of a substituent in the molecule on the stability of the nanojunction.

The E_b values summarized in Table 3 were obtained as the energy difference between the minimum energy structure for each system and the final broken structure after stretching the nanojunction. In all cases the rupture occurred at the Au - N bond. F_{rup} values were obtained using Eq. 8.

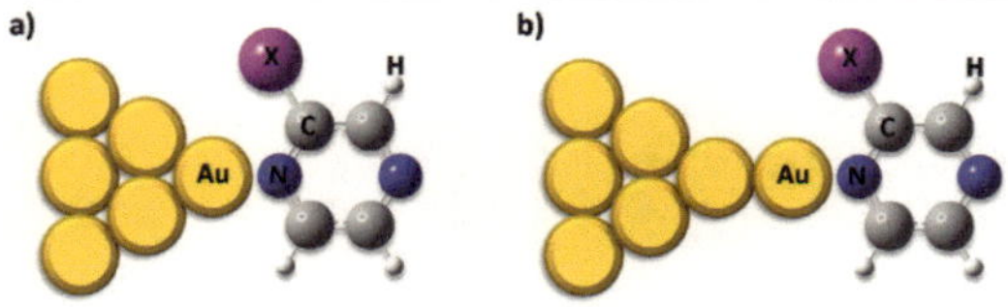

Fig. 16. Illustration of the models adopted to study the effect of ortho substituents on the mechanical stability of a Au - N nanojunction. a) Au_6 + X-pyrazine. b) Au_{6+1} + X-pyrizane.

		Au_6 + X-pyrazine		Au_{6+1} + X-pyrazine	
X	σ	E_b, eV	F_{rup}, nN	E_b, eV	F_{rup}, nN
NH_2	-0.66	0.795	1.16	0.913	1.26
CH_3	-0.17	0.731	1.07	0.828	1.13
H	0.00	0.671	1.00	0.778	1.09
Cl	0.23	0.596	0.82	0.689	0.98
$CONH_2$	0.36	0.549	0.77	---	---
CN	0.66	0.497	0.77	0.605	0.88
NO_2	0.78	0.409	0.48	0.560	0.73

Table 3. Hammett's σ (values taken from (Hansch et al., 1991)), binding energies (E_b) and rupture forces (F_{rup}) obtained from stretching experiments for Au_6 + X-pyrazines and Au_{6+1} X-pyrazines.

Using the E_b values as a *lower* limit for the activation energy to break the nanojunction, and using Eq. 6 with $\nu_u = 10^{12}$ Hz (a typical value for these systems as shown above), time constant values, τ^*, can be calculated for the different nanojunctions. The values for τ so obtained represent a *lower* limit for the lifetime of the nanojunctions. Interestingly, there is a

an excellent correlation between the logarithmic lifetime and a pure empirical parameter describing the electronic nature of the substituent such as Hammett's σ (Hansch et al., 1991), as illustrated in Fig. 17a.
Fig. 17b shows that a good correlation is also observed with respect to the calculated rupture force for the nanojunction. This is interesting since F_{rup} values can be obtained experimentally.
Note that the electronic properties of the substituent has a marked effect on the lifetime of the nanojunction, which ranges from 10^{-6} seconds for X = NO_2 to about 10^3 seconds for NH_2. On the other hand, only a slight effect of the substituent was found on the experimentally measured conductance of a structurally similar system (Venkataraman et al., 2007).
When the molecule binds to a less coordinated gold atom, such as in Au_{6+1} - X-pyrazine systems, the strength of the nanojunction increases, as reflected by the higher values obtained for both E_b and F_{rup}. This also causes an increase in the lifetime of the molecular junction, as illustrated in Fig. 17.

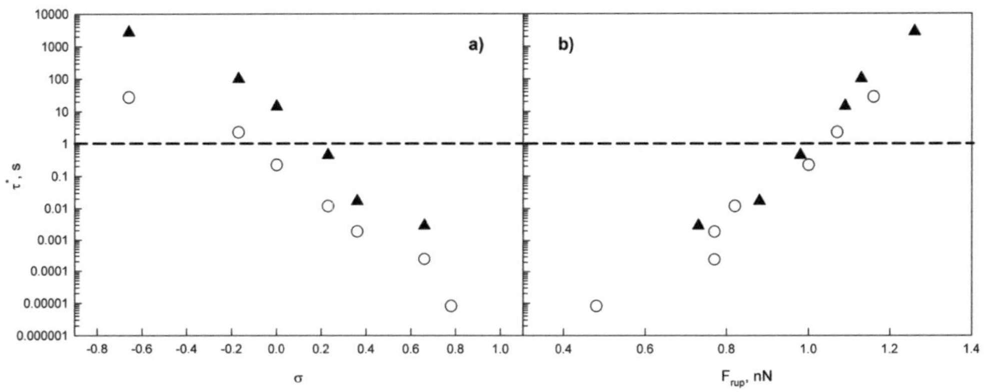

Fig. 17. Correlation between log τ, calculated as the lower limit for the nanojunction's lifetime and a) Hammett's σ and b) calculated rupture force (F_{rup}). In both plots, open circles represent the Au_6 + X-pyrazine systems, whereas filled triangles are for the Au_{6+1} - X-pyrazine systems.

6. Conclusions and perspectives

A comprehensive revision of the most recent advances with respect to the experimental techniques and computational simulations focused on the stability and mechanical properties of monatomic metallic and single molecule nanowires is presented.
We have established a simple model based on the *Transition State Theory* and using the *Minimum Energy Path* to study the long-term stability of nanowires in a time scale corresponding to that of the experimental observations. The utility of this model has been demonstrated for the quantitative evaluation of the effect of impurities in gold nanowires, as well as for a qualitative assessment of the effect exerted by a change in the electronic properties of a molecule in the temporal stability of a metal - molecule junction.
Some of the future work within this topic includes the analysis of other systems, including different metals and / or other chemical linkers for molecular nanowires. It would also be of

great interest to evaluate, if there is any, the correlation between the stability of the nanowire and its electronic properties, such as conductance.

7. References

Amorim, E. P. M.; da Silva, A. J. R. & da Silva, E. Z. (2008). Computer Simulations of Copper and Gold Nanowires and Single-wall Nanowires, *J. Phys. Chem. C*, Vol. 112, No.39, 15241-15246, ISSN: 1932-7447.

Amorim, E. P. M.; da Silva, A. J. R.; Fazzio, A. & da Silva, E. Z. (2007). Short linear atomic chains in copper nanowires, *Nanotechnology*, Vol. 18, No.14, 145701, ISSN: 0957-4484.

Anglada, E.; Torres, J. A.; Yndurain, F. & Soler, J. M. (2007). Formation of Gold Nanowires with Impurities: A First-Principles Molecular Dynamics Simulation, *Phys. Rev. Lett.*, Vol. 98, No.9, 096102-096104, ISSN: 0031-9007.

Auletta, T.; de Jong, M. R.; Mulder, A.; van Veggel, F. C. J. M.; Huskens, J.; Reinhoudt, D. N.; Zou, S.; Zapotoczny, S.; Schönherr, H.; Vancso, G. J. & Kuipers, L. (2004). β-Cyclodextrin Host–Guest Complexes Probed under Thermodynamic Equilibrium: Thermodynamics and AFM Force Spectroscopy, *J. Am. Chem. Soc.*, Vol. 126, No.5, 1577-1584, ISSN: 0002-7863.

Bahn, S. R.; Lopez, N.; Nørskov, J. K. & Jacobsen, K. W. (2002). Adsorption-induced restructuring of gold nanochains, *Phys. Rev. B*, Vol. 66, No.8, 081405 (pp 1-4), ISSN: 1098-0121.

Bahn, S. R. & Jacobsen, K. W. (2001). Chain Formation of Metal Atoms, *Phys. Rev. Lett.*, Vol. 87, No.26, 266101-1-266101-4, ISSN: 0031-9007.

Baskes, M. I.; Nelson, J. S. & Wright, A. F. (1989). Semiempirical modified embedded-atom potentials for silicon and germanium, *Phys. Rev. B*, Vol. 40, No.9, 6085-6100, ISSN: 1098-0121.

Batista, R. J. C.; Ordejon, P.; Chacham, H. & Artacho, E. (2007). Resistive and rectifying effects of pulling gold atoms at thiol-gold nanocontacts, *Phys. Rev. B*, Vol. 75, No.4, 041402-041404, ISSN: 1098-0121.

Beebe, J. M.; Engelkes, V. B.; Miller, L. L. & Frisbie, C. D. (2002). Contact Resistance in Metal-Molecule-Metal Junctions Based on Aliphatic SAMs: Effects of Surface Linker and Metal Work Function, *J. Am. Chem. Soc.*, Vol. 124, No.38, 11268-11269, ISSN: 0002-7863.

Bettini, J.; Sato, F.; Coura, P. Z.; Dantas, S. O.; Galvão, D. S. & Ugarte, D. (2006). Experimental realization of suspended atomic chains composed of different atomic species, *Nat Nano*, Vol. 1, No.3, 182-185, ISSN: 1748-3387.

Bonacic-Koutecky, V.; Burda, J.; Mitric, R.; Ge, M.; Zampella, G. & Fantucci, P. (2002). Density functional study of structural and electronic properties of bimetallic silver--gold clusters: Comparison with pure gold and silver clusters, *J. Chem. Phys.*, Vol. 117, No.7, 3120-3131, ISSN: 0021-9606.

Bürki, J.; Stafford, C. A. & Stein, D. L. (2005). Theory of Metastability in Simple Metal Nanowires, *Phys. Rev. Lett.*, Vol. 95, No.9, 090601, ISSN: 0031-9007.

Chen, F.; Li, X.; Hihath, J.; Huang, Z. & Tao, N. (2006). Effect of Anchoring Groups on Single-Molecule Conductance: Comparative Study of Thiol-, Amine-, and

Carboxylic-Acid-Terminated Molecules, *J. Am. Chem. Soc.*, Vol. 128, No.49, 15874-15881, ISSN: 0002-7863.

Cui, X. D.; Primak, A.; Zarate, X.; Tomfohr, J.; Sankey, O. F.; Moore, A. L.; Moore, T. A.; Gust, D.; Harris, G. & Lindsay, S. M. (2001). Reproducible Measurement of Single-Molecule Conductivity, *Science*, Vol. 294, No.5542, 571-574, ISSN: 0036-8075.

da Silva, E. Z.; da Silva, A. J. R. & Fazzio, A. (2001). How Do Gold Nanowires Break?, *Phys. Rev. Lett.*, Vol. 87, No.25, 256102, ISSN: 0031-9007.

da Silva, E. Z.; Novaes, F. D.; da Silva, A. J. R. & Fazzio, A. (2004). Theoretical study of the formation, evolution, and breaking of gold nanowires, *Phys. Rev. B*, Vol. 69, No.11, 115411 (pp 1-11), ISSN: 1098-0121.

De Maria, L. & Springborg, M. (2000). Electronic structure and dimerization of a single monatomic gold wire, *Chemical Physics Letters*, Vol. 323, No.3-4, 293-299, ISSN: 0009-2614.

Dreher, M.; Pauly, F.; Heurich, J.; Cuevas, J. C.; Scheer, E. & Nielaba, P. (2005). Structure and conductance histogram of atomic-sized Au contacts, *Phys. Rev. B*, Vol. 72, No.7, 075435-11, ISSN: 1098-0121.

Evans, E. & Ritchie, K. (1997). Dynamic strength of molecular adhesion bonds, *Biophysical Journal*, Vol. 72, No.4, 1541-1555, ISSN: 0006-3495.

Evans, E. (1999). Introductory Lecture Energy landscapes of biomolecular adhesion and receptor anchoring at interfaces explored with dynamic force spectroscopy, *Faraday Discussions*, Vol. 111, 1-16, ISSN: 1364-5498.

Evans, E. (2001). Probing the Relation Between Force-Lifetime and Chemistry in Single Molecular Bonds, *Annual Review of Biophysics and Biomolecular Structure*, Vol. 30, No.1, 105-128, ISSN: 1056-8700.

Eyring, H. (1935). The Activated Complex in Chemical Reactions, *J. Chem. Phys.*, Vol. 3, No.2, 107-115, ISSN: 0021-9606.

Galvão, D. S.; Rodrigues, V.; Ugarte, D. & Legoas, S. B. (2004). The role of carbon contamination in metallic nanowires, *Materials Research*, Vol. 7, 339-342, ISSN: 1516-1439.

Haiss, W.; Martín, S.; Leary, E.; Zalinge, H. v.; Higgins, S. J.; Bouffier, L. & Nichols, R. J. (2009). Impact of Junction Formation Method and Surface Roughness on Single Molecule Conductance, *The Journal of Physical Chemistry C*, Vol. 113, No.14, 5823-5833, ISSN: 1932-7447.

Haiss, W.; Nichols, R. J.; van Zalinge, H.; Higgins, S. J.; Bethel, D. & Schiffrin, D. J. (2004). Measurement of single molecule conductivity using the spontaneous formation of molecular wires, *Phys. Chem. Chem. Phys.*, Vol. 6, 4330-4337, ISSN: 14639076.

Haiss, W.; Wang, C.; Grace, I.; Batsanov, A. S.; Schiffrin, D. J.; Higgins, S. J.; Bryce, M. R.; Lambert, C. J. & Nichols, R. J. (2006). Precision control of single-molecule electrical junctions, *Nature Materials*, Vol. 5, No.12, 995-1002, ISSN: 1476-1122.

Haiss, W.; Wang, C.; Jitchati, R.; Grace, I.; Martín, S.; Batsanov, A. S.; Higgins, S.; Bryce, M. R.; Lambert, C. J.; Jensen, P. S. & Nichols, R. J. (2008). Variable contact gap single-molecule conductance determination for a series of conjugated molecular bridges, *Journal of Physics: Condensed Matter*, Vol. 20, No.37, 374119 (9pp), ISSN: 0953-8984.

Häkkinen, H.; Barnett, R. N.; Scherbakov, A. G. & Landman, U. (2000). Nanowire Gold Chains: Formation Mechanisms and Conductance, *J. Phys. Chem. B*, Vol. 104, No.39, 9063-9066, ISSN: 1520-6106.

Hansch, C.; Leo, A. & Taft, R. W. (1991). A Survey of Hammett Substituent Constants and Resonance and Field Parameters, *Chem. Rev.*, Vol. 91, No.2, 165-195, ISSN: 0009-2665.

He, H. X.; Shu, C.; Li, C. Z. & Tao, N. J. (2002). Adsorbate effect on the mechanical stability of atomically thin metallic wires, *Journal of Electroanalytical Chemistry*, Vol. 522, No.1, 26-32, ISSN: 0022-0728.

Henkelman, G. & Jonsson, H. (2000). Improved tangent estimate in the nudged elastic band method for finding minimum energy paths and saddle points, *J. Chem. Phys.*, Vol. 113, No.22, 9978-9985, ISSN: 0021-9606.

Henkelman, G.; Uberuaga, B. P. & Jonsson, H. (2000). A climbing image nudged elastic band method for finding saddle points and minimum energy paths, *J. Chem. Phys.*, Vol. 113, No.22, 9901-9904, ISSN: 0021-9606.

Hobi, E.; da Silva, A. J. R.; Novaes, F. D.; da Silva, E. Z. & Fazzio, A. (2005). Comment on "Contaminants in Suspended Gold Chains: An Ab Initio Molecular Dynamics Study", *Phys. Rev. Lett.*, Vol. 95, No.16, 169601 (p 1), ISSN: 0031-9007.

Hobi, E. Jr.; Fazzio, A. & da Silva, A. J. R. (2008). Temperature and Quantum Effects in the Stability of Pure and Doped Gold Nanowires, *Phys. Rev. Lett.*, Vol. 100, No.5, 056104 (pp 1-4), ISSN: 0031-9007.

Hou, S.; Ning, J.; Shen, Z.; Zhao, X. & Xue, Z. (2006). Influences of the molecule-electrode interface structure on the conducting characteristics of the gold-4,4 bipyridine-gold molecular junction, *Chem. Phys.*, Vol. 327, No.1, 1-9, ISSN: 0301-0104.

Hou, S.; Zhang, J.; Li, R.; Ning, J.; Han, R.; Shen, Z.; Zhao, X.; Xue, Z. & Wu, Q. (2005). First-principles calculation of the conductance of a single 4,4 bipyridine molecule, *Nanotechnology*, Vol. 16, No.2, 239-244, ISSN: 0957-4484.

Huang, Z. F.; Chen, F.; Bennett, P. A. & Tao, N. J. (2007a). Single Molecule Junctions Formed via Au-Thiol Contact: Stability and Breakdown Mechanism, *J. Am. Chem. Soc.*, Vol. 129, No.43, 13225-13231, ISSN: 0002-7863.

Huang, Z. F.; Xu, B. Q.; Chen, Y. C.; Ventra, M. D. & Tao, N. J. (2006). Measurement of Current-Induced Local Heating in a Single Molecule Junction, *Nano Lett.*, Vol. 6, No.6, 1240-1244, ISSN: 1530-6984.

Huang, Z.; Chen, F.; D'agosta, R.; Bennett, P. A.; Di Ventra, M. & Tao, N. (2007b). Local ionic and electron heating in single-molecule junctions, *Nature Nanotechnology*, Vol. 2, No.11, 698-703, ISSN: 1748-3387.

Hybertsen, M. S.; Venkataraman, L.; Klare, J. E.; Whalley, A. C.; Steigerwald, M. L. & Nuckolls, C. (2008). Amine-linked single-molecule circuits: systematic trends across molecular families, *Journal of Physics: Condensed Matter*, Vol. 20, No.37, 374115, ISSN: 0953-8984.

Jelínek, P.; Pérez, R.; Ortega, J. & Flores, F. (2008). Ab initio study of evolution of mechanical and transport properties of clean and contaminated Au nanowires along the deformation path, *Phys. Rev. B*, Vol. 77, No.11, 115447 (pp 1-12), ISSN: 1098-0121.

Kamenetska, M.; Koentopp, M.; Whalley, A. C.; Park, Y. S.; Steigerwald, M. L.; Nuckolls, C.; Hybertsen, M. S. & Venkataraman, L. (2009). Formation and Evolution of Single-Molecule Junctions, *Phys. Rev. Lett.*, Vol. 102, No.12, 126803-126804, ISSN: 0031-9007.

Ke, L.; Kotani, T.; van Schilfgaarde, M. & Bennett, P. A. (2007). Breakdown of a gold nanowire between electrodes, *Nanotechnology*, Vol. 18, No.42, 424002, ISSN: 0957-4484.

Kiguchi, M.; Miura, S.; Hara, K.; Sawamura, M. & Murakoshi, K. (2006). Conductance of a single molecule anchored by an isocyanide substituent to gold electrodes, *Appl. Phys. Lett.*, Vol. 89, No.21, 213104-3, ISSN: 0003-6951.

Kiguchi, M.; Miura, S.; Hara, K.; Sawamura, M. & Murakoshi, K. (2007). Conductance of single 1,4-disubstituted benzene molecules anchored to Pt electrodes, *Appl. Phys. Lett.*, Vol. 91, No.5, 053110-053113, ISSN: 0003-6951.

Kiguchi, M.; Miura, S.; Takahashi, T.; Hara, K.; Sawamura, M. & Murakoshi, K. (2008). Conductance of Single 1,4-Benzenediamine Molecule Bridging between Au and Pt Electrodes, *J. Phys. Chem. C*, Vol. 112, No.35, 13349-13352, ISSN: 1932-7447.

Kiguchi, M. & Murakoshi, K. (2006). Fabrication of stable Pd nanowire assisted by hydrogen in solution, *Appl. Phys. Lett.*, Vol. 88, No.25, 253112-253113, ISSN: 0003-6951.

Kim, B. S.; Beebe, J. M.; Jun, Y.; Zhu, X. Y. & Frisbie, C. D. (2006a). Correlation between HOMO Alignment and Contact Resistance in Molecular Junctions: Aromatic Thiols versus Aromatic Isocyanides, *J. Am. Chem. Soc.*, Vol. 128, No.15, 4970-4971, ISSN: 0002-7863.

Kim, Y. H.; Tahir-Kheli, J.; Schultz, P. A.; Goddard, W. A. & III (2006b). First-principles approach to the charge-transport characteristics of monolayer molecular-electronics devices: Application to hexanedithiolate devices, *Phys. Rev. B*, Vol. 73, No.23, 235419-10, ISSN: 1098-0121.

Kizuka, T. (2008). Atomic configuration and mechanical and electrical properties of stable gold wires of single-atom width, *Phys. Rev. B*, Vol. 77, No.15, 155401 (pp 1-11), ISSN: 1098-0121.

Koizumi, H.; Oshima, Y.; Kondo, Y. & Takayanagi, K. (2001). Quantitative high-resolution microscopy on a suspended chain of gold atoms, *Ultramicroscopy*, Vol. 88, No.1, 17-24, ISSN: 0304-3991.

Kondo, Y. & Takayanagi, K. (1997). Gold Nanobridge Stabilized by Surface Structure, *Phys. Rev. Lett.*, Vol. 79, No.18, 3455, ISSN: 0031-9007.

Kondo, Y. & Takayanagi, K. (2000). Synthesis and Characterization of Helical Multi-Shell Gold Nanowires, *Science*, Vol. 289, No.5479, 606-608, ISSN: 0036-8075.

Krüger, D.; Fuchs, H.; Rousseau, R.; Marx, D. & Parrinello, M. (2002). Pulling Monatomic Gold Wires with Single Molecules: An Ab Initio Simulation, *Phys. Rev. Lett.*, Vol. 89, No.18, 186402, ISSN: 0031-9007.

Krüger, D.; Rousseau, R.; Fuchs, H. & Marx, D. (2003). Towards "Mechanochemistry": Mechanically Induced Isomerizations of Thiolate-Gold Clusters, *Angew. Chem. Int. Ed. Eng.*, Vol. 42, No.20, 2251-2253, ISSN: 1433-7851.

Legoas, S. B.; Galvao, D. S.; Rodrigues, V. & Ugarte, D. (2002). Origin of Anomalously Long Interatomic Distances in Suspended Gold Chains, *Phys. Rev. Lett.*, Vol. 88, No.7, 076105 (pp 1-4), ISSN: 0031-9007.

Legoas, S. B.; Rodrigues, V.; Ugarte, D. & Galvão, D. S. (2004). Contaminants in Suspended Gold Chains: An Ab Initio Molecular Dynamics Study, *Phys. Rev. Lett.*, Vol. 93, No.21, 216103 (pp 1-4), ISSN: 0031-9007.

Li et al, C. Z. (1999). Fabrication of stable metallic nanowires with quantized conductance, *Nanotechnology*, Vol. 10, No.2, 221, ISSN: 0957-4484.

Li, C. Z. & Tao, N. J. (1998). Quantum transport in metallic nanowires fabricated by electrochemical deposition/dissolution, *Appl. Phys. Lett.*, Vol. 72, No.8, 894-896, ISSN: 0003-6951.

Li, C.; Pobelov, I.; Wandlowski, T.; Bagrets, A.; Arnold, A. & Evers, F. (2007b). Charge Transport in Single Au | Alkanedithiol | Au Junctions: Coordination Geometries and Conformational Degrees of Freedom, *J. Am. Chem. Soc.*, Vol. 130, No.1, 318-326, ISSN: 0002-7863.

Li, C.; Pobelov, I.; Wandlowski, T.; Bagrets, A.; Arnold, A. & Evers, F. (2007a). Charge Transport in Single Au | Alkanedithiol | Au Junctions: Coordination Geometries and Conformational Degrees of Freedom, *J. Am. Chem. Soc.*, Vol. 130, No.1, 318-326, ISSN: 0002-7863.

Li, Q.; Wu, X.; Huang, J. & Yang, J. (2005). Electronic transport property of 4,4'-bipyridine molecular junction, *Ultramicroscopy*, Vol. 105, No.1-4, 293-298, ISSN: 0304-3991.

Li, X.; He, J.; Hihath, J.; Xu, B.; Lindsay, S. M. & Tao, N. (2006a). Conductance of Single Alkanedithiols: Conduction Mechanism and Effect of Molecule-Electrode Contacts, *J. Am. Chem. Soc.*, Vol. 128, No.6, 2135-2141, ISSN: 0002-7863.

Li, Z. L.; Zou, B.; Wang, C. K. & Luo, Y. (2006b). Effects of Electrode Distances on Geometric Structure and Electronic Transport Properties of Molecular 4,4'-Bipyridine Junction, *Journal of Physics: Conference Series*, Vol. 29, 110-114, ISSN: 1742-6596.

Li, Z. L.; Zou, B.; Wang, C. K. & Luo, Y. (2006c). Electronic transport properties of molecular bipyridine junctions: Effects of isomer and contact structures, *Phys. Rev. B*, Vol. 73, No.7, 075326-075327, ISSN: 1098-0121.

Lorenz, R.; Georg, H.; Mathis, G.; Jean-Luc, B. & Egbert, Z. (2006). Stretching and Breaking of a Molecular Junction, *Small*, Vol. 2, No.12, 1468-1475, ISSN: 1613-6829.

Martín, S.; Haiss, W.; Higgins, S.; Cea, P.; Lopez, M. C. & Nichols, R. J. (2008). A Comprehensive Study of the Single Molecule Conductance of α,ω-Dicarboxylic Acid-Terminated Alkanes, *J. Phys. Chem. C*, Vol. 112, No.10, 3941-3948, ISSN: 1932-7447.

Merkel, R.; Nassoy, P.; Leung, A.; Ritchie, K. & Evans, E. (1999). Energy landscapes of receptor-ligand bonds explored with dynamic force spectroscopy, *Nature*, Vol. 397, No.6714, 50-53, ISSN: 0028-0836.

Michael, L. M. & Michael, P. (2007). Computing Reaction Pathways on Molecular Potential Energy Surfaces, In: *Reviews in Computational Chemistry Vol. 4*, Kenny, B. L. (Ed.), 35-65,

Mills, G. & Jónsson, H. (1994). Quantum and thermal effects in H2 dissociative adsorption: Evaluation of free energy barriers in multidimensional quantum systems, *Phys. Rev. Lett.*, Vol. 72, No.7, 1124, ISSN: 0031-9007.

Nakamura, A.; Brandbyge, M.; Hansen, L. B. & Jacobsen, K. W. (1999). Density Functional Simulation of a Breaking Nanowire, *Phys. Rev. Lett.*, Vol. 82, No.7, 1538, ISSN: 0031-9007.

Nakamura, J.; Kobayashi, N. & Aono, M. (2001). Electronic states and structural stability of gold nanowires, *RIKEN Review*, Vol. 2001, No.37, 17-20, ISSN: 0919-3405.

Novaes, F. D.; da Silva, A. J. R.; da Silva, E. Z. & Fazzio, A. (2003). Effect of Impurities in the Large Au-Au Distances in Gold Nanowires, *Phys. Rev. Lett.*, Vol. 90, No.3, 036101 (pp 1-4), ISSN: 0031-9007.

Novaes, F. D.; da Silva, A. J. R.; da Silva, E. Z. & Fazzio, A. (2006). Oxygen Clamps in Gold Nanowires, *Phys. Rev. Lett.*, Vol. 96, No.1, 016104 (pp 1-4), ISSN: 0031-9007.

O'Hanlon, J. F. (2001). *A User's Guide to Vacuum Technology*, John Wiley & Sons, Inc, ISBN: 9780471270522, Hoboken, NJ.

Ohnishi, H.; Kondo, Y. & Takayanagi, K. (1998). Quantized conductance through individual rows of suspended gold atoms, *Nature*, Vol. 395, No.6704, 780-783, ISSN: 0028-0836.

Okamoto, M. & Takayanagi, K. (1999). Structure and conductance of a gold atomic chain, *Phys. Rev. B*, Vol. 60, No.11, 7808-7811, ISSN: 1098-0121.

Park, Y. S.; Whalley, A. C.; Kamenetska, M.; Steigerwald, M. L.; Hybertsen, M. S.; Nuckolls, C. & Venkataraman, L. (2007). Contact Chemistry and Single-Molecule Conductance: A Comparison of Phosphines, Methyl Sulfides, and Amines, *J. Am. Chem. Soc.*, Vol. 129, No.51, 15768-15769, ISSN: 0002-7863.

Park, Y. S.; Widawsky, J. R.; Kamenetska, M.; Steigerwald, M. L.; Hybertsen, M. S.; Nuckolls, C. & Venkataraman, L. (2009). Frustrated Rotations in Single-Molecule Junctions, *J. Am. Chem. Soc.*, Vol. 131, No.31, 10820-10821, ISSN: 0002-7863.

Patrone, L.; Palacin, S. & Bourgoin, J. P. (2003a). Direct comparison of the electronic coupling efficiency of sulfur and selenium alligator clips for molecules adsorbed onto gold electrodes, *Applied Surface Science*, Vol. 212-213, 446-451, ISSN: 0169-4332.

Patrone, L.; Palacin, S.; Charlier, J.; Armand, F.; Bourgoin, J. P.; Tang, H. & Gauthier, S. (2003b). Evidence of the Key Role of Metal-Molecule Bonding in Metal-Molecule-Metal Transport Experiments, *Phys. Rev. Lett.*, Vol. 91, No.9, 096802, ISSN: 0031-9007.

Paulsson, M.; Krag, C.; Frederiksen, T. & Brandbyge, M. (2008). Conductance of Alkanedithiol Single-Molecule Junctions: A Molecular Dynamics Study, *Nano Lett.*, Vol. 9, No.1, 117-121, ISSN: 1530-6984.

Perez-Jimenez, A. (2005). Uncovering Transport Properties of 4,4'-Bipyridine/Gold Molecular Nanobridges, *J. Phys. Chem. B*, Vol. 109, No.20, 10052-10060, ISSN: 1520-6106.

Pu, Q.; Leng, Y.; Tsetseris, L.; Park, H. S.; Pantelides, S. T. & Cummings, P. T. (2007a). Molecular dynamics simulations of stretched gold nanowires: The relative utility of different semiempirical potentials, *J. Chem. Phys.*, Vol. 126, No.14, 144707-6, ISSN: 0021-9606.

Pu, Q.; Leng, Y.; Zhao, X. & Cummings, P. T. (2007b). Molecular simulations of stretching gold nanowires in solvents, *Nanotechnology*, Vol. 18, No.42, 424007, ISSN: 0957-4484.

Quek, S. Y.; Venkataraman, L.; Choi, H. J.; Louie, S. G.; Hybertsen, M. S. & Neaton, J. B. (2007). Amine-Gold Linked Single-Molecule Circuits: Experiment and Theory, *Nano Lett.*, Vol. 7, No.11, 3477-3482, ISSN: 1530-6984.

Quek, S. Y.; Kamenetska, M.; Steigerwald, M. L.; Choi, H. J.; Louie, S. G.; Hybertsen, M. S.; Neaton, J. B. & Venkataraman, L. (2009). Mechanically controlled binary conductance switching of a single-molecule junction, *Nat Nano*, Vol. 4, No.4, 230-234, ISSN: 1748-3387.

Quinn, J. R.; Foss, F. W.; Venkataraman, L.; Hybertsen, M. S. & Breslow, R. (2007). Single-Molecule Junction Conductance through Diaminoacenes, *J. Am. Chem. Soc.*, Vol. 129, No.21, 6714-6715, ISSN: 0002-7863.

Rodrigues, V.; Bettini, J.; Rocha, A. R.; Rego, L. G. C. & Ugarte, D. (2002). Quantum conductance in silver nanowires: Correlation between atomic structure and transport properties, *Phys. Rev. B*, Vol. 65, No.15, 153402, ISSN: 1098-0121.

Rodrigues, V. & Ugarte, D. (2001a). Real-time imaging of atomistic process in one-atom-thick metal junctions, *Phys. Rev. B*, Vol. 63, No.7, 073405 (pp 1-4), ISSN: 1098-0121.

Rodrigues, V. & Ugarte, D. (2001b). Structural and electronic properties of gold nanowires, *Eur. Phys. J. D*, Vol. 16, No.1-3, 395-398, ISSN: 1434-6060.

Rodrigues, V. & Ugarte, D. (2002). Quantum conductance properties of metal nanowires, *Materials Science and Engineering B*, Vol. 96, No.2, 188-192, ISSN: 0921-5107.

Rodrigues, V.; Bettini, J.; Silva, P. C. & Ugarte, D. (2003). Evidence for Spontaneous Spin-Polarized Transport in Magnetic Nanowires, *Phys. Rev. Lett.*, Vol. 91, No.9, 096801, ISSN: 0031-9007.

Rodrigues, V.; Fuhrer, T. & Ugarte, D. (2000). Signature of Atomic Structure in the Quantum Conductance of Gold Nanowires, *Phys. Rev. Lett.*, Vol. 85, No.19, 4124-4127, ISSN: 0031-9007.

Rubio, G.; Agraït, N. & Vieira, S. (1996). Atomic-Sized Metallic Contacts: Mechanical Properties and Electronic Transport, *Phys. Rev. Lett.*, Vol. 76, No.13, 2302-2305, ISSN: 0031-9007.

Rubio-Bollinger, G.; Bahn, S. R.; Agraït, N.; Jacobsen, K. W. & Vieira, S. (2001). Mechanical Properties and Formation Mechanisms of a Wire of Single Gold Atoms, *Phys. Rev. Lett.*, Vol. 87, No.2, 026101 (pp 1-4), ISSN: 0031-9007.

Sánchez-Portal, D.; Artacho, E.; Junquera, J.; Ordejón, P.; García, A. & Soler, J. M. (1999). Stiff Monatomic Gold Wires with a Spinning Zigzag Geometry, *Phys. Rev. Lett.*, Vol. 83, No.19, 3884-3887, ISSN: 0031-9007.

Sato, F.; Moreira, A. S.; Bettini, J.; Coura, P. Z.; Dantas, S. O.; Ugarte, D. & Galvao, D. S. (2006a). On the Formation of Copper Linear Atomic Suspended Chains, *Condensed Matter*, arXiv:cond-mat/0602092v1.

Sato, F.; Moreira, A. S.; Bettini, J.; Coura, P. Z.; Dantas, S. O.; Ugarte, D. & Galvao, D. S. (2006b). Transmission electron microscopy and molecular dynamics study of the formation of suspended copper linear atomic chains, *Phys. Rev. B*, Vol. 74, No.19, 193401-193404, ISSN: 1098-0121.

Schönherr, H.; Beulen, M. W. J.; Bügler, J.; Huskens, J.; van Veggel, F. C. J. M.; Reinhoudt, D. N. & Vancso, G. J. (2000). Individual Supramolecular Host–Guest Interactions Studied by Dynamic Single Molecule Force Spectroscopy, *J. Am. Chem. Soc.*, Vol. 122, No.20, 4963-4967, ISSN: 0002-7863.

Skorodumova, N. V. & Simak, S. I. (2003). Stability of gold nanowires at large Au-Au separations, *Phys. Rev. B*, Vol. 67, No.12, 121404 (pp 1-4), ISSN: 1098-0121.

Skorodumova, N. V. & Simak, S. I. (2004). Stabilization of monoatomic gold wires by carbon impurities, *Solid State Communications*, Vol. 130, No.11, 755-757, ISSN: 0038-1098.

Skorodumova, N. V.; Simak, S. I.; Kochetov, A. E. & Johansson, B. (2007). Ab initio study of electronic and structural properties of gold nanowires with light-element impurities, *Phys. Rev. B*, Vol. 75, No.23, 235440 (pp 1-4), ISSN: 1098-0121.

Sørensen, M. R.; Brandbyge, M. & Jacobsen, K. W. (1998). Mechanical deformation of atomic-scale metallic contacts: Structure and mechanisms, *Phys. Rev. B*, Vol. 57, No.6, 3283, ISSN: 1098-0121.

Sørensen, M. R.; Jacobsen, K. W. & Jónsson, H. (1996). Thermal Diffusion Processes in Metal-Tip-Surface Interactions: Contact Formation and Adatom Mobility, *Phys. Rev. Lett.*, Vol. 77, No.25, 5067, ISSN: 0031-9007.

Stadler, R.; Thygesen, K. S. & Jacobsen, K. W. (2005). Forces and conductances in a single-molecule bipyridine junction, *Phys. Rev. B*, Vol. 72, No.24, 241401-241404, ISSN: 1098-0121.

Stolberg, L.; Lipkowski, J. & Irish, D. E. (1990). Adsorption of pyridine at the Au(110) - solution interface, *Journal of Electroanalytical Chemistry*, Vol. 296, No.1, 171-189, ISSN: 0022-0728.

Takai, Y.; Kawasaki, T.; Kimura, Y.; Ikuta, T. & Shimizu, R. (2001). Dynamic Observation of an Atom-Sized Gold Wire by Phase Electron Microscopy, *Phys. Rev. Lett.*, Vol. 87, No.10, 106105, ISSN: 0031-9007.

Todorov, T. N.; Hoekstra, J. & Sutton, A. P. (2001). Current-Induced Embrittlement of Atomic Wires, *Phys. Rev. Lett.*, Vol. 86, No.16, 3606, ISSN: 0031-9007.

Torres, J. A.; Tosatti, E.; Dal Corso, A.; Ercolessi, F.; Kohanoff, J. J.; Di Tolla, F. D. & Soler, J. M. (1999). The puzzling stability of monatomic gold wires, *Surface Science*, Vol. 426, No.3, L441-L446, ISSN: 0039-6028.

Ulrich, J.; Esrail, D.; Pontius, W.; Venkataraman, L.; Millar, D. & Doerrer, L. H. (2006). Variability of Conductance in Molecular Junctions, *J. Phys. Chem. B*, Vol. 110, No.6, 2462-2466, ISSN: 1520-6106.

van Ruitenbeek, J.; Scheer, E. & Weber, H. (2005). Contacting Individual Molecules Using Mechanically Controllable Break Junctions, In: *Introducing Molecular Electronics*, Cuniberti, G., Fagas, G., and Richter, K. (Ed.), 253-274, Springer Berlin Heidelberg, ISBN: 1616-6361.

Vélez, P.; Dassie, S. A. & Leiva, E. P. M. (2005). First Principles Calculations of Mechanical Properties of 4,4'-Bipyridine Attached to Au Nanowires, *Phys. Rev. Lett.*, Vol. 95, No.4, 045503-045504, ISSN: 0031-9007.

Vélez, P.; Dassie, S. A. & Leiva, E. P. M. (2008). When do nanowires break? A model for the theoretical study of the long-term stability of monoatomic nanowires, *Chem. Phys. Lett.*, Vol. 460, No.1-3, 261-265, ISSN: 0009-2614.

Vélez, P.; Dassie, S. A. & Leiva, E. P. M. (2007). Understanding the properties of nanocontacts. Recent advances in experiment and theory, In: *Recent Advances in Nanoscience*, Mariscal, M. M. and Dassie, S. A. (Ed.), 1-38, Research Signpost, ISBN: 978-81-308-0207-7, Kerala, India.

Venkataraman, L.; Klare, J. E.; Tam, I. W.; Nuckolls, C.; Hybertsen, M. S. & Steigerwald, M. L. (2006a). Single-Molecule Circuits with Well-Defined Molecular Conductance, *Nano Lett.*, Vol. 6, No.3, 458-462, ISSN: 1530-6984.

Venkataraman, L.; Park, Y. S.; Whalley, A. C.; Nuckolls, C.; Hybertsen, M. S. & Steigerwald, M. L. (2007). Electronics and Chemistry: Varying Single-Molecule Junction Conductance Using Chemical Substituents, *Nano Lett.*, Vol. 7, No.2, 502-506, ISSN: 1530-6984.

Venkataraman, L.; Klare, J. E.; Nuckolls, C.; Hybertsen, M. S. & Steigerwald, M. L. (2006b). Dependence of single-molecule junction conductance on molecular conformation, *Nature*, Vol. 442, No.7105, 904-907, ISSN: 0028-0836.

Villarba, M. & Jónsson, H. (1994). Diffusion mechanisms relevant to metal crystal growth: Pt/Pt(111), *Surface Science*, Vol. 317, No.1-2, 15-36, ISSN: 0039-6028.

Vineyard, G. H. (1957). Frequency factors and isotope effects in solid state rate processes, *Journal of Physics and Chemistry of Solids*, Vol. 3, No.1-2, 121-127, ISSN: 0022-3697.

Wu, X.; Li, Q.; Huang, J. & Yang, J. (2005). Nonequilibrium electronic transport of 4,4'-bipyridine molecular junction, *J. Chem. Phys.*, Vol. 123, No.18, 184712-184716, ISSN:

Xu, B.; Xiao, X. & Tao, N. J. (2003a). Measurements of Single-Molecule Electromechanical Properties, *J. Am. Chem. Soc.*, Vol. 125, No.52, 16164-16165, ISSN: 0002-7863.

Xu, B. Q.; Li, X. L.; Xiao, X. Y.; Sakaguchi, H. & Tao, N. J. (2005). Electromechanical and Conductance Switching Properties of Single Oligothiophene Molecules, *Nano Lett.*, Vol. 5, No.7, 1491-1495, ISSN: 1530-6984.

Xu, B.; He, H.; Boussaad, S. & Tao, N. J. (2003b). Electrochemical properties of atomic-scale metal wires, *Electrochimica Acta*, Vol. 48, No.20-22, 3085-3091, ISSN: 0013-4686.

Xu, B. & Tao, N. J. (2003). Measurement of Single-Molecule Resistance by Repeated Formation of Molecular Junctions, *Science*, Vol. 301, No.5637, 1221-1223, ISSN: 0021-9606.

Yanson, A. I.; Bollinger, G. R.; van den Brom, H. E.; Agraït, N. & van Ruitenbeek, J. M. (1998). Formation and manipulation of a metallic wire of single gold atoms, *Nature*, Vol. 395, No.6704, 783-785, ISSN: 0028-0836.

Yasuda, S.; Yoshida, S.; Sasaki, J.; Okutsu, Y.; Nakamura, T.; Taninaka, A.; Takeuchi, O. & Shigekawa, H. (2006). Bond Fluctuation of S/Se Anchoring Observed in Single-Molecule Conductance Measurements using the Point Contact Method with Scanning Tunneling Microscopy, *J. Am. Chem. Soc.*, Vol. 128, No.24, 7746-7747, ISSN: 0002-7863.

Yoshida, M.; Kurui, Y.; Oshima, Y. & Takayanagi, K. (2007). In situ Observation of the Electrical Conductance of the Bridged Single Fullerene using TEM-STM, *Microscopy and Microanalysis*, Vol. 13, No.Supplement S02, 748-749, ISSN: 1431-9276.

Zhang, C.; Barnett, R. N. & Landman, U. (2008). Bonding, Conductance, and Magnetization of Oxygenated Au Nanowires, *Phys. Rev. Lett.*, Vol. 100, No.4, 046801-046804, ISSN: 0031-9007.

Zoloff Michoff, M. E.; Vélez, P. & Leiva, E. P. M. (2009). Substituent Effect on the Mechanical Properties of Au–N Nanojunctions, *J. Phys. Chem. C*, Vol. 113, No.9, 3850-3854, ISSN: 1932-7447.

Zou, S.; Schonherr, H. & Vancso, G. J. (2005). Force Spectroscopy of Quadruple H-Bonded Dimers by AFM: Dynamic Bond Rupture and Molecular Time–Temperature Superposition, *J. Am. Chem. Soc.*, Vol. 127, No.32, 11230-11231, ISSN: 0002-7863.

2

Modelling Metallic Nanowires Breakage for Statistical Studies: Ni Case as Example

Samuel Peláez[1], Carlo Guerrero[2,3], Ricardo Paredes[2],
Pedro A. Serena[1], and Pedro García-Mochales[4]
[1]ICMM, Consejo Superior de Investigaciones Científicas;
[2]Instituto Venezolano de Investigaciones Científicas;
[3]La Universidad del Zulia;
[4]Universidad Autónoma de Madrid
[1,4]Spain;
[2,3]Venezuela

1. Introduction

During the last two decades, the study of the properties of nanowires has been one of the keystones for the development of nanotechnology since these nanometric size objects exhibit electrical and mechanical properties of interest in fundamental knowledge as well as for the future development of technological applications. In particular, many experimental studies have afforded the creation, stabilization and deformation of metallic nanowires in order to describe their mechanical properties and their electronic transport quantum features appearing due to electron transversal confinement (Serena & Garcia, 1997; Agraït et al., 2003). The standard approximation for the experimental study of such metallic nanowires includes the formation, elongation and breakage of ultranarrow nanocontacts, as for instance, those formed between an STM tip and a metallic surface. Since the atomic arrangement formed during every indentation-breakage cycle of the STM tip is different, these experimental studies require the accumulation and analysis of a huge amount of data to characterize the electron transport properties of the resulting nanocontacts.

With the advent of powerful computational resources and the achievement of realistic descriptions of the atomic interactions, it has been possible to reproduce "in silicon" many of such formation-breaking experiments. An important part of these simulation studies has been done using Molecular Dynamics (MD) algorithms, allowing to elucidate how this formation-elongation-breakage occurs. Furthermore, MD simulations allow the accurate determination of the different structures that appear during the final stages of the breaking process. Getting insight of such structures is a crucial matter since they control the electron transport through the nanowire, allowing a comparison with the experimental data.

However, the comparison between experimental results and MD computational simulations requires an extra ingredient: the use of statistics. In general, the study of nanoscale systems demands a statistical approach. This becomes more evident in electron transport studies, since little modifications of the atomic positions results in large conductance variations. It is necessary to statistically address the study of many nanowire breaking events, mimicking the experimental indentation-retraction cycles. In these cases, semiclassical approximations based in accurate interatomic potential are usually used since they are much less demanding of computational resources than sophisticated but time consuming *ab initio* methods.

MD simulations have not only proved successful in accounting for complex properties of many nanowires during the breaking process, allowing to detect the presence of ultranarrow structures as monomer, dimers and long atomic chains (Bahn & Jacobsen, 2001) that govern the electrical properties of the nanowire just before its breaking. MD techniques have also used to demonstrate the formation of new non-crystalline structures, as helical or weird structures (Gülseren et al., 1998), not corresponding to the initial FCC or BCC crystallographic disposition of the metal under study. This is also the case of the icosahedral or pentagonal nanowires formed by subsequent staggered parallel pentagonal rings (with a relative rotation of $\pi/5$) connected with single atoms (Mehrez & Ciraci, 1997; García-Mochales et al., 2008a; Peláez et al., 2009), showing a characteristic -5-1-5-1 ordering that have been confirmed experimentally for some metallic species (González et al., 2004).

In this chapter we will describe how to simulate the breakage of nanowires using MD techniques with Embedded Atom Method potentials, although the MD approach can be used with any other semiclassical potential (Section 2). We will describe the methodology used to extract different physical magnitudes of interest from the simulation, in particular the minimum cross-section that allows monitor the evolution of the breaking nanowire (Section 3). Also we shall describe the computational tools and algorithms applied to identify and characterize the different atomic structures formed that are relevant to the mechanical and transport properties of these nanocontacts: (i) monomers and dimers (and their surrounding environments) just before the breakage occurs (Section 4); and (ii) pentagonal nanowires formed during last stages of the elongation process (Section 5). We will show the procedure to calculate the atomic stress and apply it to investigate the pentagonal nanowires formed during the stretching process (Section 6).

Finally we will show how the statistical results corresponding to different stretching directions can be merged in order to compare with experimental findings (Section 7). This merging procedure is needed since experimental data usually correspond to the statistical average of nanowire breaking events involving random stretching directions. Although the method and a brief discussion will be given at Section 7, results expected for random stretching orientations will be included in its corresponding figures throughout the chapter for an easy comparison with those obtained for well determined stretching directions.

All these simulation and characterization methods will be focussed on the study of the breaking mechanisms of nickel nanowires, as a study case. The study will account for three stretching (crystallographic) directions, two initial sizes and a large range of temperatures. The statistical MD approach will describe hundreds of breaking events. The dependence of the breakage behaviour with several parameters (that can be controlled in experiments) will permit to determinate the optimal conditions for the production of some specific structures (monomer and dimer point-contacts, icosahedral nanowires) for this metallic species.

2. Computational methods

Our approach is based on the statistical study of the structural evolution of many nanowires under stretching, using standard Molecular Dynamics (MD) simulations. Each breakage event, even starting from identical initial conditions, evolves showing its own features. Therefore, the only possibility to carry out an accurate study of the mechanical and electrical properties of the breakage process is performing a statistical analysis. We follow a similar strategy to the experimental one, simulating hundreds of independent breaking events, in order to determine the presence of preferred configurations. In particular, we have developed a methodology (Hasmy et al. 2001; Medina et al., 2003; Hasmy et al., 2005; García-Mochales et al., 2005) able to generate hundreds of computational breaking events and analyse them statistically to determine the different structures appearing at the breaking processes. We will describe this methodology along next sections.

In general, there are many computational simulations focused on the description of single nanowire breaking events, neglecting the study of statistical effects due to their high computational cost (Olesen et al., 1994; Bratkovsky et al., 1995; Barnett & Landman, 1997; Sørensen et al., 1998; Ikeda et al., 1999; Branício & Rino, 2000; Bahn and Jacobsen, 2001; Heemskerk et al., 2003; Sutrakar & Mahapatra, 2009), but several MD studies have statistically established that there exist a correlation between preferred atomic configurations and experimental conductance histograms at different temperatures (Hasmy et al. 2001; Medina et al., 2003; Hasmy et al., 2005; García-Mochales et al., 2005; Pauly et al., 2006; García-Mochales et al., 2008a; García-Mochales et al., 2008b).

Of course the price to pay for carrying out these statistical studies is the need to use "cheap" computational potentials for describing the atomic interactions. In spite of their accuracy, the large computational resources required by *ab initio* based methods make them inappropriate to determine computationally statistical features of these systems. A high computational effort is also required when other approaches are used, as tight-binding molecular dynamics simulations (González et al., 2004) or hybrid algorithms (Hasmy et al., 2005; Pauly et al., 2006). This means that semiclassical potentials, due to their relatively low resources requirements, are nowadays the most appropriate for statistical studies. On the other hand, "cheap" not mean "unaccurate", since we need to rely on obtained results of breaking nanowires. In consequence, the chosen potential must be complex enough to accurately reproduce low coordination situations as those appearing during the stretching process and structures with high surface/volume ratios.

We have simulated the nanowire atoms evolution using a MD scheme where atomic interaction is represented by Embedded Atom Method (EAM) potentials (Daw & Baskes, 1983; Foiles, 1985). On the EAM approach, the potential energy function for the system reads

$$E = \frac{1}{2}\sum_{ij}\varphi(r_{ij}) + \sum_{i} F(\overline{\rho}_i) \qquad (1)$$

where i and j run over the number of atoms. In the first term, $\varphi(r_{ij})$ corresponds to a pair potential depending only on the distance r_{ij} between every pair of "different" atoms i and j. The second term is the so-called embedding energy, which depends on the mean electronic density $\overline{\rho}_i$ at atom i's location. This electron density is approximated in EAM as the sum of the contributions due to the surrounding atoms, $\overline{\rho}_i = \sum_{i\neq j}\rho(r_{ij})$. The embedding energy is calculated by evaluating and summing the embedding function $F(\rho)$ at each atom's position. Depending on the material and the specific physical properties to be studied, different pair potential $\varphi(r_{ij})$, embedding energy $F(\rho)$, and density $\rho(r_{ij})$ functions can be defined.

The computational technique here described is a general one and can be used with any other semiclassical potential. However, in order to reach accurate results, it is important that the chosen one provides a good description of low dimensional situations (monomers, dimers, chains, open surfaces, edges, regions with vacancies, etc). We have used the EAM parameterization proposed by Mishin et al. (1999). This parameterization is a state-of-the-art potential able to describe high and low coordinated systems as those found during the last nanowire breaking stages. It has been constructed by fitting almost 30 different properties obtained from experimental measurements or accurate *ab initio* calculations: bulk properties as cohesive energy, lattice constants, elastic constants, phonon dispersion curves, vacancy formation and migration energies, surface energies, ... This EAM potential has been used to describe $Ni_{x=1,150}$ clusters (Grigoryan & Springborg, 2003) showing a good agreement with *ab initio* calculations and experimental data. These small clusters contain configurations with very different coordination numbers (ranging from 1 to 12), therefore including the typical coordination numbers we have found during the nanowire breakage processes.

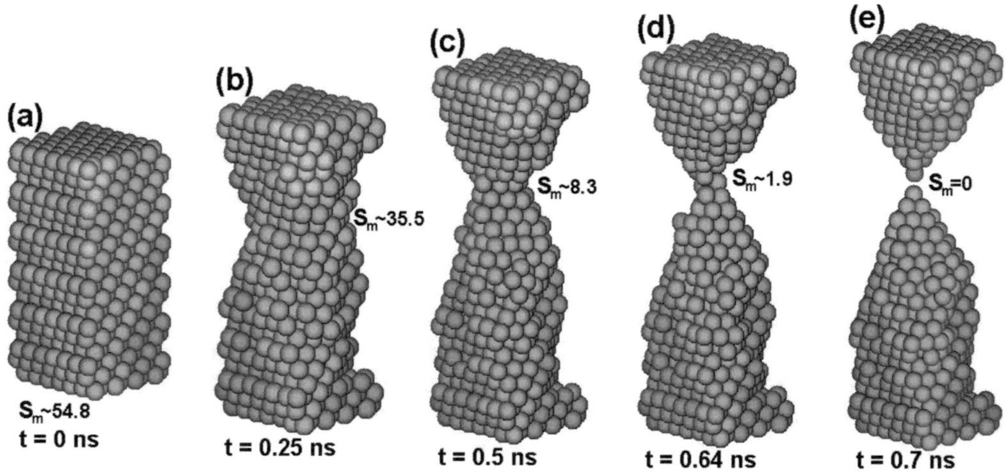

Fig. 1. Five different snapshots of a [111] Ni nanowire during stretching at T = 300 K from its initial parallelepiped shape. Each snapshot includes information on the simulation time and the minimum cross-section S_m at that time.

Nanowire dynamics has been studied at constant temperature T using a standard velocities scaling algorithm at every MD step (Rapaport, 1995). Nine temperatures ranging from 4 to 865 K (half of the Ni bulk melting temperature) have been considered in this work in order to describe the nanowire behaviour with the temperature. The time interval used for the integration of atomic trajectories (MD time step) is $\Delta t = 10^{-2}$ ps. Atomic trajectories and velocities were determined using conventional Verlet velocity integration algorithms. We have checked that results from this time step are equivalent to those obtained with shorter time intervals, ensuring the quality of the results with the smallest computational cost.
The simulation of a single nanowire breaking event consists of three stages. The first stage corresponds to the definition of the initial unrelaxed structure. We consider as initial nanowire a bulk super-cell with parallelepiped shape, containing hundreds of atoms ordered according to a FCC structure with bulk Ni lattice parameter (a = 3.52 Å). The initial parallelepiped height coincides with the stretching direction and is larger than the base edges. We define the z axis as the stretching (pulling) direction. In these studies we have considered three different stretching directions corresponding to the [100], [110] and [111] crystalline directions and two different parallelepiped sizes (large and small) with similar

aspect ratio for each direction. The initial sizes used were (layers × atoms per layer): for [100] case, 21×49=1029 atoms and 12×16=192 atoms; for the [110] case, 29×35=1015 atoms and 17×12=204 atoms; for the [111] case, 18×56=1008 atoms and 10×16=160 atoms. Fig. 1a shows the initial configuration corresponding to a large-type [111] nanowire. At the beginning of the simulation, each atom velocity is assigned at random according to the Maxwellian distribution corresponding to the simulation temperature (Rapaport, 1995).

The second stage corresponds to the relaxation of the bulk-like initial structure. Firstly, we define two supporting bilayers at the top and bottom of the supercell. Atomic x and y coordinates within these bilayers will be kept frozen during the simulation. The nanowire will remain attached to these two bulk-like supporting bilayers during the relaxation stage. This stage lasts for 3000 MD steps in order to optimize the geometry of the isolated parallelepiped-like nanowire. The role of these "frozen" bilayers is very important, since they represent the connection with the macroscopic world. If free or periodic boundary conditions were taken into consideration, the resulting nanowire evolution would present different patterns, including phase transformations upon loading as it has been noticed for narrow Ni nanowires (Liang & Zhou, 2006; 2007). During the third stage (stretching process), the z coordinate of those atoms forming the top (down) frozen bilayer is forced to increase (decrease) a quantity $\Delta z = 10^{-4}$ Å after every MD step. This incremental process simulates the separation of the supporting bilayers in opposite directions at constant velocity of 2 m/s, giving rise to the subsequent nanowire fracture. Those atoms located between the frozen regions move following the forces derived from their EAM-like interaction with the surrounding atoms. Notice that the stretching velocity is much larger than that used in experiments. However, our computational description of the nanowire breaking is comparable to that of actual experimental traces since the stretching velocity is smaller than the sound speed in nickel. We consider that the nanowire breaking process is completed when the minimum cross-section of the nanowire is zero (in the next section is explained the way this magnitude is calculated).

In Fig. 1 we depict five snapshots corresponding to different stages of a representative nanowire breaking process (a large nanowire stretched along the [111] orientation at 300K), starting with the initial (unrelaxed) configuration and finishing with the configuration just after the breakage. During stretching the nanowire shape develops through typical slip processes leading to inelastic deformations. Notice that during the last stages of the breaking process the nanowire structure evolves, forming a bipyramidal structure, each pyramid formed by three energetically favourable (111) facets.

3. Characterization of breaking nanowires: Cross-section histogram

During the stretching stage, the accurate knowledge of the atomic coordinates and velocities allows the full determination of the minimum cross-section S_m. This quantity provides relevant information on the favourable configurations appearing at the narrowest part of the nanowire during its evolution under stretching. Furthermore, S_m provides a first-order approximation of the conductance G (Sharvin, 1965). The minimum cross-section S_m is calculated in units of atoms following standard procedures that have been successfully used in previous studies (Bratkovsky et al., 1995; Sørensen et al., 1998). In our case, we define the atomic radius r_0 to be equal to half the FCC (111) interplanar distance ($r_0 = d_{111}/2$).

In order to calculate the cross-section S_i at a given z_i position, we firstly compute the total atomic volume $V_{\mathrm{tot},i}$ inside a "detecting cursor" width Δz. We have used $\Delta z = d_{111}$. We assign

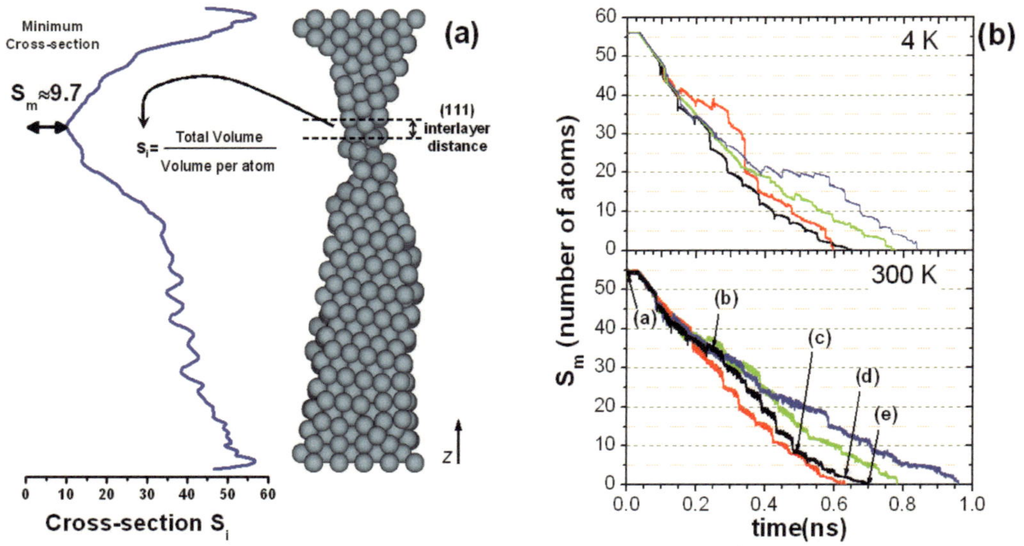

Fig. 2. (a) Example of how cross-section S_i is calculated along a [110] nanowire and its minimum value S_m (b) Evolution of the minimum cross-section S_m (measured in units of the number of atoms) as a function of time for four [111] Ni nanowires during their breaking process at two temperatures T = 4 K and T = 300 K. Labels (a)–(e) pointing at the black line correspond to the five different snapshots depicted in Fig 1.

a volume $V_0 = 4\pi r_0^3/3$ to each atom and, therefore, the quantity $S_i = V_{\text{tot},i}/V_0$ corresponds to the nanowire section (in number of atoms) at the z_i position. In Fig. 2a is depicted a cartoon of the method used to measured S_i. The detecting cursor moves along the z axis between the two frozen bilayers, using a step equal to $0.1\times d_{111}$. This allows calculate the cross-section S_i along the nanowire. Finally, from the set of collected S_i values, we determine the minimum cross-section value S_m. Note that the cursor size Δz is kept fixed independently on the nanowire crystalline direction along the z axis. This allows a true comparison between histograms obtained for different orientations, especially at the last breaking stages. In our study S_m is calculated every 10 MD steps. In Fig. 2a we illustrate how the S_i and S_m are calculated with an example obtained from a [110] Ni nanowire simulation at 4K.

The evolution of S_m versus time shows a typical staircase trace. In Fig. 2b some examples of $S_m(t)$ illustrate this behaviour. $S_m(t)$ traces show a stepped profile with well-marked jumps associated to atomic rearrangements that take place within the nanowire. We have verified that these jumps are correlated with jumps in the force acting on the supporting slabs. In general, S_m decreases monotonically between two subsequent jumps reflecting the existence of elastic stages. These elastic stages have been associated to the experimentally observed conductance plateaus. As it is expected, the quantity S_m presents larger fluctuations when temperature increases, as it is shown when comparing the traces obtained at 4 and 300 K. These behaviours have been observed for the three stretching orientations.

Following the same methodology used in experiments, we summarize the information of many breaking events in histograms. In particular, the minimum cross-section histograms $H(S_m)$ are built by accumulating $S_m(t)$ traces acquired during the simulation of hundreds of nanowire stretching processes. We have show that 100 independent nanowire breakages can be enough to built a cross-section histogram, but more samples could be necessary for performing the statistical analysis of other quantities (García-Mochales et al., 2005).

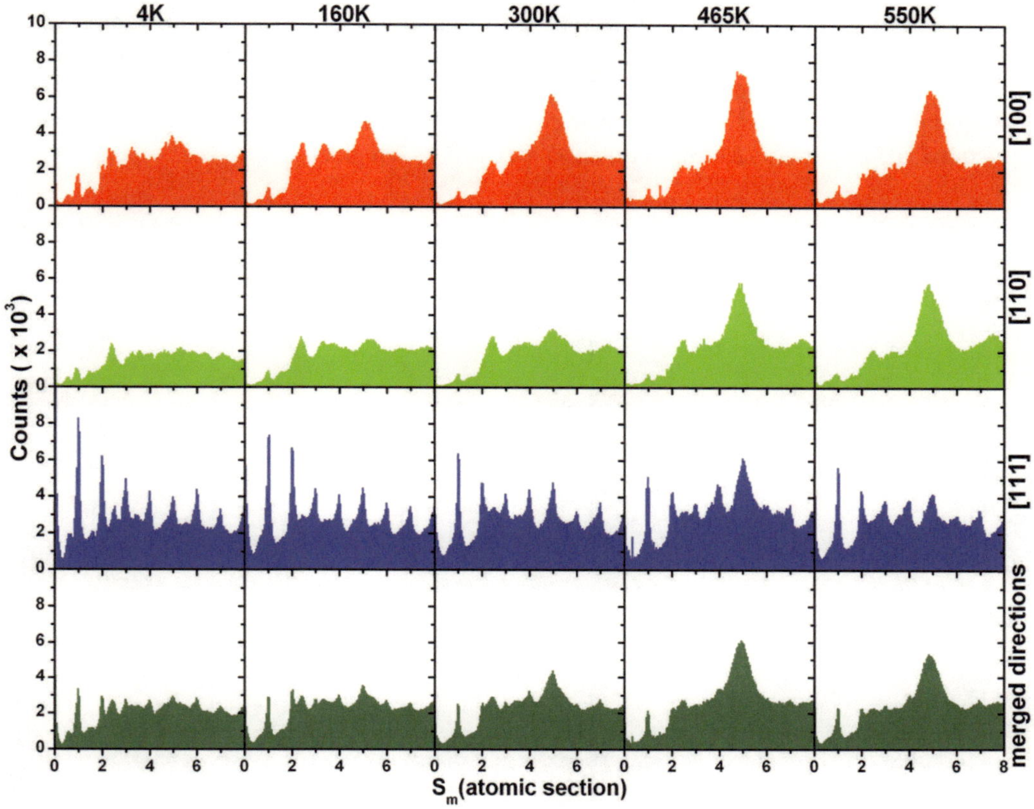

Fig. 3. Minimum cross-section histograms $H(S_m)$ of large Ni nanowires at T=4, 160, 300, 465 and 550K along the [100], [110] and [111] crystallographic directions. Last row show the expected results if nanowires would be stretched along random directions (see section 6).

In Figs. 3 and 4 (for large and small initial size nanowires, respectively) we show the histograms $H(S_m)$ for five temperatures (T=4, 160 300, 465 and 550 K), constructed with 300 independent breakages for the [100], [110] and [111] stretching directions. The different histograms are depicted in the range $0 < S_m < 8$. A first inspection of these figures reveals the existence of well-defined peaks associated to preferred nanowire configurations as it has been shown in previous works (Hasmy et al. 2001; Medina et al., 2003; Hasmy et al., 2005; García-Mochales et al., 2005).

Our main finding is that $H(S_m)$ histograms are very dependent on the stretching (i.e., the nanowire axis) direction. The $H(S_m)$ histograms associated to the [111] stretching direction present a well-defined peaked structure for both sizes at the whole range of temperatures depicted. The main difference we have found between both sizes corresponds to the small increase of the peak located at $S_m \sim 5$ above room temperature (RT) that only occurs for large size nanowires. For the [100] and [110] cases, very low temperature histograms present for both sizes a noisy structure. As temperature increases, histograms peaks show rounded shapes and the general structure presents less noise. The more prominent feature of [100] and [110] $H(S_m)$ histograms is the appearance and growth of a $S_m \sim 5$ peak as T increases (especially for small initial size nanowires). We will discuss later how this protruding peak, which is not developed at low temperatures and fade out at T > 650 K, is related with the existence of (non-crystalline) icosahedral nanowires and it became the dominant structure at temperatures above RT. Apart the $S_m \sim 5$ peak, there are differences in the low S_m region between the [111] histograms and [100] and [110] ones. The [111] direction provides histograms with well-defined decreasing peaked structure whereas for [100] and [110] directions $H(S_m)$ presents a clear 'depletion' in the $S_m < 2$ region.

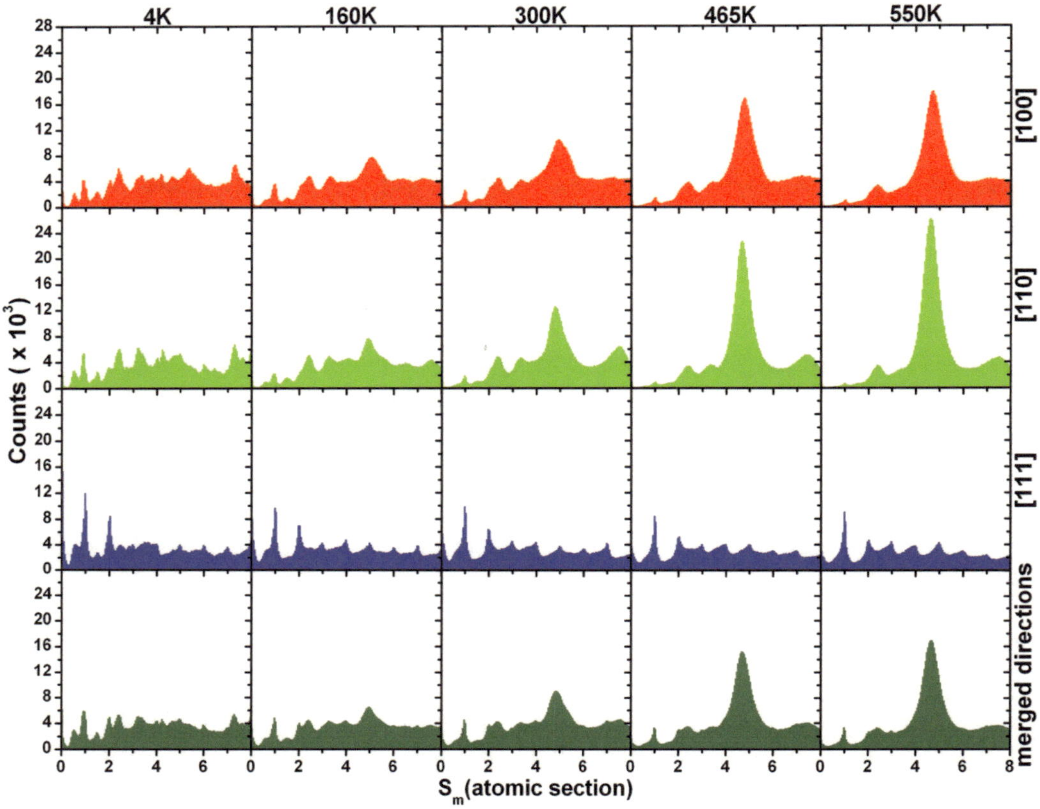

Fig. 4. Minimum cross-section histograms $H(S_m)$ of small Ni nanowires at T=4, 160, 300, 465 and 550K. Different rows correspond to nanowires stretched along [100], [110], [111] and random crystallographic directions.

The thermal evolution of histograms can be seen on the three nanowire orientations, but they are more remarkable for [100] and [110] stretching directions. Small peaks at T=4K correspond to metastable configurations with slightly higher cohesive energies with respect to other metastable configurations. The increase of temperature allows the exploration of more configurations during the stretching process, and, in this way, those metastable configurations with local minimum energy are easily accessible, leading to a better definition of their associated $H(S_m)$ peaks. For the three stretching directions, higher temperatures than those showed in Figs. 3 and 4 present a progressive vanishing of the peak structure. It is clear that at high temperatures the nanowire structures formation from stretching competes against nanowire melting processes (taking place at smaller temperatures than the bulk melting temperature). Nanowire atoms in a pre-melting state made progressively more difficult the formation of any stable structure.

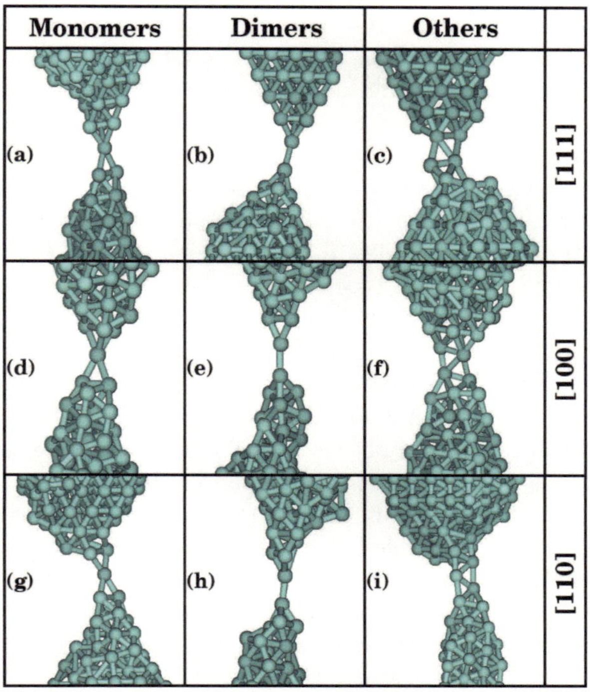

Fig. 5. Examples illustrating the type of nanowire configurations just before breakage at 4K. This classification scheme has been used for all the studied temperatures, since nanowire always break following these evolution patterns.

4. Characterisation of the nanowire before the breakage

4.1 Identification of monomers and dimers

Lowest conductance peaks observed in conductance histograms are associated with those configurations involving few atoms and appearing during the last breakage stages. We have found that final stages can be classified according to three different categories: monomers, dimers and "others" (as illustrated in Fig. 5). The monomer structure is characterized by a central atom standing between two "pyramids". In the dimer structure, the apex atoms of two opposite pyramidal configurations form a two atoms chain. Final configurations that do not match these two categories, presenting more complex structures, have been labelled as "others". As expected for Ni (Bahn & Jacobsen, 2001), we have not found any atomic chain formed by three or more atoms for the three studied stretching directions.

Complex structures (those labelled as "others") generally show an abrupt jump from $S_m \geq 2$ (i.e., structures formed by two or more atoms) to S_m=0. This feature could be used as a signature of such type of breaking pattern. On the other hand, monomers and dimers are characterized by similar S_m values (S_m~1), being difficult to distinguish among them if we only take into account the minimum cross-section value. We require an additional method to discriminate between monomer and dimer structures.

The procedure used is based on the "burning" algorithm introduced by Herrmann & Stanley (1984a; 1984b) to study the internal structure of percolating cluster at the percolation threshold. A simple generalization of this algorithm to disordered lattices was made to solve the problem that we are concerned. If an atom belonging to a monomer or dimer structure is suppressed from the nanocontact, no connection will be between both sides of the wire. For this reason these objects (also called 'red' bonds) are the links in percolation theory. In a nanowire configuration, two atoms are identified as neighbours (i.e., they are linked) if its separation is smaller than a predefined distance d_{Cond}. This distance was chosen between the first and second nearest FCC neighbours distance.

The steps of the actual algorithm are: (i) Two atoms from both frozen sides of the wire are chosen. This atoms are called *a* and *b*. (ii) From atom *a* the structure is "burned" for the first time. This means that atom *a* is labelled *1*. All the neighbours of this atom are labelled *2*. Next, all the neighbours of atoms of type 2 are labelled *3* except atoms that were already labelled. This procedure is followed until the atom *b* is labelled as *n*. (iii) From atom *b* the structure is "burned" again. Atom *b* with label *n* and all its neighbours whose label is strictly lower than the label of atom *b* are separated in a new object. The neighbours with label *n*-2 from these atoms with labels *n*-1 are added to the new object, and so on. The procedure finish when atom *a* is added to the object. This object, called the elastic backbone, is the union of all the minimal paths between atoms *a* and *b*. It contains all atoms belonged to monomers and dimmers. (iv) To finally obtain monomers and dimmers, each atom for the elastic backbone is temporally separated from it and a burning procedure is running on this structure. This atom belongs to a monomer or a dimmer if no connection is observed. (v) In order to understand the structure of monomers and dimers, we also identify the neighbour atoms of those atoms forming the monomer or the dimer. To distinguish if they are connected to *a* or *b*, burning procedures are used.

We consider that two atoms are connected when the distance between their centres is lower than the distance d_{Cond}=3Å. Results using distances closer to the Ni nearest neighbours distance do not show relevant differences. Using this d_{Cond}, we have determined for each stretching nanowire the type of breaking pattern and analysed the statistical weights of monomers, dimers and complex structures appearing around the S_m~1 histogram peak for different simulation parameters (temperature, stretching direction, and initial size). In Fig. 6 the fractions of monomers, dimers and complex structures found in the range $0.25 < S_m < 1.75$ are depicted as function of the temperature for the two initial sizes of Ni nanowires.

These figures denote that the temperature have little influence on the production ratio of final structures. The main change is a slight decrease of other structures fraction with the temperature, especially for large size nanowires. Therefore, the decrease of $S_m \approx 1$ and $S_m \approx 0.5$ peaks with the temperature must be attributed to a decrease of the mean lifetime due to instabilities of these structures at high temperatures. This lifetime shortening is more remarkable for the dimer structure.

For the [111] stretching direction the joined proportion of monomer and dimers (~80%) is larger than the fraction of other structures (~20%). This fraction seems to be rather independent of the temperature as well as the nanowire size, except for small ones near the nanowire melting temperature, where monomers fraction increases and dimers fraction decreases. For the [100] and [110] cases, there is a large fraction of complex structures (30-40%). The temperature behaviour of this fraction shows a dependence on the nanowire size: it keeps constant for small size nanowires but decreases for the large size ones. The fraction of monomers is of the same order but slightly lower than that found for the [111] case, meanwhile the dimers fraction takes values below of the [111] case. On the contrary of the [111] case, dimers fraction increases with temperature for the large size nanowires set, especially if the stretching occurs along the [110] direction.

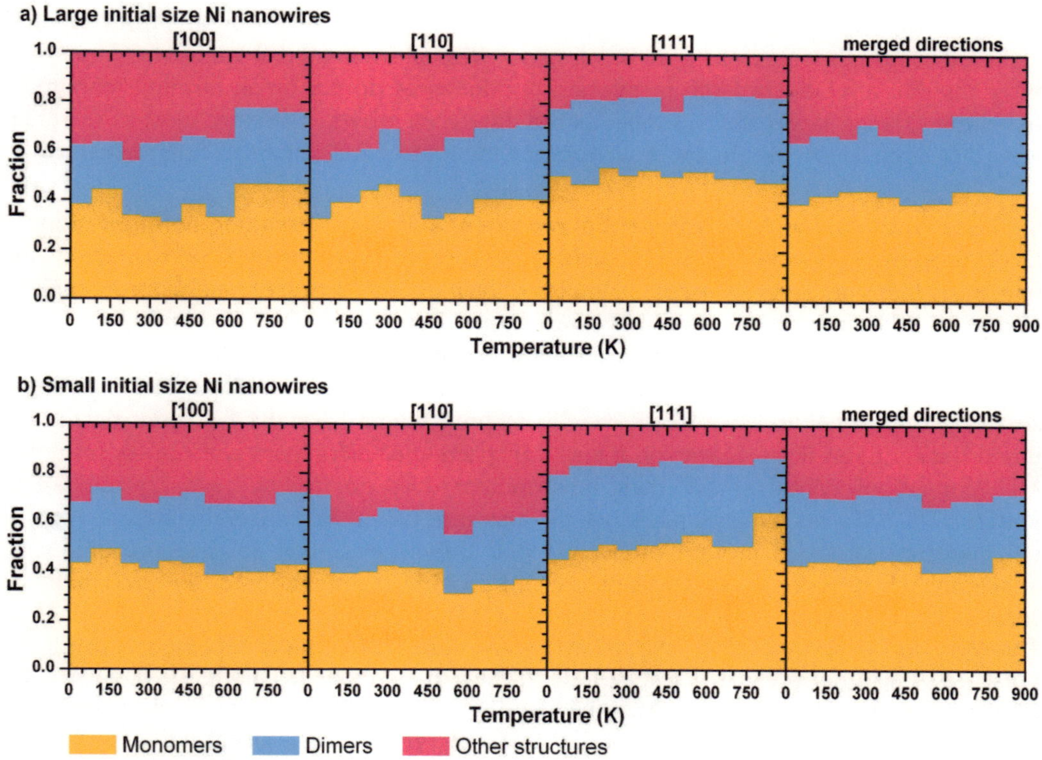

Fig. 6. Fraction of monomers, dimers and more complex structures appearing during the Ni nanowire breaking process as functions of the temperature for Ni nanowires of large (a) and small (b) initial size. Different columns correspond to nanowires stretched along [100], [110], [111] and random (merged) crystallographic directions.

Also the final structures of the nanowire at the breaking moment can be related with its whole evolution and the full histogram $H(S_m)$ building partial histogram that only include traces that present monomer and dimer configurations (García-Mochales et al., 2008). For the [111] stretching direction, that monomer and dimer based partial histogram is similar to the full histogram, indicating that the nanowire generally evolves from its initial structure to a monomer and/or dimer configuration before the breakage. This breaking scheme seems to be independent on the initial nanowire size and temperature. However, the behaviour of partial histogram for the [100] and [110] stretching directions depend on the initial nanowire size and temperature. For small nanowires with [100] and [110] orientation and at low temperatures the breaking pattern obtained for the [111] case is recovered: most of the nanowires present monomer or dimer configurations during the last stages of the elongation process and full and partial histograms are quite similar. As the temperature increases (and the S_m~5 peak appears) the histograms start to differ although the general shape is conserved. Nanowires forming monomers and dimers are responsible of the $0.25 < S_m < 1.75$ $H(S_m)$ region, but for larger sections they show a significantly lower contribution. For the largest nanowire case for [100] and [110] orientations, independently of the temperature, nanowires with monomer and dimer structures at the final breaking stage are solely responsible of those smaller peaks at the low S_m region, but do not recover the full peaked structure appearing in $H(S_m)$. These divergences of the partial histogram from the full histogram (at high temperature for small nanowires, at any temperature for the large one) indicates that a relatively large proportion of nanowires do not break forming monomers nor dimers, being larger their number as the difference grows. Therefore most of the [100] and [110] large nanowires do not break forming monomers and dimers as final structure.

4.2 Local and non-local environment of monomer and dimer breaking configurations

Electron transport through a given breaking nanowire mainly depends on the size of the narrowest nanocontact cross-section. However, the atomic structure of neighbouring regions also plays a key role in the electronic conductance (Pauly et al., 2006; Hasmy et al., 2005). The study of the electronic transport through monomer-like or dimer-like Ni nanocontact has centred on static structures with surrounding configurations obtained from crystalline structures (Sirvent et al., 1996; Jacob et al., 2005), proposing 3-1-3 and 4-1-4 as the most likely monomer configurations. However, a visual inspection of monomers and dimers from our MD simulations confirms that their environment does not follow such crystalline-like pattern. The MD calculations allow us the accurate determination of the actual type of neighbourhood formed around monomers and dimers and their relative probability of appearance. In addition, this information could provide an estimation of the probability of conductance associated to the lower peaks of the S_m histograms.

The electron transport through the nanowire also depends on more factors, including the presence of disorder (defects, vacancies, impurities, dislocations ...) along the nanowire. Its effect on the electrical conductance can be very important at the latest stages of the breaking process: if the scattering due to the disorder becomes very large, the electron transport could leave the ballistic regime and quantum diffusive transport features should appear.

Once the atom that forms the monomer is identify, it is trivial to determine its neighbour atoms (those closer than the distance d_{Cond}) and classify them with respect to the monomer z coordinate. This gives us a configuration of type "n-1-m", i.e., a configuration where there are n atoms at one side of the monomer and m atoms at the other side. For dimer structures the procedure is similar, providing configurations of type "n-1-1-m".

In Fig. 7 we present the occurrence probability of different atomic configurations around monomers and dimers, showing that the atomic structures (and its probability of occurrence) around monomers and dimers are roughly the same regardless of the size of the initial nanowire, the stretching direction or the temperature (although only T=4 and 300K are showed). We notice that the most likely monomer structure presents the configuration 2-1-3. Around 40% of the monomers exhibit this configuration. Besides, we found that 2-1-4, 3-1-3, 3-1-4 and 2-1-2 configurations appear with a probability ranging between ~20% and ~10%. Some small deviations from this rule can be found for the [110] and small size nanowires: at low temperature 2-1-2 configuration presents a probability larger than 30% but 3-1-3 structure have less than a 5%, whereas at RT probability of the 3-1-3 configuration is higher than the 20% (growing at the expense of the 2-1-3 structure). Comparing the actual monomer surrounding structures with those proposed on previous works, the non-trivial 2-1-3 structure (the most probable, as we found) was not even considered. Moreover, we have not found evidences of the 4-1-4 crystalline-like configuration in our simulations.

In Figs. 7 the local environment for the dimer configurations show that the most common structures are 3-1-1-4, 3-1-1-3 with occurrence probability between ~50% and ~20%. The most likely configuration depends slightly on the temperature, size and stretching direction of the nanowire. The 4-1-1-4 configuration has also a relative high probability (~10%-20%),

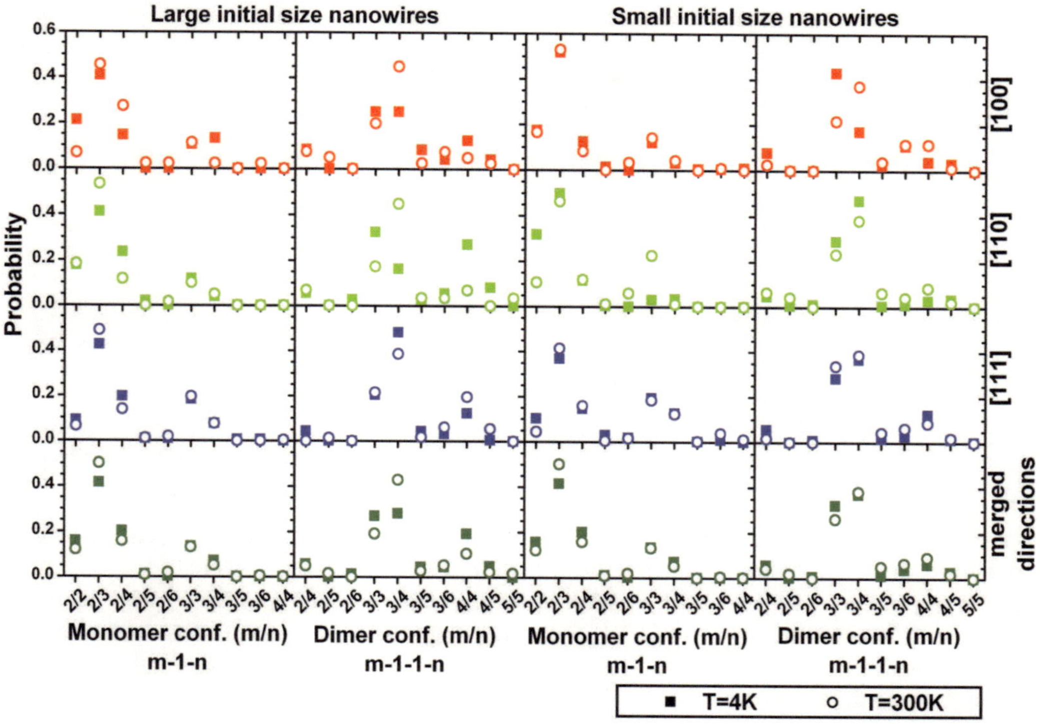

Fig. 7. Probability of the occurrence of different atomic environments around a Ni monomer and a Ni dimer for different stretching directions and temperature T=4 and 300K. A "m/n" configuration means that the monomer (dimer) configuration is of the type "*m*-1-*n*" or "*n*-1-*m*" (type "*m*-1-1-*n*" or "*n*-1-1-*m*").

being for the [110] stretching direction, low temperature and large initial size nanowire more probable (~30%) than the 3-1-1-4 configuration (that decreases below 20%). Other configurations show probabilities lower than 5%, reflecting that each dimer atom has to be connected with a stable base formed by 3 or 4 atoms.

As previously mentioned, the electron transport through the nanowire is also dependent on the existence of disorder along the nanowire. To determine the existence of a long-range structure (order) on the monomer configuration at T=4K, in Fig. 8 we plot the average number of atoms $\langle N(z)\rangle$ located at a distance z from its position. This approximation will detect if the atomic structure keeps its crystalline character as the z distance from the monomer (or dimer) increases. We have calculated $\langle N(z)\rangle$ for the three stretching directions and two initial sizes. There is not significant difference between the two sizes, both show similar $\langle N(z)\rangle$ values up to 8 Å and the decrease observed from this distance for smaller nanowires is expected due to their size. A similar behaviour is observed around dimers (not shown).

On one hand, for the [111] case, we note clear peaks separated a distance close to the typical separation of the [111] planes d_{111}=2.03Å, indicating that the system try to keep the same crystallographic structure from both fixed slabs towards the monomer position. On the other hand, for the [100] and [110] stretching directions, the situation is quite different, since for increasing z we could not distinguish a well defined peaked structure. For these orientations, the fixed slabs try to maintain the [100] or [110] structure but nanowires present dislocations, disorder and the formation of large non-crystalline areas leading to

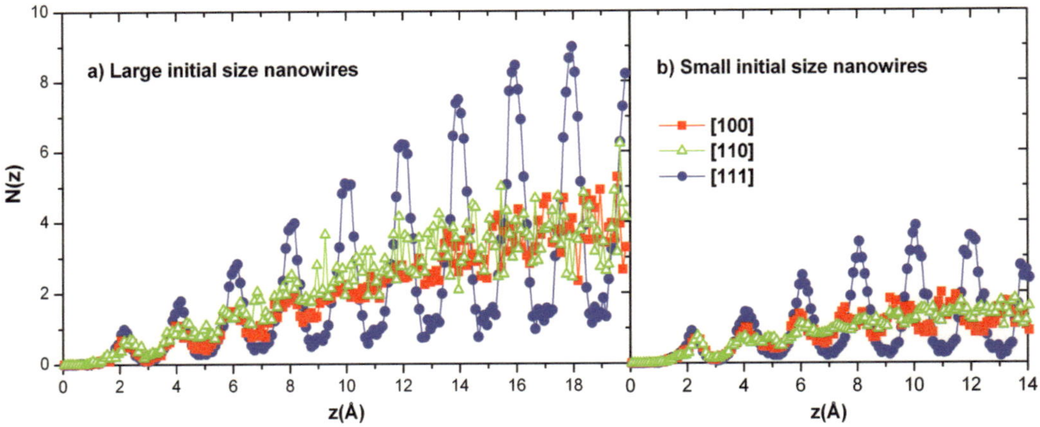

Fig. 8. Average number of atoms $\langle N(z) \rangle$ at a distance z from the monomer position for the [100], [110] and [111 directions at T=4K: large (a) and small (b) initial size nanowires.

amorphization of the nanowire. Similar behaviour has been previously noticed for infinite nanowires under high strain rates (Branicio & Rino, 1999; Ikeda et al., 1999).
As pointed previously, the amorphization of the structure can be important in the electronic transport during latest stages of the nanowire breakage: the presence of disorder on a large nanowire region will lead to a conductance decrease. But in this case is especially important because we could observe two transport behaviours depending on the stretching direction. In the [111] case the electron moves inside a well ordered system (keeping the ballistic transport features), meanwhile for the [100] and [110] orientations the electron suffers many scattering events due to the disorder (leading to a diffusive transport regime).

5. Icosahedral or Pentagonal nanowires: Identification and characterisation

In previous subsections we showed that $H(S_m)$ presents a huge peak around S_m~5 for the [100] and [110] cases and that the peak increases with the temperature. We have analyzed the types of configurations which contribute to this peak. We have depicted in Fig. 9 different time snapshots of one of those structures, for a [110] nanowire at T=550K, showing its formation, growth and breakage. It presents along its length pentagonal rings (responsible for the S_m~5 cross-section value) that name these structures: pentagonal nanowires. We found that these structures (for the Ni case) are very common for the [100] and [110] stretching directions, whereas they rarely occur for the [111] case.
Icosahedral or pentagonal nanowires are formed by subsequent staggered parallel pentagonal rings (with a relative rotation of $\pi/5$) connected with single atoms, showing a characteristic -5-1-5-1- ordering (see as example Fig. 9). The atomic sequence -5-1-5-1- presents a fivefold symmetry with respect to the nanowire axis. This symmetry does not correspond to any crystallographic FCC nor BCC structures. The -5-1-5-1- staggered nanowire configuration may be understood in terms of a sequence of interpenetrated icosahedra. This icosahedral symmetry is very common in very small systems due to the large stability and high coordination characterizing such geometry (Bulienkov & Tytik, 2001). Contrary to monoatomic chains, pentagonal nanowires are rather robust structures at relatively high temperatures and, therefore, they may consider as a promising candidate for being used as nanodevice components.

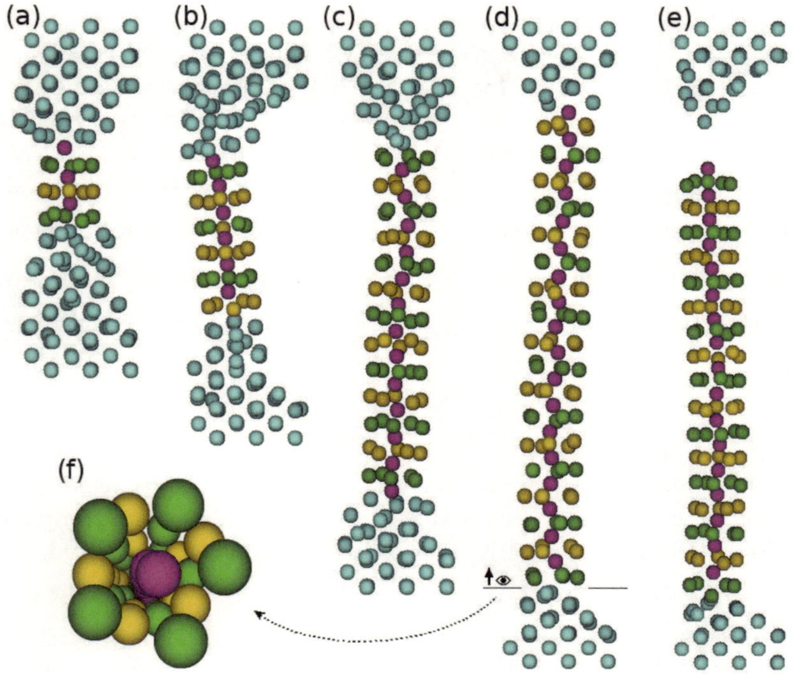

Fig. 9. (a-e) Formation, growth and breakage of an icosahedral nanowire from a small [110] Ni nanowire at T=550K; (f) shows the cross-sectional perspective view as seen from the position indicate on (d). Colours are used to identified the -5-1-5-1- structure.

Different computational works during the last decade have shown the formation of these structures on different metallic species. The formation of staggered pentagonal configurations during the stretching process has been already reported for Na (Barnet & Landman, 1997) using first principles methods, and for Cu nanowires with different Molecular Dynamic (MD) approaches (Mehrez & Ciraci, 1997; Gülseren et al., 1998; González et al., 2004; Sutrakar & Mahapatra, 2009). In particular the high stability of the Cu nanowire was confirmed with ab-initio calculations. Pentagonal motives also appear in infinite Al and Pb nanowires obtained from MD simulated annealing methods (Sen et al., 2002). More recently we have reported such structures for stretched Ni nanowires with different crystallographic orientations (Garcia-Mochales et al, 2008a; 2008b; 2008c), being very stable and presenting a high plastic deformation under strain.
The formation of pentagonal nanowires can be seen in the curve S_m vs. time as a plateau around S_m~5. Its large statistical occurrence is reflected in histograms of minimum cross-section $H(S_m)$ as a huge peak centred at S_m~5. These quantities (the length of those cross-section plateaus, the relative height or area of that peak) were proposed as parameters to classify the conditions and probability of formation of the icosahedral nanowires. But these magnitudes provide little information about the actual lengths and the length distribution of such pentagonal chains and cannot determine either the number of pentagonal rings that form the tubular structure. To overcome these limitations, we developed an algorithm that identifies the pentagonal structures that form the icosahedral nanowire (Pelaez et al., 2009). This algorithm also allows us to define its length L_p as the distance between the outermost pentagonal rings and to count the number of pentagonal rings n_p that form it.

The algorithm is based in the determination of the angular distribution of the nearest neighbour atoms and provides a parameter ($\alpha(z)$) which compares the angular distribution of the projected nanowire atomic coordinates with that corresponding to a perfect pentagonal nanowire. For a given z coordinate we consider a slice perpendicular to the z (stretching) direction with a thickness of 2 Å and centred on the z value. The N_t atoms inside such slice are projected onto the xy plane, each one getting new 2D coordinate $\vec{\eta}_i$; then the centroid of this structure is calculated $\vec{\eta}_0 = \sum_i \vec{\eta}_i / N_t$. The angular distribution is calculated from the angles $\theta_{i,j}$ between the pairs of vectors $\vec{\eta}_i'$ and $\vec{\eta}_j'$ defining the projected atomic coordinates with respect to the centroid ($\vec{\eta}_i' = \vec{\eta}_i - \vec{\eta}_0$). The parameter α is calculated as

$$\alpha = \frac{2}{N_a} \sum_{i,j}^{N_a} \frac{|\theta_{i,j} - m\,\theta_0|}{\theta_0} \tag{2}$$

where N_a is the number of pair of atoms considered, $\theta_0=\pi/5$ is the reference angle of a perfect staggered pentagonal structure and m is the integer that minimizes the expression $|\theta_{i,j} - m\,\theta_0|$. To avoid spurious contributions from atoms near the centroid, only vectors satisfying $|\vec{\eta}_j'| > a/4$ are considered, i.e., centre atoms are excluded from the calculation of α. This algorithm is applied along the z-coordinate of the nanowire, displacing the imaginary slab δz=0.1 Å at a time. This results in a $\alpha(z)$ profile of the nanowire, showing where pentagonal structures appear.

In order to minimize artifacts, the $\alpha(z)$ curve is softened over a Δz interval. This softened curve $\langle\alpha\rangle(z)$ is defined as

$$\langle\alpha\rangle(z) = \frac{1}{\Delta z} \int_{z-\Delta z/2}^{z+\Delta z/2} \alpha(z')\,dz' \tag{3}$$

where a value of Δz=1Å has been found appropriate. This average of α ($\langle\alpha\rangle$) over a 1Å interval provides a quantity that distinguishes between pentagonal and non-pentagonal structures through the nanowire. We have observed that if the parameter $\langle\alpha\rangle(z) < 0.5$, the set of atoms around z forms a structure similar to that of a pentagonal ring. On the contrary, if $\langle\alpha\rangle(z) > 0.5$ the set of atoms presents another structure (bulk like -FCC-, helical or disordered).

The ability of the algorithm to discriminate between different structures was checked using different test structures (square, pentagonal, hexagonal and heptagonal nanowires) with increasing amount of disorder (Peláez et al., 2009).

In order to further check the ability of the algorithm to identify pentagonal regions, we tested its performance over simulated nanowires showing icosahedral structure. Fig. 9 shows several snapshots of a nanowire breaking process where an icosahedral structure is observed. For these snapshots, Fig. 10 shows the $\langle\alpha\rangle(z)$ profile curves, as well as the radial and angular distribution functions $g(r)$ and $g(\theta)$ through different sections of the nanowire at snapshot (b). As illustrated in the figure, the algorithm returns value near to 1 when is applied to the ordered regions of the nanowire, and values below 1 for the pentagonal region. Minima of $\langle\alpha\rangle$ correspond to the position of the pentagonal rings; as they are not perfect ordered structures (though still can be recognized as pentagons) their $\langle\alpha\rangle$ values are a bit greater than zero. We have chosen the value of $\langle\alpha\rangle$=0.5 as the limit value to recognize a pentagonal structure. This value $\langle\alpha\rangle$=0.5 discriminates between pentagonal and non-pentagonal structures. We define the pentagonal nanotube length $L_p(t)$, observed during stretching at a given time t, as the distance between the maximum and minimum z coordinates with $\langle\alpha\rangle$=0.5. We also define n_p as the number of pentagonal rings forming the icosahedral nanowire at its late stage (equivalent to the number of $\langle\alpha\rangle$ minima below 0.5).

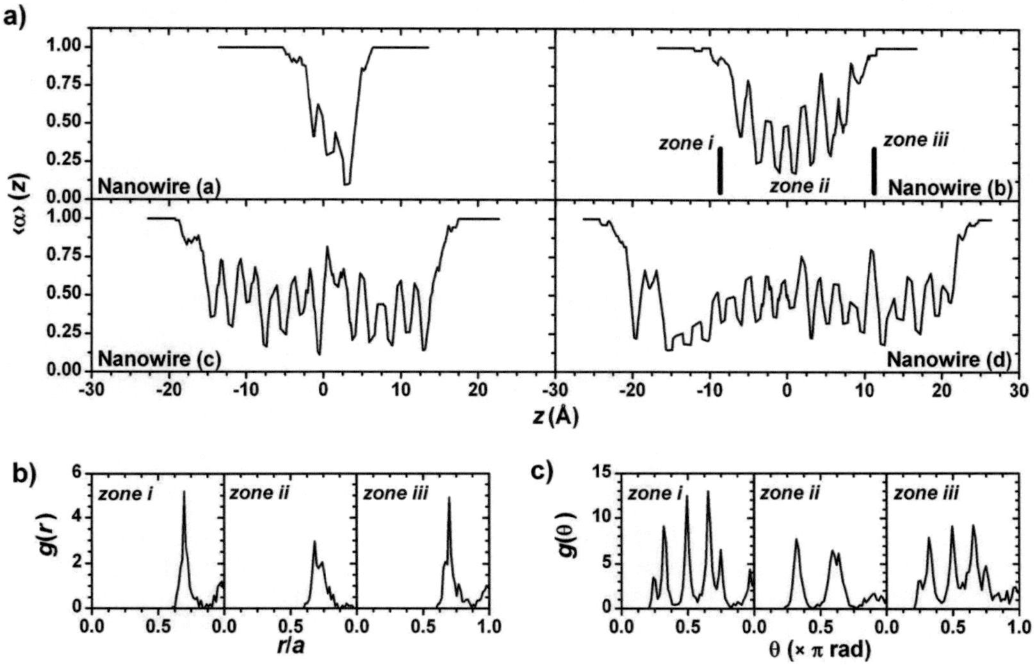

Fig. 10. a) Average α-parameter $\langle\alpha\rangle(z)$ along the Ni [110] simulated nanowire showed in Fig. 9 before breakage. Minimum values of $\langle\alpha\rangle$ below 0.5 correspond to the position of pentagonal rings forming the icosahedral nanowire. b-c) Distance $g(r)$ and radial $g(\theta)$ distributions for the three marked region of the nanowire labelled as (b).

In Fig. 10 we have also divided the nanowire in three regions that seem to have different structures from visual inspection. For each one of these three regions we show the radial and angular distribution functions $g(r)$ and $g(\theta)$. We can see that the outermost regions exhibit a mostly FCC structure, with the FCC typical interatomic distances and angles. The central region corresponds to the pentagonal structure according to both visual inspection and the $\langle\alpha\rangle$ profile. Inspecting the angular distribution function in this region we can see how the central peak of the FCC structure has disappeared and the peak at the right has displaced a bit towards the left. This is the typical structure of the pentagonal angular distribution function. Regarding the interatomic distances, not a big change is appreciated in this region from the radial distribution function. As it can be seen, the parameter $\langle\alpha\rangle$ proves to be a very efficient parameter to identify the pentagonal regions in a breaking nanowire.

We use this algorithm as the standard tool to characterize the pentagonal nanowires produced during the stretching simulations and extract information about probabilities of formation of such structures, distributions of their maximum lengths L_p^m (the length L_p of icosahedral nanowire just before its breakage) and the number of pentagonal rings n_p on the nanotube just before breaking. Since our simulation were carry out over a broad range of

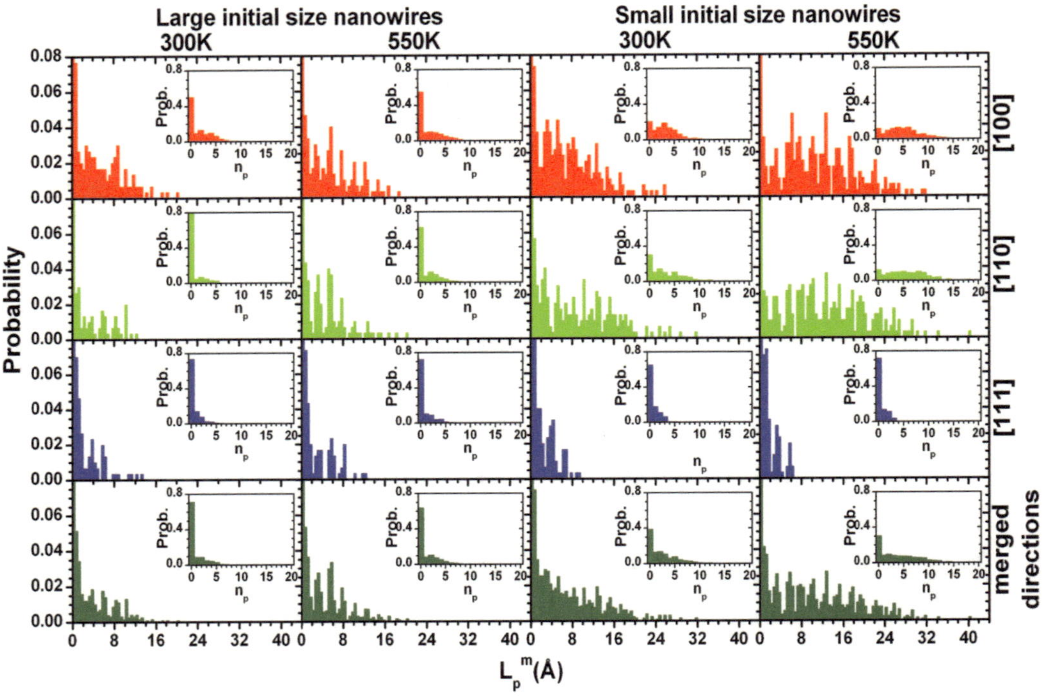

Fig. 11. Distribution function $D(L_p^m)$ for the maximum pentagonal nanowire length L_p^m and (as insets) distribution function $D(n_p)$ for the maximum number of pentagonal rings n_p for the two initial size nanowires at T=300 and 550K for [100], [110], [111] and random stretching directions.

temperatures, this studies will give the optimal temperature T_{Opt} required for maximizing the occurrence probability and/or length distribution of pentagonal nanowires.

In Fig. 11, for the three orientations and two sizes of nanowire considered, we have plotted the probability distribution $D(L_p^m)$ of obtaining a pentagonal nanowire of maximum length L_p^m. Also it is depicted, as insets, the probability distributions $D(n_p)$ of the maximum number of pentagonal rings. We only show results for two temperatures (T=300 and 550K). Distributions of L_p^m present a clearly peaked shape. It indicates that pentagonal chains have well defined favourable lengths. The peaks are separated approximately by integer values of the distance $d_{5\text{-}5}$=2.22 Å, the calculated equilibrium separation between successive staggered pentagonal rings (García-Mochales et al, 2008b; 2008c).

The distribution value for L_p^m=0 corresponds to those cases where the nanowire does not form any pentagonal structure (n_p=0). Those cases with $0 < L_p^m < 2$ Å correspond to nanowires showing a unique pentagonal ring (n_p=1). We have considered that an icosahedral wire is formed only if the chain satisfies $n_p \geq 2$. The shortest pentagonal chain (n_p=2) presents a unique icosahedron in its structure. The L_p^m distributions of T=300 and 550K shown on Fig. 11 exhibit large tails corresponding to those pentagonal nanowires including a large number of pentagonal rings. Distributions from low temperatures (not presented here) show also a peaked structure, but they have a shorter tail and larger probability on the $L_p^m < 2$ Å region (Peláez et al., 2009). The same behaviour is observed for very high temperatures (as it approximates to the nanowire melting temperature).

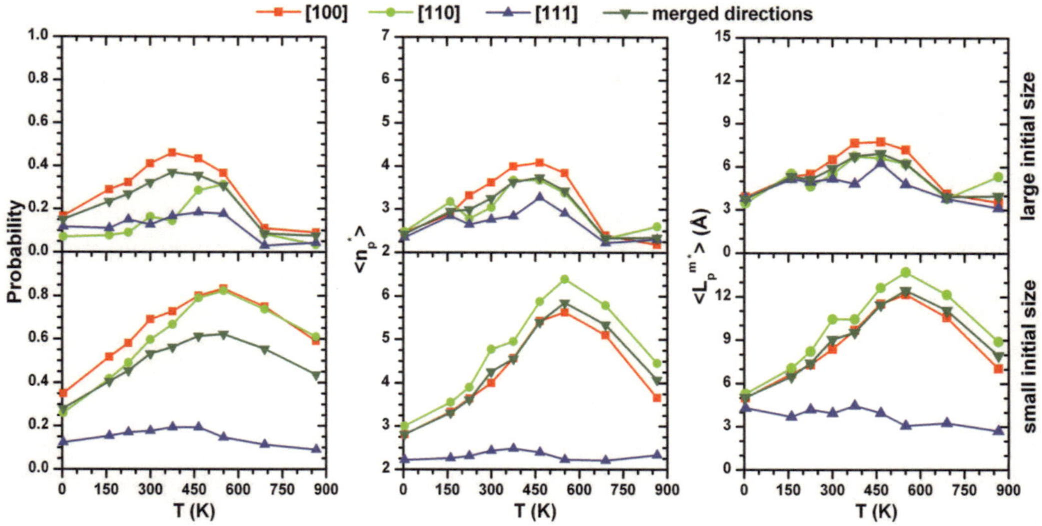

Fig. 12. Temperature dependence for the two initial sizes of the probability p_i of finding pentagonal chains (with number of pentagonal rings $n_p \geq 2$) and mean number of pentagonal rings $\langle n_p^* \rangle$ and mean length $\langle L_p^{m*} \rangle$ of such icosahedra chains detected.

$D(n_p)$ distributions are shown in Fig.11 for the same simulations sets. Usually $D(n_p)$ presents two maxima. The first one at $n_p=0$ identifies the breakages that do not produce icosahedral chains. It is the most probable case at all temperatures. The second maximum in a zone with $n_p > 2$ indicates that, once a seed is produced to nucleate a pentagonal wire, it will grow with high probability. This two-peak feature of the distribution occurs at any temperature except at very low or high temperatures. The position of the second maximum depends on breakage simulation set parameters (temperature, size, orientation), but it is correlated with the probability of formation of an icosahedral nanowire. And, as it is expected, the length of the $D(n_p)$ tail goes along with the $D(L_p^m)$ tail. Ideally every structure with $n_p \geq 2$ would give a sharp peak in the $D(L_p^m)$ distribution. Since real structures show tilted pentagonal rings, there is not perfect alignment with z-axis and the pentagonal rings separation is longer than that of the equilibrium, producing broad peaks in the $D(L_p^m)$ figure.

$D(n_p)$ determines the pentagonal wires production efficiency using the percentage p_i of breakages that produce an icosahedral chain, i.e, $p_i = \sum_{p\geq 2} D(n_p)$, elucidating which conditions favour the appearance of icosahedral nanowires. This is shown in left panels of Fig. 12 as function of the temperature for the three stretching directions and the two sizes. For same temperature and stretching orientation, it is easier to form a pentagonal tube if we start with a small size nanowire. The probability of getting icosahedral chains is much lower when breaking along the [111] direction than from the [100] or [110] directions. For the [111] breaking orientation this probability is below the 20% of the simulations, while for the other orientations the probability can be even over 80% for the small initial size nanowires.

In central and right panels of Fig. 12 are represented respectively the average number of pentagonal rings $\langle n_p^* \rangle$ and the average icosahedral tube length $\langle L_p^{m*} \rangle$ of those simulations that present a pentagonal nanowire during their stretching (i.e., the average is performed only over simulations with $n_p \geq 2$). Again the [111] orientation produces pentagonal nanowires of average length lower that those produced on the [100] and [110] orientation.

But in this case there are differences between the two initial nanowire sizes. For the small size, in average, the [111] orientation produces pentagonal tubes of bit more than 2 pentagonal ring while the other orientations produce much longer nanowires. For large size nanowires, average lengths of all orientations are more similar: $\langle n_p^* \rangle$ and $\langle L_p^{m*} \rangle$ of the [111] orientation grows with respect of the small size, and concurrently these quantities for the [100] and [110] directions decreases. Combining the information of these lengths with the information of the probability p_i, it can be seen that for small size nanowires it is highly probable to form an icosahedral nanowire for the [100] and [110] orientation and it would have a relative large length. For large nanowires these orientation would produce pentagonal nanowires at a quite good ratio, but their average size will be smaller. For the [111] orientation, both sizes would produce pentagonal chains at a low proportion, but if a large size initial nanowire is used, the length of the few created chains should be comparable with those produced from the [100] and [110] orientations.
Finally, from curves depicted in Fig. 12 we can extract information about the optimal temperature of creation of icosahedral nanowires. Maxima of these quantities indicate the optimal temperatures T_{Opt}. For each simulation set parameters (size and orientation) there can be two possible values of T_{Opt} depending on whether we are looking for the maximum probability of nanowire formation or the highest length of the pentagonal tube. Although slightly differences on the simulation parameters and the chosen criterion, T_{Opt} is around 550 K for parameters investigated. As it can be seen in Fig. 11, in this T_{Opt} temperature $D(L_p^m)$ and $D(n_p)$ show also the broadest distributions.

6. Atomic stress calculations

One of the advantages of the MD methodology is the possibility of analyzing mechanical properties of the systems. I.e., we are able to monitor the time evolution of the tensile atomic stress along the stretching direction of the nanowires. This measurement gives us information about the stress distribution on the nanowire at every stage of the breaking process. Fig. 13 shows an example of the time evolution of the atomic stress. When a

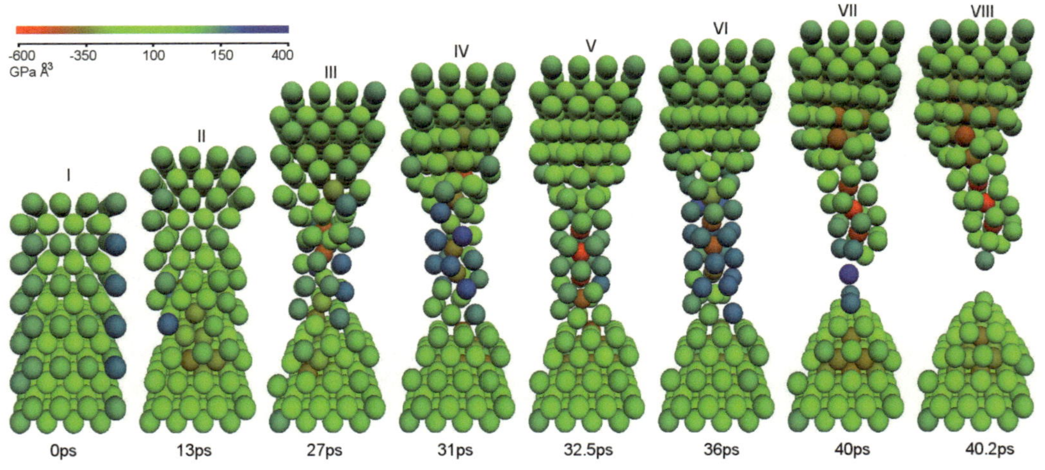

Fig. 13. Snapshots of the breaking process of a Ni nanowire that forms a pentagonal structure. Colours account for the atomic tensile stress.

simulation starts, the nanowire shape is that of a parallelepiped (snapshot I of Fig. 13). In these conditions those atoms at the edges of the structure are in a more stressed situation than those on the surfaces. And these atoms on the surfaces are more stressed than bulk ones. This stress distribution is related to the atomic coordination. Atoms at low coordination situations feel more stressed and unstable, while high coordination atoms are in a more comfortable condition.

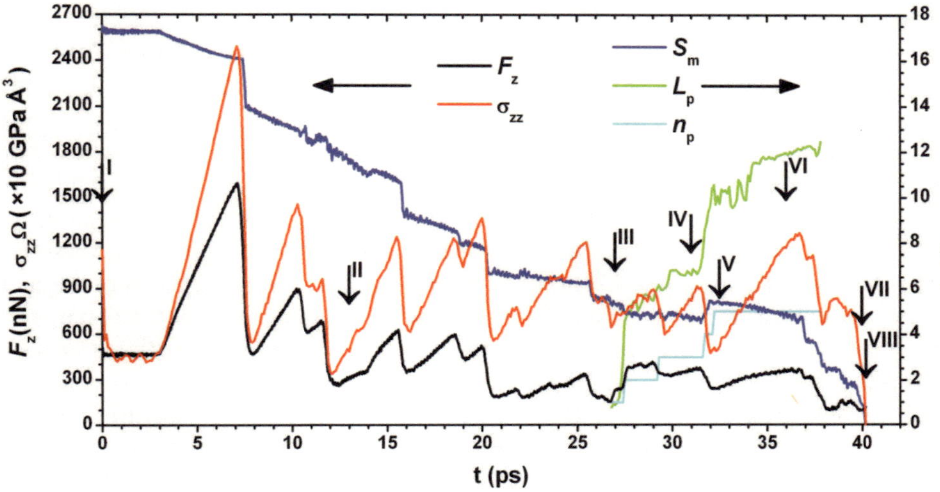

Fig. 14. Time evolution of several quantities from the simulation shown in Fig. 13: force F_z on the extremes of the nanowire; atomic stress σ_{zz} is shown here multiplied by the system volume Ω (constant during the simulation); minimum cross-section S_m; the pentagonal length $L_p(t)$ and the number of pentagonal rings $n_p(t)$ that form the pentagonal structure starting at $t \sim 26$ns. Vertical arrows indicate the time position of snapshots shown in Fig. 13.

It is well known that icosahedral is one of the preferred structures for small clusters and icosahedral arrangements are quite robust (Bulienkov & Tytik, 2001). However, the stability of pentagonal nanowires has not been studied so far. There is a large difference in the atomic coordination between those atoms in the core of the pentagonal nanowire and those on the surface. So we are interested in analyzing how the stress is distributed along this icosahedral configuration (within and around the pentagonal structure) during its formation, growth and breakage as the stretching process occurs.

As it can be seen in Fig. 13, an interesting behaviour of the atomic tensile stress is observed in the narrowest region of the nanowire (the central region) as the stretching process goes on. Since snapshot III, a clear distinction can be made between surface and core atoms. Atoms in the surface are much more stressed than core atoms. This means that surface atoms fell strained (they are bluer), while core atoms feel compressed (they are redder) compared to those atoms (green) out of the pentagonal structure. This can be understood the following way: the system tends to keep a uniform overall stress along all the regions of the nanowire. In the central region this means that the sum of the surface and core atoms tensile stress (weighted by the number of atoms of each kind) should be approximately equal to the overall average tensile stress of the nanowire. In this central region most of the atoms are on the surface. Having so few core atoms makes in compensation their compressive stress much stronger than the strained stress of surface atoms.

The uniaxial tensile stress σ_{zz} of the nanowire showed on Fig. 13 is depicted in Fig. 14. It is calculated from the virial theorem for static systems using the formula (Tsai, 1979)

$$\sigma_{zz} = \frac{1}{2V}\sum_{i,j} F_{ij}\frac{(r_{iz}-r_{jz})^2}{r_{ij}} \tag{4}$$

where V is the whole volume of the nanowire, $F_{i,j}$ is the force exerted by atom j over atom i, r_{ij} is the distance between them and r_{iz} is the z-coordinate of atom i.
Fig. 14 shows the time evolution of $\sigma_{zz}(t)$ and the force $F_z(t)$ on the frozen slabs of the nanowire. Both curves exhibit a sawtooth structure, with a similar behaviour. A linear increase with time is associated to an elastic strain, while an abrupt drop is related to plastic atomic rearrangements to release the elastic energy. Some of these F_z or stress drops coincide with a drop in the minimum cross section S_m, also shown in Fig. 14. In these cases the atomic rearrangements have driven the system to have a thinner cross section in its narrowest region. Notice, nonetheless, the particular case of the abrupt increase of S_m between snapshots IV and V. Fig. 13 shows that at snapshot IV the section where the central region and the bottom reservoir join is formed by a thin 4-atom structure. At snapshot V, however, this region has a more robust pentagonal structure that increases the cross section and relaxes the tensile stress.
Starting from snapshot III, our algorithm for detection of pentagonal structures shows the presence of pentagonal rings. In Fig. 14 we also plot the length of this pentagonal structure and the number of pentagonal rings detected. We can see that the path from having 2 to 3 pentagonal rings is accompanied by a drop in the tensile stress and the force F_z. Similar drops in S_m and F_z are observed when the system changes from 3 to 5 pentagonal rings. The variable L_p, also plotted in Fig. 14, illustrates how the pentagonal region elongates until a plastic rearrangement takes place and a new pentagonal ring is formed.

7. Comparing with experimental results

In previous sections of this chapter we have showed how, using powerful computational resources and accurate description of the atomic interactions, it is possible to reproduce many of the formation-breaking experiments "in silicon". And indeed, that it is possible to analyse situations and structures that experimentally are difficult to study. We have stressed the importance of statistical studies of these phenomena because, given a particular initial geometrical configuration and temperature, each breakage event evolves differently. Of course this is the standard approximation for the experimental study of metallic nanowires, where statistical data is accumulated during many indentation-retraction cycles.
However, the comparison between experimental results and the numerical simulations (as they have been presented up to here) requires an additional element. Experimental data usually correspond to the statistical average of nanowire breaking events involving random stretching directions. In a standard experiment there is no reason that indicates that any particular orientation would be preferred during the nanowire formation and breakage. Therefore, to accomplish a complete statistical analysis equivalent to the experimental one, computational calculations must simulate breaking events on random stretching directions. This could be a potential problem since it is not easy to perform simulations with arbitrary (random) initial stretching directions. Fortunately, it is not necessary to compute "every" stretching orientation to get the statistical behaviour of the breaking nanowires.

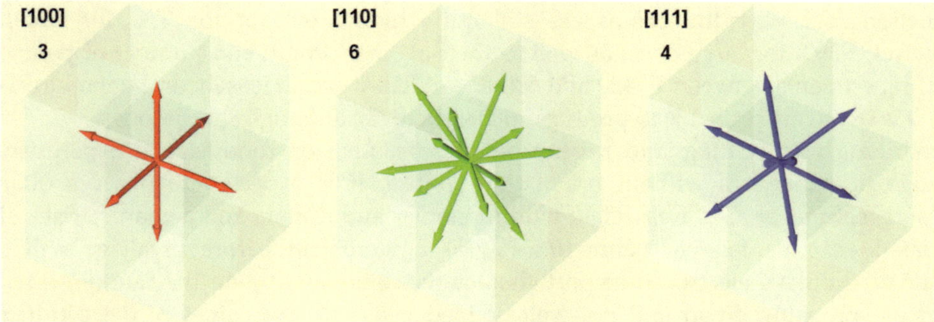

Fig. 15. FCC main crystalline directions ([100], [110], [111]) and their multiplicities (3, 6 and 4 respectively).

In an experiment, the final stages and behaviour of a nanowire during its breakage are leaded by the type of crystalline structure closest to its elongation direction. So it should be in simulations. Therefore, we need only to analyse main crystalline orientations, those that show different structural type, and average their results according to their occurrences. I.e., to achieve orientation statistics in computational simulations, results from the main crystalline directions have to be merged with the appropriate weights. These weights will be proportional to the zone axis multiplicity (González et al., 2004). Of course, each main direction must have an accurate statistics to guarantee the correctness of the global results.

Nickel has a FCC crystalline structure. In Fig. 15 we show FCC main crystalline directions ([100], [110] and [111]) and their multiplicities (3, 6 and 4 respectively). Their relative occurrence at random stretching orientation should be 3/13, 6/13 and 4/13. These are the weights we have used to merge the statistical results from the three main directions and obtain the average statistical results that can be compared with the experimental ones.

In figures showed in previous sections we have included these results corresponding to this weighted average. In Figs. 2 and 3, expected minimum cross-sections histograms for random stretching orientation are shown for large and small nanowires. Merged histograms conserve main features of [111] direction (well defined peaks for low S_m values) and [100] and [110] directions (huge peak at S_m~5 that increases with the temperature).

The merged results for the monomer and dimer production are shown in Fig. 6. For small sizes the proportion of monomer (~45%), dimers (~25%) and more complex structures (~30%) is constant with the temperature; for large systems, monomer ratio keeps constant (~45%) but that of dimers increases (~25 to ~30%) slightly with temperature. Their surrounding structures (Fig. 7, bottom panels) show the main characteristics already described. For monomer, the 2-1-3 structure is the most probable (~50%) followed by 2-1-4, 2-1-2, 3-1-3 and 3-1-4, independently of the temperature. No signal found from a 4-1-4 structure. For the dimer structure, the 3-1-1-4 configuration is the most probable one followed by the 3-1-1-3 and 4-1-1-4; their relative probabilities depend on the nanowire size and the temperature. With respect the non local environment, two different behaviours (that will be mixed in standard experiments) are expected if stretching occurs along the [111] direction (ordered structure from the monomer/dimer) or along the [100] and [110] orientations (amorphization around the monomer/dimer). This can lead to a mix of two transport regimes (ballistic and diffusive) in experiments were the conductance is measured. The pentagonal nanowires production for the [100] and [110] directions is much higher than for the [111]. When results from the three directions are merged (Fig. 12), probabilities of

production of icosahedral chains are still quite high (~60% at the T_{Opt} for small size nanowires; ~30% for large ones) as well as their average length and number of pentagonal rings. Experiments between T=450 and 650 K will likely obtain icosahedral nanowires; even at RT it is relatively probable to produce such structures, although shorter ones.
Summarizing and putting into relation all this results, for conductance experiments of breakage of nanowires without orientation control, it is expected to observe quantum transport features at $S_m \leq 1$ associate with monomer and dimers and a characteristic signal associated with pentagonal nanowires (S_m~5). System just before breakage will show ballistic or diffusive electron transport characteristics depending on the random direction, producing probably broad and not well defined peaks at low values of the conductance histogram. To identify the formation of icosahedral nanowires, they should produce a broad peak at relative high conductance value, easily seen at temperatures above RT.

8. Conclusions and future work

In this chapter we have described the huge potential and reliability of MD techniques within the EAM framework as a tool for the description of the mechanical properties and atomic configurations of metallic nanowires. In particular, we apply this methodology to the study of nanowires subjected to a tensile stress leading to fracture, and studied the final stages of the breaking events when the nanowire narrowest region is formed by low coordination configurations. This computational approach goes beyond the description of isolated single breaking events and it is able to provide a statistical view of a large set of breaking events, mimicking the methods established in the laboratories long time ago.
We have presented several techniques to analyse the average behaviour of different physical magnitudes of interest provide by the simulation, as the minimum cross-section, the tensile stress, the local and non-local geometry of formed monomers, dimers and pentagonal chains. In particular we have introduced a procedure to merge results obtained from simulations with well defined stretching direction in order to reproduce more realistic experimental conditions where geometry of stretched nanowire is not so well defined.
The statistical approach and the analysis techniques has been used to compute and analyse hundreds of breaking events of Ni nanowires taking into account several parameters as the nanowire size, temperature, and crystallographic stretching direction. Results cover by this large set of variables would help to the interpretation of the experimental findings.
This powerful tool provides a new approach to understand many physical problems related with the behaviour of metallic nanowires. In particular, future work will address the study of formation of monomer, dimer, monoatomic chain and icosahedral structures for different metallic species. However, more open questions must be addressed, as the role of the initial nanowire shape and size especially if the initial cross-section shape corresponds to a magic number configuration. Or the responsibility of indentation-retraction processes in the formation of histograms: histograms presented in this chapter are formed by adding independent breaking events whereas in the experimental situation is slightly different, since all the contacts are formed in the same region. The technique should also be useful to determinate the optimal condition and procedures to produce nanostructures of interest as the pentagonal nanowires. These are few examples that indicate that our methodology will be intensively exploited in future studies for answering the interesting questions raised in the fascinating and useful domain of the nanowires.

9. Acknowledgment

We would like to thank A Hasmy, J J Sáenz, M Díaz, J L Costa-Krämer and E Medina for the interesting discussions and ideas. This work has been partially supported by the Spanish MICINN through Projects FIS2006-11170-C02-01 CSD2007-00046, and FIS2009-14320-C02 and by the Madrid Regional Government through the S-0505/MAT/0202 (NanoObjetos-CM) project. PG-M acknowledges the Spanish MEC "Ramón y Cajal" Programme support.

10. References

Agraït, N.; Levy Yeyati, A. & Van Ruitenbeek, J. M. (2003). *Physics Report*, 377, 81-279, ISSN: 0370-1573.

Bahn, S. R. & Jacobsen, K. W. (2001). *Phys. Rev. Lett.*, 87, 2661101, ISSN: 0031-9007.

Barnet, R. N. & Landman, U. (1997). *Nature*, 387, 788-791, ISSN: 0028-0836.

Branício, P. S. & Rino, J. P. (2000). *Phys. Rev. B*, 62, 16950-16955, ISSN: 1098-0121.

Bratkovsky, A. M.; Sutton, A. P. & Todorov, T. N. (1995). *Phys. Rev. B*, 52, 5036-5051, ISSN: 1098-0121.

Bulienkov, N. A. & Tytik, D. L. (2001). *Russ. Chem. Bull. Int. Ed.*, 50, 1-19, ISNN: 1066-5285.

Daw, M. S. & Baskes, M. I. (1983). *Phys. Rev. Lett.*, 50, 1285-1288, ISSN: 0031-9007.

Foiles, S. M. (1985). *Phys. Rev. B*, 32, 3409-3415, ISSN: 1098-0121.

García-Mochales, P.; Peláez, S.; Serena, P. A.; Medina, E. & Hasmy, A. (2005). *Appl. Phys. A*, 81, 1545-1549, ISSN: 0947-8396.

García-Mochales, P.; Paredes, R.; Peláez, S. & Serena, P. A. (2008a). *Nanotechnology*, 19, 225704, ISSN: 0957-4484.

García-Mochales, P.; Paredes, R.; Peláez, S. & Serena, P. A. (2008b). *Journal of Nanomaterials*, 2008, 361464, ISSN: 1687-4110.

García-Mochales, P.; Paredes, R.; Peláez, S. & Serena, P. A. (2008c). *Phys. Status Solidi A*, 205, 1317-1323, ISSN: 1862-6300.

González, J. C.; Rodrigues, V.; Bettini, J.; Rego, L. G. C.; Rocha, A. R.; Coura, P. Z.; Dantas, S. O.; Sato, F.; Galvao, D. S. & Ugarte, D. (2004). *Phys. Rev. Lett.*, 92, 126102, ISSN: 0031-9007.

Grigoryan, V. & Springborg, M. (2003). *Chem. Phys. Lett.*, 375, 219-226, ISSN: 0009-2614.

Gülseren, O.; Ercolessi, F. & Tosatti, E. (1998). *Phys. Rev. Lett.*, 80, 3775-3778, ISSN: 0031-9007.

Hasmy, A.; Medina, E. & Serena, P. A. (2001). *Phys. Rev. Lett.*, 86, 5574-5577, ISSN: 0031-9007.

Hasmy, A.; Pérez-Jiménez, A. J.; Palacios, J. J.; García-Mochales, P.; Costa-Krämer, J. L.; Díaz, M.; Medina, E. & Serena, P. A. (2005). *Phys. Rev. B*, 72, 245405, ISSN: 1098-0121.

Heemskerk, J. W. T.; Noat, Y.; Bakker, D. J.; Van Ruitenbeek, J. M.; Thijsse, B. J. & Klaver, P. (2003). *Phys. Rev. B*, 67, 115416, ISSN: 1098-0121.

Herrman, H. J. & Stanley, H. E. (1984a). *Phys. Rev. Lett.*, 53, 1121-1124, ISSN: 0031-9007.

Herrman, H. J. & Stanley, H. E. (1984b). *J. Phys. A: Math. Gen.*, 17, L261-266, ISSN: 1751-8113.

Ikeda, H.; Qi, Y.; Cagin, T.; Samwer, K.; Johnson, W. L. & Goddard III, W. A. (1999). *Phys. Rev. Lett.*, 82, 2900-2903, ISSN: 0031-9007.

Jacob, D.; Fernández-Rossier, J. & Palacios, J. J. (2005). *Phys. Rev. B*, 71, 220403, ISSN: 1098-0121.

Liang, W. & Zhou, M. (2006). *Phys. Rev. B*, 73, 115409, ISSN: 1098-0121.

Liang, W. & Zhou, M. (2007). *Philos. Mag.*, 87, 2191-2220, ISSN: 0141-8610.

Medina, E.; Díaz, M.; León, N.; Guerrero, C.; Hasmy, A.; Serena, P. A. & Krämer, J. L. (2003). *Phys. Rev. Lett.*, 91, 026802, ISSN: 0031-9007.
Mehrez, H. & Ciraci, S. (1997). *Phys. Rev. B*, 56, 12632-12642, ISSN: 1098-0121.
Mishin, Y.; Farkas, D.; Mehl, M. J. & Papaconstantopoulos, D. A. (1999). *Phys. Rev. B*, 59, 3393-3407, ISSN: 1098-0121.
Olesen, L.; Laegsgaard, E.; Stensgaard, I.; Besenbacher, F.; Schiøtz, J.; Stoltze, P.; Jacobsen, K. W. & Nørskov, J. K. (1994). *Phys. Rev. Lett.*, 72, 2251-2254, ISSN: 0031-9007.
Pauly, F.; Dreher, M.; Viljas, K. J.; Häfner, M.; Cuevas, J. C. & Nielaba, P. (2006). *Phys. Rev. B*, 74, 235106, ISSN: 1098-0121.
Peláez, S.; Guerrero, C.; Paredes, R.; Serena, P. A. & García-Mochales P. (2009). *Phys. Status Solidi C*, 6, 2133-2138, ISSN: 1610-1634.
Rapaport, D. C. (1995). *The Art of Molecular Dynamics Simulation*, Cambridge University Press, ISBN: 0521599423, Cambridge, United Kingdom.
Serena, P. A. & García, N. (Ed.) (1997). *Nanowires* (*NATO ASI Series E* vol 340), Kluwer Academic Publishers, ISBN: 0792346270, Dordrecht, The Netherlands
Sen, P.; Gülseren, O.; Yildirim, T.; Batra, I. P. & Ciraci, S. (2002). *Phys. Rev. B*, 65, 235433, ISSN: 1098-0121.
Sharvin, Y. V. (1965). *Sov. Phys. JEPT*, 21, 655-656, ISSN: 0038-5646.
Sirvent, C.; Rodrigo, J. G.; Vieira, S.; Jurczyszyn, L.; Mingo, N. & Flores, F. (1996). *Phys. Rev. B*, 53, 16086-16090, ISSN: 1098-0121.
Sørensen, M. R.; Brandbyge, M. & Jacobsen, K. W. (1998). *Phys. Rev. B*, 57, 3283-3294, ISSN: 1098-0121.
Sutrakar, V. K. & Mahapatra, D. R. (2009). *Nanotechnology*, 20, 045701, ISSN: 0957-4484.
Tsai, D. H. (1979). *J. Chem. Phys.*, 70, 1375-1382, ISSN: 0021-9606.

3

Fabrication of Functional Metallic Nanowires Using Electrodeposition Technique

Takeshi Ohgai
Nagasaki University
Japan

1. Introduction

The array of numerous metallic nanowires with straight shape has much attention due to their shape anisotropy and extremely large surface area. This unique structure can be applied to develop the novel functional nano-materials such as electronic, magnetic and optical nano-scale devices. Metallic nanowires can be fabricated by manipulating metallic atoms one by one using a scanning tunneling microscope (STM) probe, while they can be also prepared by electrochemically depositing metallic atoms into a nanoporous template with numerous cylindrical nanopores. In a fabrication technique using electrodeposition of nanowires, nanoporous templates such as anodized aluminum oxide films with high density of nanopores (about 10^8~10^{10} pores•cm^{-2}) have been used so far.

In 1975, using an anodic oxide coating film on aluminum, magnetic properties of electrodeposited Co nanowires and Co-Ni alloy nanowires were investigated (Kawai & Ueda, 1975). In 1986, as an application to perpendicular magnetic recording medium, alumite films containing Fe nanowires were fabricated (Tsuya et al., 1986). In 1988, magnetization curling process in perpendicular direction was studied using Fe nanowire arrays in alumite media (Huysmans et al., 1988). In 1990, influence of the packing density on the magnetic behaviour was investigated using alumite media containg magnetic nanowires (Zhang et al., 1990).

On the other hand, in 1991, using the polymer membrane template with numerous nanopores, polymeric and metal microtubules were synthesized (Martin, 1991). In 1993, arrays of Ni and Co nanowires were electrodeposited in polymer templates with the nanometer-sized pores prepared by nuclear track etching technique (Whitney et al., 1993). They found that the preferred magnetization direction is perpendicular to the film plane and enhanced coercivities as high as 680 Oe. Remnant magnetization of the nanowires up to 90% had been reported. In 1994, a membrane-based synthetic approach to obtain nano-materials was introduced to materials science field (Martin, 1994). He suggested that the template synthesis method is useful to prepare polymers, metals, semiconductors, and other materials on a nano-scopic scale.

In 1994, giant magnetoresistance (GMR) properties were found in Co/Cu multilayered nanowires electrodeposited in nanoporous polymer template (Piraux et al., 1994). In the report, magnetoresistance measurements with the current perpendicular to the layers were performed on the array of parallel nanowires. They observed that the 15% of GMR at room temperature on Co/Cu multilayered nanowires. In the almost same time, GMR of nanowires with Co/Cu and Ni-Fe/Cu multilayers was demonstrated (Blondel et al., 1994). They made the multilayered nanowires with 6 μm long, 80 nm in diameter and each layer thickness of 5–10 nm in nanoporous polymer template. At ambient temperature, GMR of 14% for Co/Cu and of 10% for Ni-Fe/Cu was demonstrated in the current perpendicular to the layers.

On the contrary, in 2000, current perpendicular to plane giant magnetoresistance (CPP-GMR) was found using the multilayered nanowires electrodeposited in a commercially available anodic aluminum oxide membrane filter (Evans et al., 2000). In the report, Co–Ni–Cu/Cu multilayered nanowires were synthesized and extremely large values (55% at room temperature and 115% at 77 K) of CPP-GMR were demonstrated. In 2003, Co/Cu multilayered nanowires with spin-valve effect and GMR response were synthesized in self-organized anodized aluminum oxide nanopores grown at the surface of bulk aluminum (Ohgai et al., 2003). In the report, the aluminum oxide template with pore-length as short as 2000 nm was used and the pore bottom oxide layer was removed by a chemical etching technique. 20% of GMR was demonstrated in Co/Cu multilayered nanowires at room temperature, while the typical resistance switching of spin-valves was observed in Co/Cu/Co tri-layered nanowires.

Compound semiconductor nanowires were also fabricated as well as ferromagnetic metal nanowires using an electrodeposition technique. Bi_2Te_3 nanowires (Wang et al., 2004), CdTe nanowires (Ohgai et al., 2005), InSb nanowires (Zhang et al., 2005) and ZnSe nanowires (Katkar & Tait, 2007) have been synthesized so far. They reported that these semiconductor nanowires can be applied to light emission diodes, photovoltaic devices, solar cells and thermovoltaic devices.

If the template synthesis technique on the basis of fabricating nanopores and electrodepositing nanowires as mentioned above is applied to produce micro- and nano-scale devices in ultra large scale integration (ULSI), position of each nanopore, inter-pore distance, pore diameter and pore shape should be controlled precisely. To fabricate novel electronic devices in ULSI, lithographic galvanoforming (LIGA) process using laser, UV, X-ray, electrons and ions can be used.

In this chapter, fabrication process such as electrodeposition behavior, morphology, crystal structure and magnetic properties (magnetization and magnetoresistance) of metallic nanowires such as Ni, Co, Fe (in section 2), Ni-Fe alloy (in section 3) and Co/Cu multilayer (in section 4) were discussed for the application to novel functional micro- and nano-scale devices.

2. Ni, Co and Fe nanowires

2.1 Electrodeposition process of Ni, Co and Fe nanowires

Ion track-etched polycarbonate membrane filters with pore-diameter of 40, 80, 160 and 300 nm, pore-length of 6000 nm and pore-density of 10^8 pore•cm^{-2} were used as a template for growing metallic nanowires as shown in Fig.1. On a surface of the membrane filter, a gold layer was sputter-deposited to cover the pores and make a cathode. Aqueous solution containing metal sulfate was used as electrolyte.

To determine the optimum deposition potential for growing nanowires, cathodic polarizarion behavior was investigated in a wide range of cathode potential. Figure 2 shows cathodic polarization curves obtained from the solutions containing each iron-group metal ions (Ni^{2+}, Co^{2+} or Fe^{2+}).

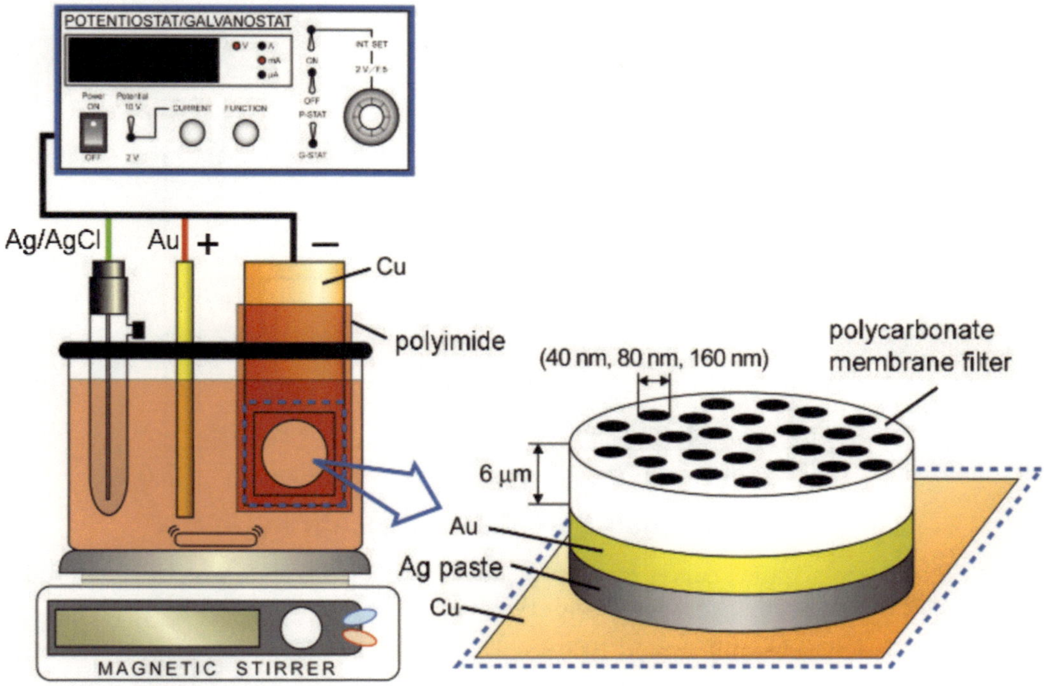

Fig. 1. Experimental apparatus for electrodepositing metallic nanowires in polycarbonate membrane filter with numerous cylindrical nanochannels.

The cathodic current begins to occur at ca. 0 V vs.Ag/AgCl, which is more-noble than the equilibrium potential of iron-group metals. It is well known that the standard electrode potentials for Ni, Co and Fe are -0.46, -0.48 and -0.64 V vs.Ag/AgCl. Therefore, this cathodic current is presumed to be reduction current of H^+ ions. With increasing the cathodic current, at around 10^{-3} A, the potential polarizes to be around -0.8 V due to the diffusion limit of H^+ ions.

At around -0.8 V, the cathodic current increases again. It is also well known that the electrodeposition of iron-group metals proceeds accompanying some over potential. Therefore, this increase in cathodic current is mainly caused by an increase in deposition current of iron-group metal ions. At the potential region less than -1.2 V, with increasing cathodic current, the potential polarizes significantly to be less-noble region due to the diffusion limit of iron-group metal ions.

Consequently, the optimum cathode potential region for growing iron-group metal nanowires is determined to be from -0.9 to -1.2 V according to the cathodic polarization curves as shown in Fig.2.

To investigate the growing process of nanowires, time dependence of cathodic current was monitored during the electrodeposition. Ni, Co and Fe nanowires were potentio-statically electrodeposited at room temperature. Figure 3 shows the effect of cathode potential on the time-dependence of cathodic current in the solution containing Ni^{2+} ions. During the electrodepostion of Ni, cathode potentials were fixed to -0.9, -1.0, -1.1 and -1.2 V. If the potential was kept to -1.0 V, the cathodic current reached up to ca. 1 mA at the beginning of electrolysis within several tens minutes. Then, the current rapidly decresed to be ca. 0.4 mA

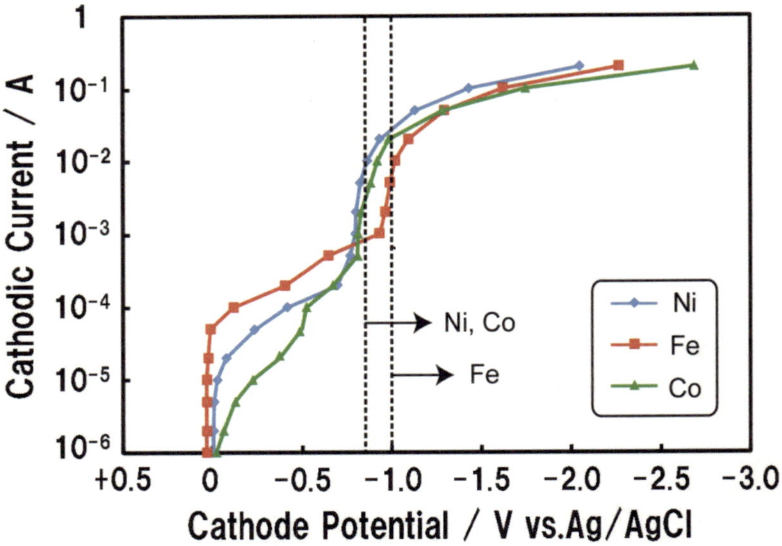

Fig. 2. Cathodic polarization curves for electrodeposition of Ni, Co and Fe on Cu sheet from aqueous solutions containing Ni^{2+} or Co^{2+} or Fe^{2+} ions.

and kept the magnitude until around 1000 sec. During this process, electrodeposition of Ni proceeds in the nanopores.

At the initial stages of the electrodepostion, large cathodic current was observed in every cathode potential. The concentration of metal ions in the nanopores will decrease with increasing electrodeposition time due to the cathodic reduction, while the metal ions will be provided from the bulk solution to the nanopore, where the metal ions are consumed by the electrodeposition.

Finally, as shown in Fig.3, the cathodic current rapidly increases at the deposition time more than 1000 sec. At this stage, electrodeposited nanowires reach the surface of the membranes and large hemispheric Ni deposits are formed. Growth rate of Ni nanowires can be estimated as ca. 6 $nm \cdot sec^{-1}$ at the cathode potential of -1.0 V. Time-dependence of cathodic current for Co and Fe deposition also showed similar behavior as well as Ni deposition. Growth rate of the nanowires increases up to around 30 $nm \cdot sec^{-1}$ with increasing cathode potential up to -1.2 V.

2.2 Crystal structure of Ni, Co and Fe nanowires

After the growing nanowires, polycarbonate membrane filters were dissolved in organic solvent (dichloromethane and chloroform) and the remains consisted of nanowires and a gold layer was served as a sample for scanning electron microscope (SEM) observation. On the other hand, using ion-milling technique, as-deposited nanowires in a membrane filter were thinned and a cross-sectional area of the nanowires was observed using transmission electron microscope (TEM).

Figure 4 shows SEM images of Co nanowires separated from the polycarbonate templates. Diameter (40, 80, 160 and 300 nm) and length (6000 nm) of the nanowires corresponds well to that of nanopores and the cylindrical shape was precisely transferred from the nanopores to the nanowires. Aspect ratio of the nanowires reaches up to around 150.

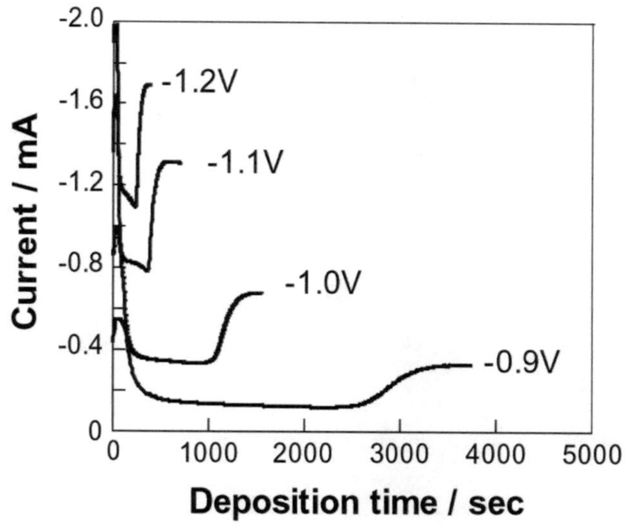

Fig. 3. Effect of cathode potential on the time-dependence of cathodic current during electrodeposition of Ni nanowires.

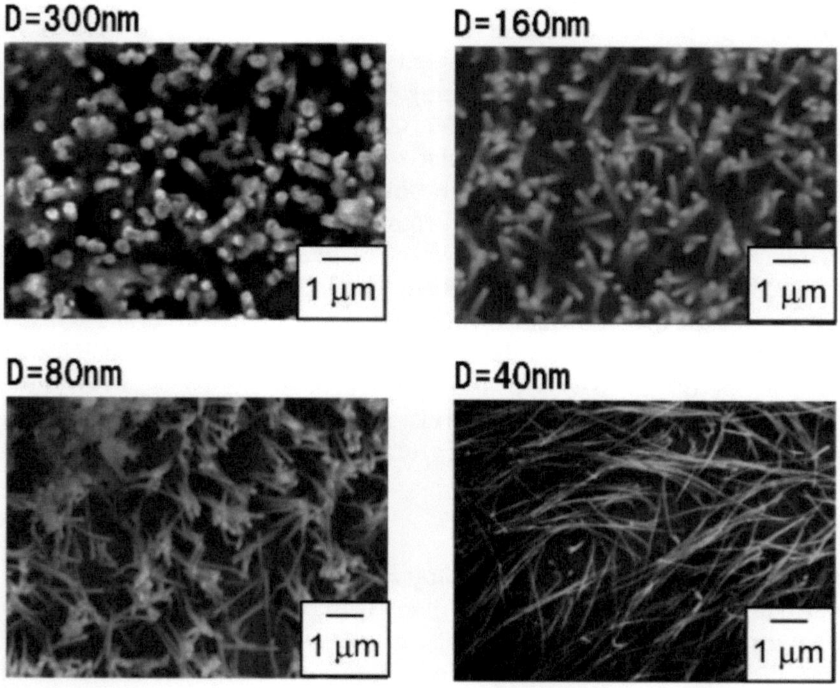

Fig. 4. SEM images of electrodeposited Co nanowires separated from polycarbonate membrane filters.

TEM bright images and diffraction patterns of electrodeposited Ni, Co and Fe nanowires were also investigated as shown in Fig.5. According to TEM bright images, cross-sectional area of the nanowire was almost to be round shape, while the diffraction patterns are composed of spots, which means a nanowire consists of a single crystal domain.

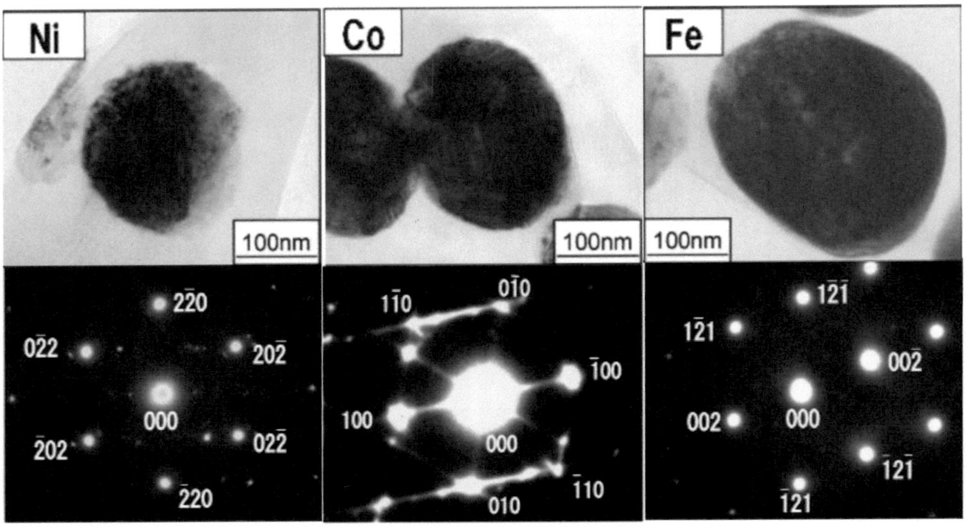

Fig. 5. Cross-sectional TEM images of Ni, Co and Fe nanowires electrodeposited in polycarbonate membrane filter.

2.3 Magnetic properties of Ni, Co and Fe nanowires

Magnetic hysteresis loops of electrodeposited nanowires were obtained using vibrating sample magnetometer (VSM) with increasing the magnetic field up to 10 kOe. Figure 6 shows the magnetic hysteresis loops of Ni, Co and Fe nanowires electrodeposited into polycarbonate templates with pore-diameter of 80, 160 and 300 nm. Magnetic field was applied to perpendicular directions to the film plan, which corresponds to the parallel direction to the long axis of nanowires. Coercive force of the nanowires was increased in decreasing the pore-diameter. Figure 7 shows the magnetic hysteresis loops of Ni, Co and Fe nanowires electrodeposited into polycarbonate templates with pore-diameter of 40 nm. As shown in these figures, magnetization switching was observed at around 1 kOe and residual magnetization was almost equal to the saturated magnetization. The coercive force of Co nanowires reaches up to 1084 Oe and the magnetic hysteresis loop shows typical perpendicular magnetization behavior. This is resulting from the uni-axial magnetic anisotropy and single magnetic domain structure of the nanowires with large aspect ratio.

3. Ni-Fe alloy nanowires

3.1 Electrodeposition process of Ni-Fe alloy nanowires

Ion track-etched polycarbonate membrane filters with pore-diameter of 80, 160, 300 and 800 nm, pore-length of 6000 nm and pore-density of 10^8 pore•cm^{-2} were used as a template for growing metallic nanowires while copper foils were used as a cathode for electrodepositing metallic films. On a surface of the membrane filter, a gold layer was sputter-deposited to cover the pores and make a cathode. Aqueous solution containing metal sulfate was used as electrolyte. The solution compositions are shown in Table 1.

To determine the optimum deposition potential for growing nanowires, cathodic polarizarion behavior was investigated in a wide range of cathode potential. Ni and Ni-Fe alloy nanowires were potentio-statically electrodeposited at 313 K. Figure 8 shows cathodic polarization curves for the electrodeposition of Ni and Ni-Fe alloy. For the Ni-Fe alloy deposition, the solutions containing 1.5 % of Fe^{2+} ions concentration ratio ($R^{Fe/bath}$=1.5%) was used as shown in Table 1.

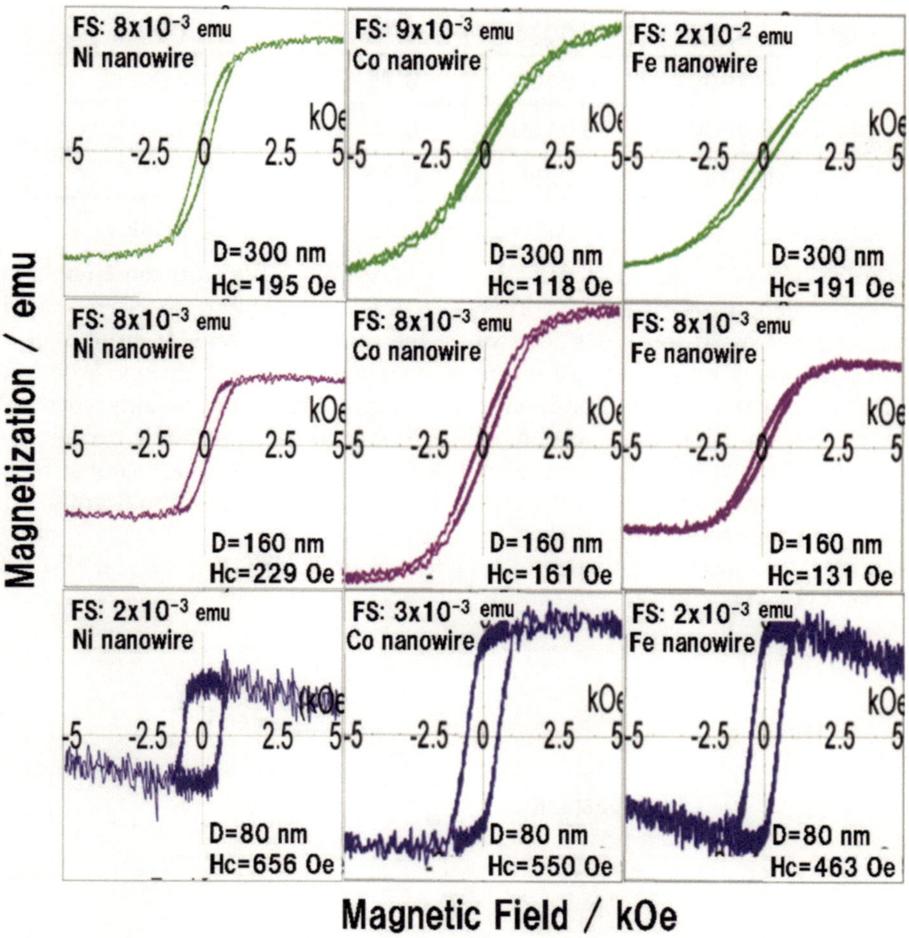

Fig. 6. Magnetic hysteresis loops of Ni, Co and Fe nanowires electrodeposited into polycarbonate templates with channel-diameter of 80, 160 and 300 nm.

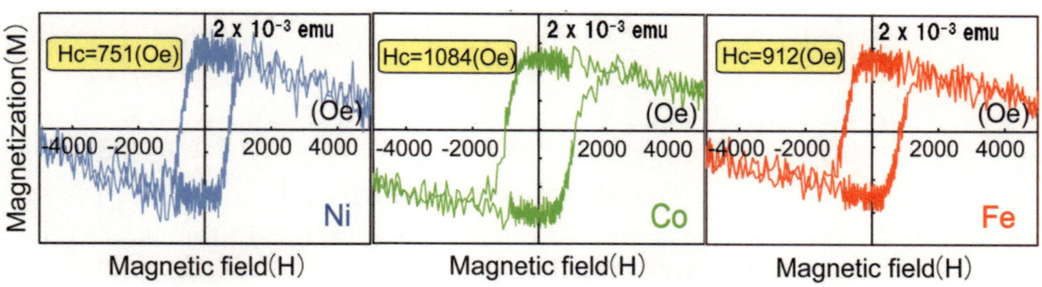

Fig. 7. Magnetic hysteresis loops of Ni, Co and Fe nanowires electrodeposited into polycarbonate templates with channel-diameter of 40 nm.

	$NiSO_4 \cdot 7H_2O$ (mol/L)	$FeSO_4 \cdot 7H_2O$ (mol/L)	H_3BO_3 (mol/L)	$[Fe^{2+}] / ([Fe^{2+}] + [Ni^{2+}])$ (%)
①	0.4950	0.0050	0.75	1.0
②	0.4925	0.0075	0.75	1.5
③	0.4900	0.0100	0.75	2.0
④	0.4850	0.0150	0.75	3.0
⑤	0.4350	0.0650	0.75	13.0

Table 1. Electrolytic solution composition for electrodeposition of Ni-Fe alloys.

The cathodic current begins to occur at ca. 0 V vs.Ag/AgCl, which is more-noble than the equilibrium potential of Ni and Fe. Therefore, this cathodic current is presumed to be reduction current of H^+ ions. With increasing the cathodic current, at around 10^{-3} A, the potential polarizes to be around -0.7 V due to the diffusion limit of H^+ ions. At around -0.7 V, the cathodic current for Ni deposition increases again, while the cathodic current for Ni-Fe alloy deposition increases at around -0.8 V. This increase in cathodic current is mainly caused by an increase in deposition current of Ni and Ni-Fe. At the potential region less than -1.2 V, with increasing cathodic current, the potential polarizes significantly to be less-noble region due to the diffusion limit of Ni^{2+} and Fe^{2+} ions.
Consequently, the optimum cathode potential for electrodeposition of Ni and Ni-Fe alloys is determined to be -1.0 V.

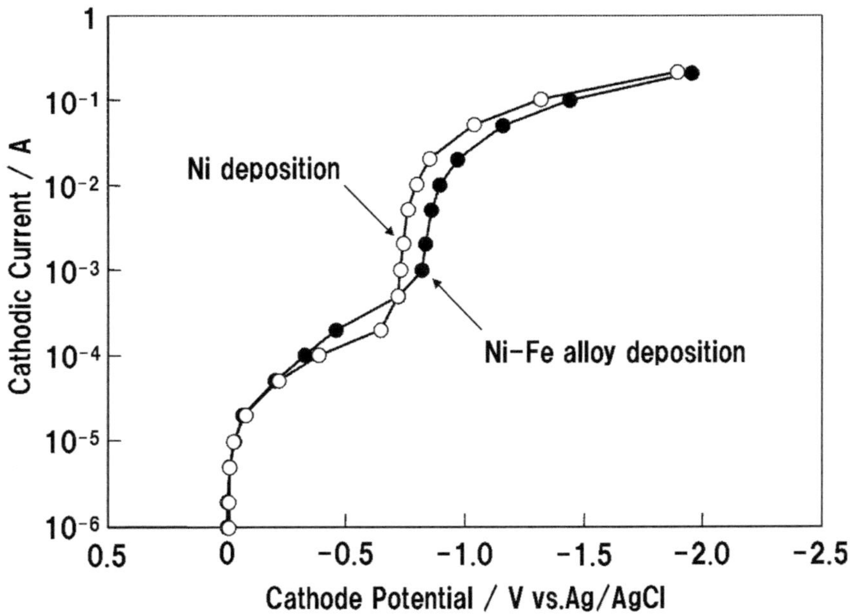

Fig. 8. Cathodic polarization curves for the electrodeposition of Ni and Ni-Fe alloy on Cu sheet from aqueous solutions containing Ni^{2+} and Fe^{2+} ions.

To investigate the relationship between Fe^{2+} concentration ratio ($R^{Fe/bath}$=$[Fe^{2+}]/([Fe^{2+}]+[Ni^{2+}])$) in baths and Fe content ($R^{Fe/depo}$ =[Fe]/([Fe]+[Ni])) in Ni-Fe alloy deposits, the composition of electrodeposits were determined using EDX analysis. Figure 9 shows the relationship between Fe^{2+} concentration ratio ($R^{Fe/bath}$) in baths and Fe content ($R^{Fe/depo}$) in Ni-Fe alloy deposits. With increase in $R^{Fe/bath}$, $R^{Fe/depo}$ also increases. In the figure, composition reference line (C.R.L) means $R^{Fe/depo}$ corresponds to $R^{Fe/bath}$. In this study, less-noble Fe preferentially electrodeposited rather than Ni. For example, Ni-22at.%Fe alloy deposit ($R^{Fe/depo}$=22%) was obtained from the solution containing 1.5 % of Fe^{2+} ions ($R^{Fe/bath}$=1.5%). Fe ratio in deposit was condensed ca. 10 times higher than Fe^{2+} ratio in bath. This result can be explained by the anomalous codeposition mechanism due to the formation and adhesion of $Fe(OH)_2$ on cathode. In this mechanism, $Fe(OH)_2$ on cathode would suppress and inhibit the electrodeposition of Ni.

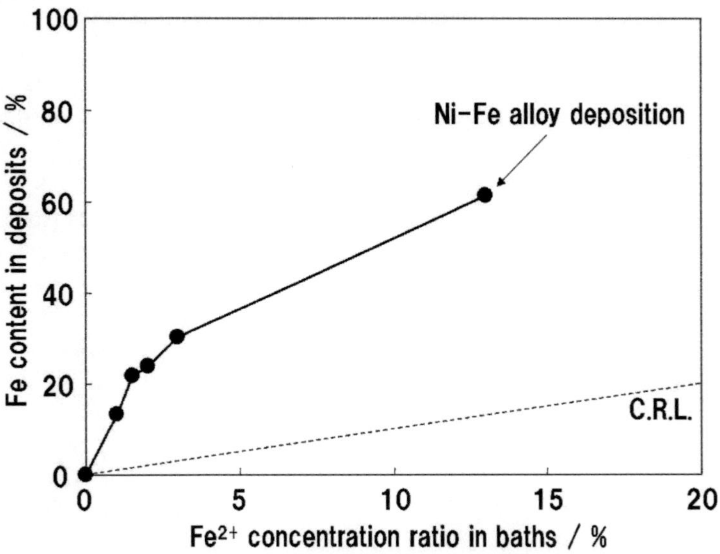

Fig. 9. Relationship between Fe^{2+} concentration ratio ($[Fe^{2+}]/([Fe^{2+}]+[Ni^{2+}])$) in baths and Fe content in Ni-Fe alloy deposits.

Fig. 10. SEM image of electrodeposited Ni-22at.%Fe alloy nanowires separated from polycarbonate membrane filter.

3.2 Crystal structure of Ni-Fe alloy nanowires

After the growing nanowires, polycarbonate membrane filters were dissolved in organic solvent (dichloromethane and chloroform) and the remains consisted of nanowires and a gold layer was served as a sample for SEM observation. Figure 10 shows SEM image of Ni-22at.%Fe alloy nanowires separated from polycarbonate template. Diameter and length of the nanowires corresponds well to that of nanochannels and the cylindrical shape was precisely transferred from the nanochannels to the nanowires as well as the case of pure Ni and pure Fe nanowires as shown in Fig.4.

3.3 Magnetic properties of Ni-Fe alloy nanowires

Magnetic hysteresis loops of electrodeposited films and nanowires were obtained using VSM with increasing the magnetic field up to 10 kOe. Figure 11 shows the magnetic hysteresis loops of Ni film and Ni-22at.%Fe alloy film electrodeposited on a copper foil. Magnetic field was applied to in-plan direction (red line) and perpendicular direction (blue line) to the film plan. As shown in these figures, in perpendicular direction, the films were hardly magnetized. On the other hand, in-plan direction, the films were easily magnetized and the magnetization reached to saturation at less than 1 kOe. Coercive force of Ni-22at.Fe alloy film was only ca. 1 Oe, which is quite smaller than that of Ni film (ca. 110 Oe). This is resulting from decreasing the magnetic anisotropy constant and magnetostriction constant of Ni with increase in Fe content in Ni-Fe alloy.

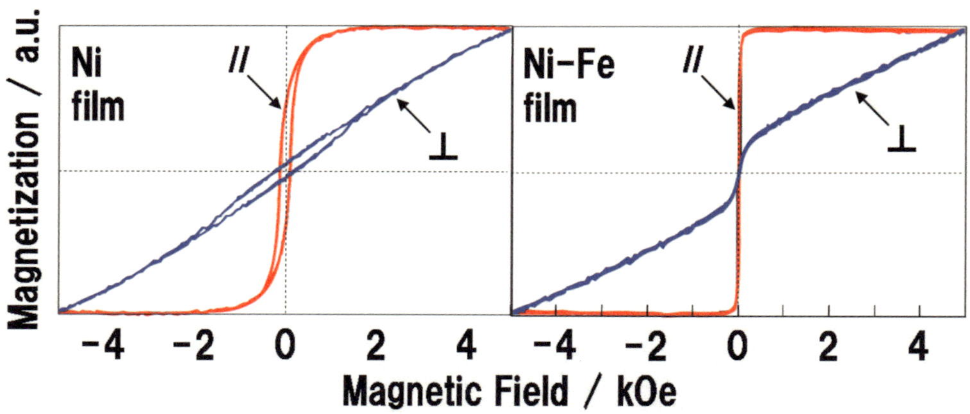

Fig. 11. Magnetic hysteresis loops of Ni film and Ni-22at.%Fe alloy film electrodeposited from aqueous solutions. Magnetic field was applied to in-plan direction (red line) and perpendicular direction (blue line) to the film plan.

Figure 12 shows the relationship between Fe content in Ni-Fe alloy deposits ($R^{Fe/depo}$) and the coercive force obtained from magnetic hysteresis loops with the applied magnetic field of in-plan direction. With increase in Fe content in deposits, the coercive force decreased down to ca. several Oe level. It is well known that coercive force, H_c is expressed by the following equation if the rotation process is dominant in magnetization.

$$H_c = 2K/M_s \quad (1)$$

Here, K and M_s mean magnetic anisotropy constant and saturated magnetization. Magnetic anisotropy constants of Ni and Fe are followings. K_{Ni} = -4.5 kJ/m^3 (-562.5 kGOe) and K_{Fe} =

+48 kJ/m³ (+6000 kGOe). Saturated magnetization of Ni and Fe are followings. M_{Ni} = 0.61 T (6100 G) and M_{Fe} = 2.16 T (21600 G). Therefore, the theoretical coercive force of Ni, H_{Ni} is estimated to be around 184 Oe, which is larger than the value (ca. 110 Oe) obtained in this study. If the magnetic anisotropy constant and saturated magnetization of Ni-Fe alloy can be expressed by the following equation,

$$K_{Ni\text{-}Fe} = K_{Ni} \times (1- R^{Fe/depo}) + K_{Fe} \times R^{Fe/depo} \quad (2)$$

$$M_{Ni\text{-}Fe} = M_{Ni} \times (1- R^{Fe/depo}) + M_{Fe} \times R^{Fe/depo} \quad (3)$$

The theoretical coercive force of Ni-Fe alloy, $H_{Ni\text{-}Fe}$ will be expressed by the following equation,

$$H_{Ni\text{-}Fe} = 2K_{Ni\text{-}Fe}/M_{Ni\text{-}Fe} \quad (4)$$

In this theory, Ni-9at.%Fe alloy will show minimum coercive force, which is almost zero.
On the other hand, if the domain wall process is dominant in magnetization, H_c is expressed by the following equation.

$$H_c \propto \lambda\sigma/M_s \quad (5)$$

Magnetostriction constant of Ni and Fe are followings. $\lambda_{Ni} = -2.0 \times 10^{-5}$ and $\lambda_{Fe} = +2.0 \times 10^{-5}$. In this theory, Ni-50at.%Fe alloy will show minimum coercive force, which is almost zero. In this study, as shown in Figure 4, the Ni-Fe alloy films with low coercive force were obtained over the wide range of Fe content from 20 % to 60 %.

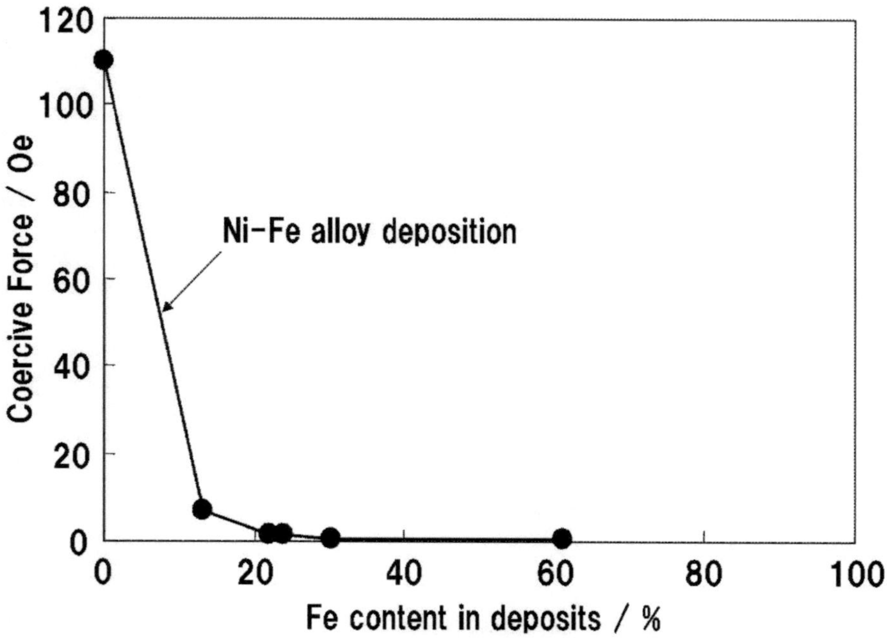

Fig. 12. Relationship between Fe content in Ni-Fe alloy deposits and the coercive force obtained from magnetic hysteresis loops of the alloy.

Figure 13 shows the magnetic hysteresis loops of Ni and Ni-22at.%Fe alloy nanowires electrodeposited into polycarbonate templates with channel-diameter of 160 nm. Magnetic field was applied to in-plan direction (red line) and perpendicular direction (blue line) to the membrane film plan. The perpendicular direction to the membrane film plan corresponds to the parallel direction to the long axis of nanowires.

As shown in these figures, in-plan direction, the nanowires were hardly magnetized. On the other hand, in perpendicular direction, the nanowires were easily magnetized and the magnetization reached to saturation at less than 2 kOe. Coercive force of Ni-22at.Fe alloy nanowires was ca. 100 Oe, which is almost half value of Ni nanowires (ca. 200 Oe) and the magnetic hysteresis loops show unique soft magnetic behavior. This is resulting from decreasing the uni-axial magnetic anisotropy due to the alloying effect of Fe. According to the equation (2), $K_{Ni\text{-}Fe}$ will decrease with increasing the content of Fe. On the other hand, due to the equation (3), $M_{Ni\text{-}Fe}$ will increase with increasing the content of Fe. Consequently, $H_{Ni\text{-}Fe}$ will decrease with increasing the content of Fe according to the equation (4).

Figure 14 shows the magnetic hysteresis loops of Ni nanowires and Ni-22at.%Fe alloy nanowires electrodeposited into polycarbonate templates with channel-diameter of 80, 160, 300 and 800 nm. Magnetic field was applied to perpendicular directions to the film plan, which corresponds to the parallel direction to the long axis of nanowires.

According to the figures, coercive force of Ni-22at.%Fe alloy nanowires was decreased with increasing the channel-diameter as well as the case of Ni nanowires. The coercive force of Ni-22at.%Fe alloy nanowires with channel-diameter of 800 nm reached down to 58 Oe. The magnetic hysteresis loop of the alloy nanowires shows unique linear dependence over the wide range of magnetic field.

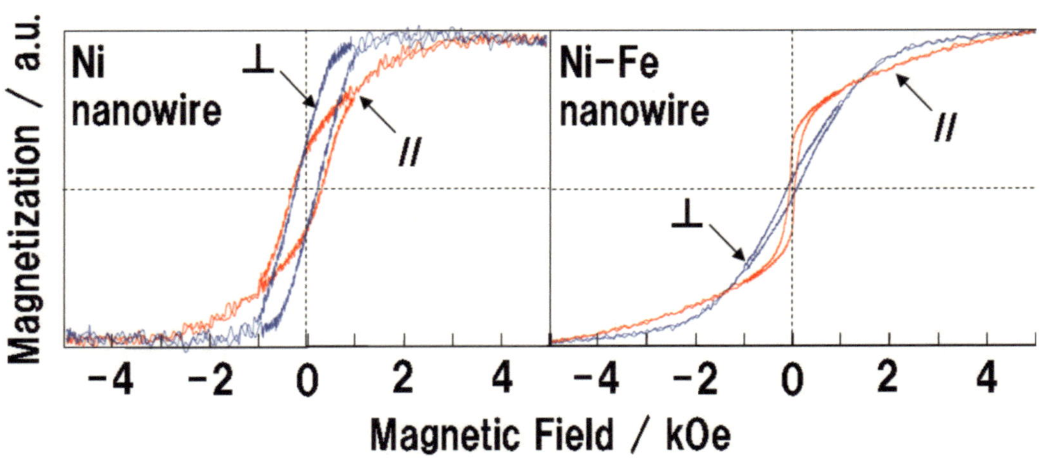

Fig. 13. Magnetic hysteresis loops of Ni nanowires and Ni-22at.%Fe alloy nanowires electrodeposited into polycarbonate templates with channel-diameter of 160 nm. Magnetic field was applied to in-plan direction (red line) and perpendicular direction (blue line) to the membrane film plan. The perpendicular direction to the membrane film plan corresponds to the parallel direction to the long axis of nanowires.

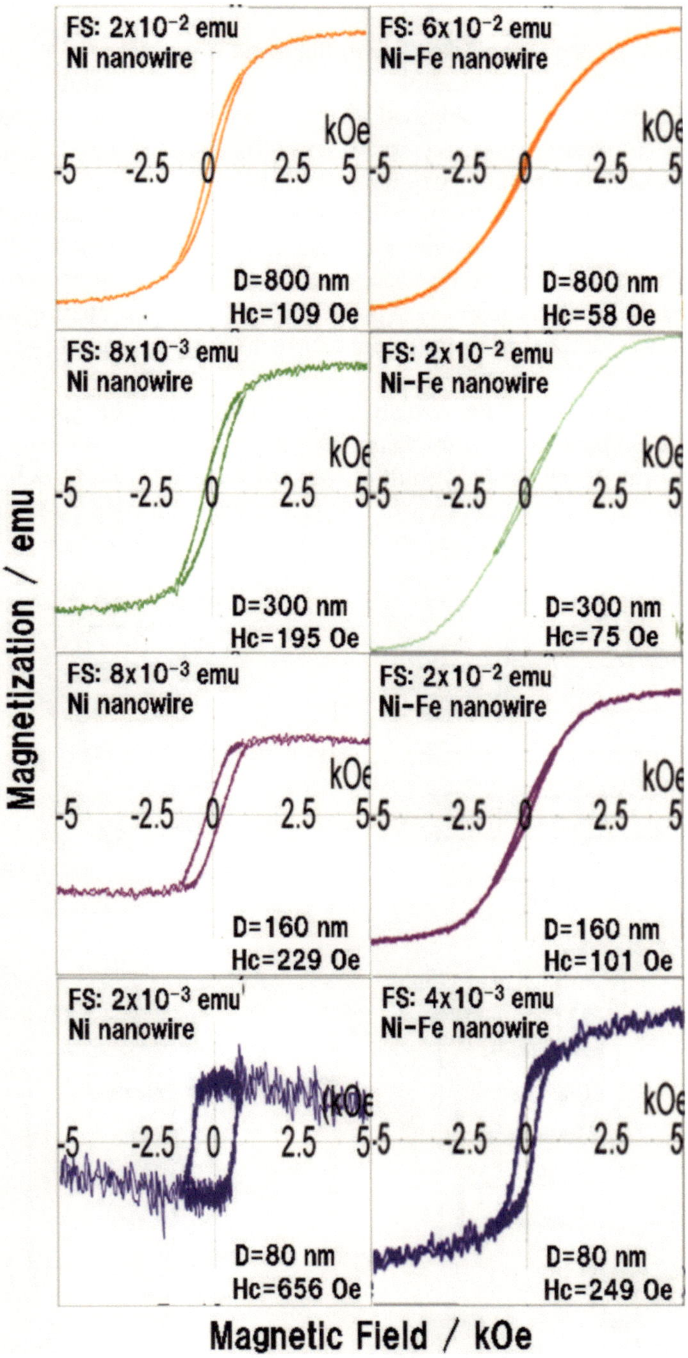

Fig. 14. Magnetic hysteresis loops of Ni nanowires and Ni-22at.%Fe alloy nanowires electrodeposited into polycarbonate templates with channel-diameter of 80, 160, 300 and 800 nm. Magnetic field was applied to perpendicular directions to the film plan, which corresponds to the parallel direction to the long axis of nanowires.

4. Co/Cu multilayered nanowires

4.1 Electrodeposition process of Co/Cu multilayered nanowires

Figure 15 illustrates the fabrication process of Co/Cu multilayered nanowires electrodeposited in nanochannels obtained by the heavy-ion track etching technique applied in this study: (a) recording of heavy ion tracks, (b) etching of tracks, (c) formation of electrodes, (d) electrodeposition of Co/Cu nanowires.

First, by irradiating the sample with an ion beam from a linear accelerator, straight ion tracks with a density of 10^8 tracks cm^{-2} were created perpendicular to the film surface as shown in Fig.15-(a). The tracks were selectively etched in a 6 M aqueous solution of NaOH to produce cylindrical nanochannels as depicted in Fig.15-(b). The etching process following the irradiation was optimized to produce a uniform cross section all along the channel with very small roughness.

A copper layer of thickness 1 μm, which acts as cathode for the nanowire growth, was electrodeposited on a gold layer of thickness 100 nm sputtered on the membrane as shown in Fig.15-(c). An aqueous electrolytic solution was synthesized from $CoSO_4 \bullet 7H_2O$ 120 g L^{-1}, $CuSO_4 \bullet 5H_2O$ 1.6 g L^{-1} and H_3BO_3 45 g L^{-1} for electrodeposition of Co-Cu alloy nanowires and Co/Cu multilayered nanowires as shown in Fig.15-(d).

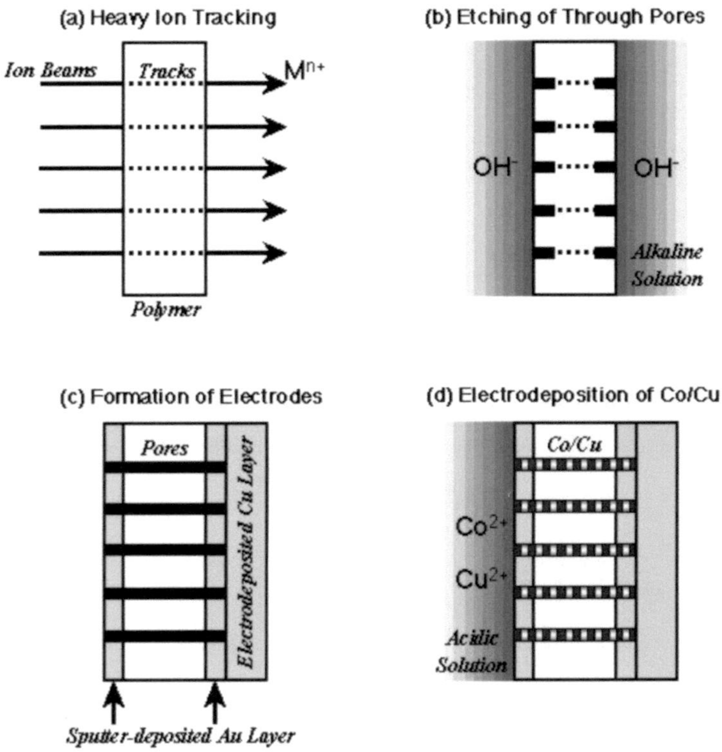

Fig. 15. Schematic of fabrication process of Co/Cu multilayered nanowires electrodeposited in nanochannels obtained by heavy-ion-track etching: (a) preparing latent tracks using heavy-ion beams, (b) etching of latent tracks leading to channels, (c) sputtering a gold layer on both sides of the surface and electrodeposition of Cu layer on the sputtered gold layer, (d) electrodeposition of Co/Cu nanowires in the channels.

A cathodic polarization curve was measured over a wide range of cathode potential to determine the optimum potential for Cu and Co deposition.

Figure 16 shows a cathodic polarization curve of polycarbonate template with channel length 30 μm and diameter 200 nm. The equilibrium potentials of Cu and Co are estimated to be around +0.05 V and -0.48 V (vs.Ag/AgCl) on the basis of the Nernst equation as follows.

$$E^{eq} = E^0 + RT\,(nF)^{-1}\ln C_M \qquad (6)$$

Here, E^0 is the standard electrode potential, and C_M denotes the concentration of metal ions (Cu^{2+} or Co^{2+}).

The cathodic current occurs at the potential region close to the equilibrium potential of Cu as shown in Fig.16. It is well-known that Cu^{2+} ions begin to electrodeposit without an accompanying overpotential from the aqueous solution. Therefore, this cathodic current corresponds to the deposition current of Cu. With increasing cathodic current, at around 10^{-5}A, the potential significantly polarizes to the less-noble region. This phenomenon seems to be caused by the diffusion control of Cu^{2+} ions. In the potential region less-noble than the equilibrium potential of Co, the cathodic current increases again at circa -0.8 V. It is also well-known that the electrodeposition of iron-group metals such as Ni, Co, and Fe is accompanied by the overpotential due to the rate determining multi-step reduction mechanism. Therefore, this increase in cathodic current is mainly caused by the deposition current of Co. It is well known that the potential dependence of growth rate for the nanowires corresponds well to the polarization curve for Cu and Co deposition.

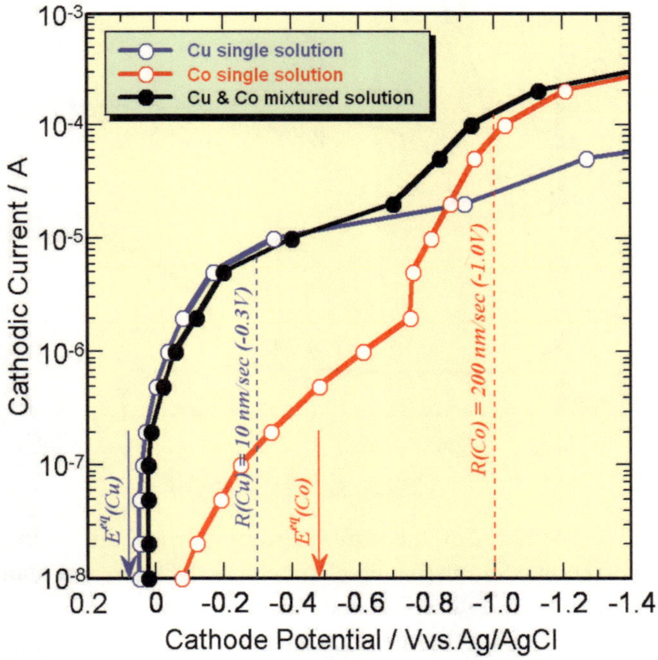

Fig. 16. Cathodic polarization curves for Cu (blue line), Co (red line) and Co-Cu alloy (black line) nanowires, both as a function of applied cathode potential during electrodeposition in a polycarbonate template with channel length 30 μm and diameter 200 nm.

Growth rates of nanowires were estimated by the channel filling time, which was determined from the time - dependence of deposition current at each potential.
Figure 17 shows the effect of the cathode potential on the time-dependence of the cathodic current during Co-Cu alloy nanowire deposition in the polycarbonate template with channel length 30 μm and diameter 200 nm. The cathode potentials were fixed to -0.7, -0.9, -1.1, and -1.3 V. To determine the wire growth rate, the channel-filling time was estimated by monitoring the deposition current. When the wires reach the membrane surface, the current will increase drastically due to the formation of hemispherical caps. If the radius of a hemispherical cap increases linearly with increase in deposition time, the current increases in proportion to the square of deposition time because the current linearly depends on the surface area of a hemispherical cap. Deposition rates were estimated by dividing channel length by channel-filling time. At –0.7 V, this time is around 600 s and the deposition rate is estimated to be about 50 nm s^{-1}, while the filling time is close to 60 s at –1.3 V and the deposition rate is estimated to be around 500 nm s^{-1}.

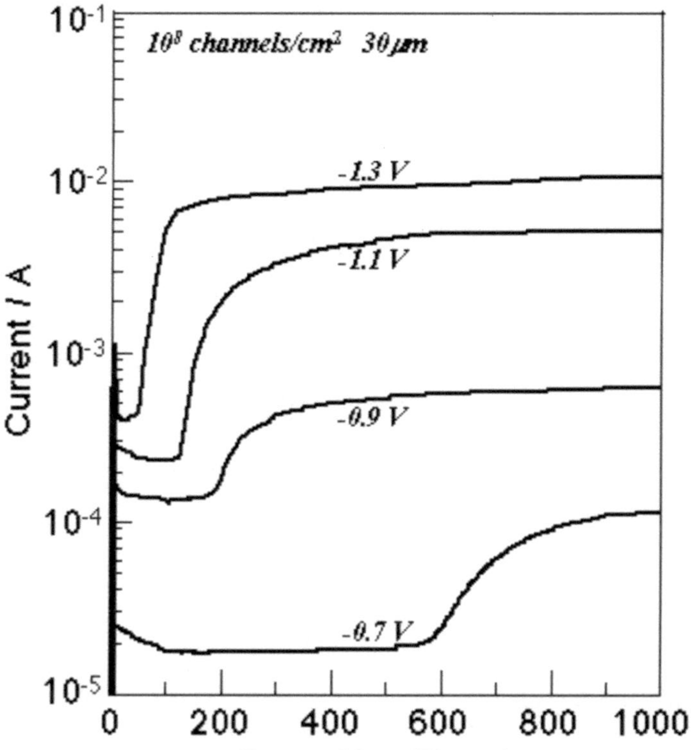

Fig. 17. Effect of cathode potential on measured current during Co-Cu alloy nanowire deposition in a polycarbonate template with channel length 30 μm and diameter 200 nm. The cathode potentials were -0.7, -0.9, -1.1, and -1.3 V.

On the basis of the results shown in Fig.16 and Fig.17, the optimum deposition potentials of Cu and Co are determined to be about –0.3 and –1.0 V (vs. Ag/AgCl), that is, at potentials nobler than the diffusion limit region of each metal ion. Typical deposition rates of Cu and Co were roughly 10 nm s^{-1} (at –0.3 V) and 200 nm s^{-1} (at –1.0 V).

Co/Cu multilayered nanowires were electrodeposited by alternatingly changing the cathode potential from -0.3 V vs. Ag/AgCl (for Cu layer) to -1.0 V (for Co layer) as shown in Fig.18. According to this figure, when the potential is switched from -1.0 V to -0.3 V, anodic current is observed. This is resulting from the dissolution of electrodeposited Co, because -0.3V is more noble than the equilibrium potential of Co. At this potential, it is estimated that the Cu deposition and Co dissolution will proceed simultaneously.

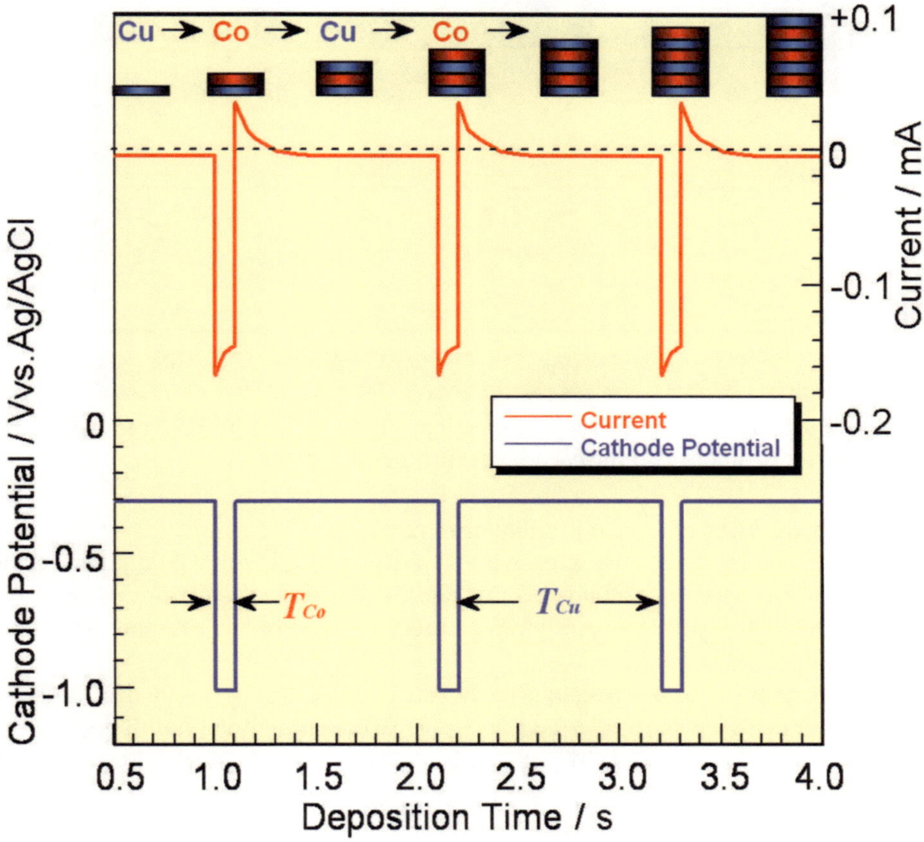

Fig. 18. Time dependence of observed-current and applied cathode potential during electrodeposition of Co/Cu multilayered nanowires.

4.2 Crystal structure of Co/Cu multilayered nanowires

To determine the alloy composition of the wires, a sample was prepared as follows. First, nanowires were electrodeposited at constant cathode potential in polycarbonate channels with length 6 μm and diameter 40 nm. Then, the polycarbonate was dissolved in an organic solvent to expose the wires. Finally, the wire composition was determined by EDX. Co-Cu alloy nanowires electrodeposited at –1.0 V contained 81%-Co and 19%-Cu.

Figure 19 shows the SEM image and EDX profiles of electrodeposited Co/Cu multilayered nanowires with the diameter of 40 nm and the each layer thickness of 30nm. According to the SEM image, multilayered structure is clearly observed and periodical EDX profiles also well correspond to the multilayered structure of the nanowires.

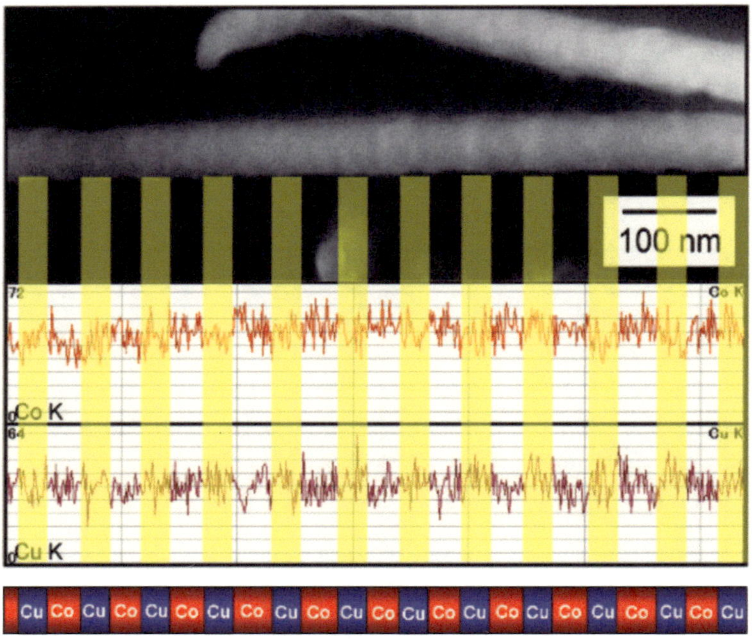

Fig. 19. SEM image and EDX profiles of electrodeposited Co/Cu multilayered nanowires with the diameter of 40 nm and the each layer thickness of 30nm.

4.3 Magnetic properties of Co/Cu multilayered nanowires

To measure the wire resistance the wires were in-situ contacted with a gold layer during the electrodeposition process as illustrated in Fig.15-(d). Magneto-resistance curves were measured at room temperature applying a direct current of 10 μA and changing the magnetic field up to 10 kOe.

Figure 20 exhibits the magnetoresistive hysteresis of Co-Cu alloy nanowires and Co/Cu multilayered nanowires electrodeposited in a polycarbonate template with channel length 30 μm and diameter 200 nm. Here, θ is defined as the angle between the magnetic field and the nanowire axis (current direction). The MR ratio is defined by the following equation

$$\text{MR ratio } (\%) = 100(R_0 - R_{10})/R_{10} \quad (7)$$

Here, R_0 and R_{10} are the resistance of zero field and 10 kOe, respectively. AMR curves of Co-Cu alloy nanowires show the usual dependence on the direction of the applied magnetic field. In the direction parallel to the wire (0 deg), the effect of magnetic field on the resistance was very small and the MR ratio was almost zero, while in the direction perpendicular to the wire (90 deg), the MR effect was maximum. An AMR ratio of 0.6 % was observed for $Co_{81}Cu_{19}$ alloy wires, which is typical of this system.

The structure of Co/Cu nanowires electrodeposited in polymer templates is illustrated schematically in Fig.20. The layer thickness of Co and Cu is around to 10 nm. GMR strongly depends on Co and Cu layer thickness, and a maximum effect could be observed at about 10 nm. Therefore, a thickness of 10 nm was realized for each layer by controlling the deposition time. The GMR curves of Co/Cu multilayered nanowires possess only a small angular dependence, and the MR ratio reaches up to 8.0 %. For these samples, the saturation field seems to be around 5 kOe, which is much smaller than the field from which the AMR curves of Co-Cu alloy nanowires result. We have already reported that in porous alumina membranes Co/Cu nanowires containing 100 and 300 bi-layers showed 20% and 12% of GMR ratio, respectively, while the Co/Cu nanowires containing 1,500 bi-layers showed 8.0% in this study. It is not surprising that the GMR ratio of Co/Cu nanowires will decrease when increasing the number of bi-layers. For example, this GMR decrease could be caused by the enhancement of Co/Cu interfacial roughness with increasing the number of bi-layers.

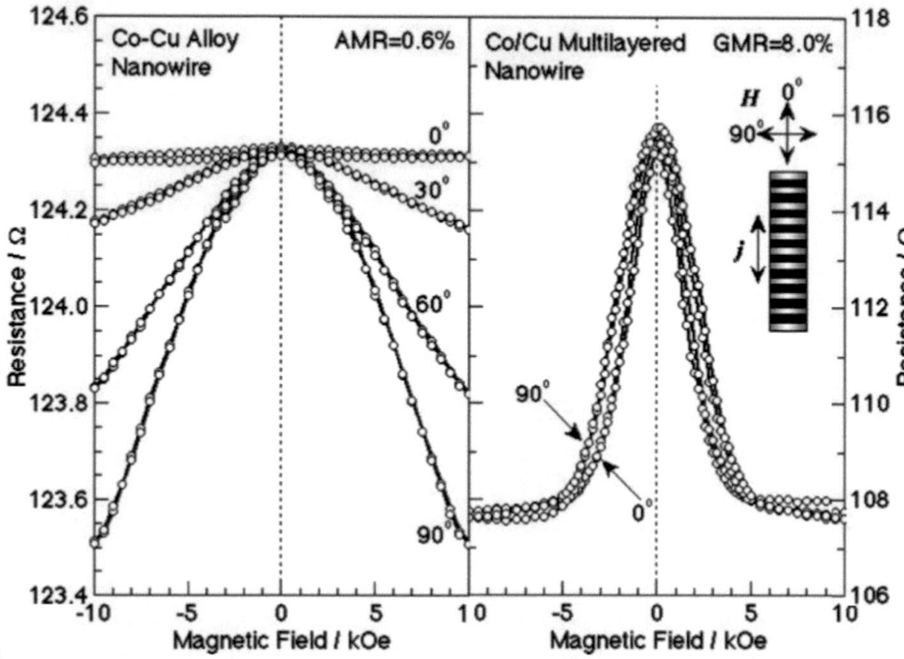

Fig. 20. Magnetoresistance of Co-Cu alloy nanowires and Co/Cu multilayered nanowires with length 30 μm and diameter 200 nm electrodeposited in polycarbonate nanochannels. The layer thickness of Co and Cu is 10 nm.

Figure 21 shows the magnetoresistance of Co/Cu multilayered nanowires with length 6 μm and diameter 60 nm electrodeposited in polycarbonate nanochannels. The each layer thickness of Co and Cu is adjusted to 10 nm. The GMR curves of Co/Cu multilayered nanowires with length of 6 μm possess only a small angular dependence as well as the case of multilayered nanowires with length of 30 μm as shown in Fig.20. MR ratio of multilayered nanowires with length of 6 μm reaches up to 12 % as shown in Fig.21, which is much larger than that of multilayered nanowires with length of 30 μm. This is resulting from the difference of layered structure in nanowires. Layered structure will depend on the number of layers. With increasing number of layers, interlayer structure will be disappeared and each layer will form alloy. Therefore, GMR of multilayered nanowires with short length would show larger value than that of the multilayered nanowires with long length. Furthermore, GMR value will also depend on the magnetic domain structure in nanowires. It is well known that the magnetic domain structure depends on the diameter size of nanowires. With decreasing the diameter, the domain structure will be changed from multi-domain to single domain. The domain wall in ferromagnetic layer will decrease GMR value. The domain wall area of nanowires with diameter 60 nm is smaller than that of nanowires with diameter 200 nm. Consequently, GMR of multilayered nanowires with short diameter would show larger value than that of the multilayered nanowires with long diameter.

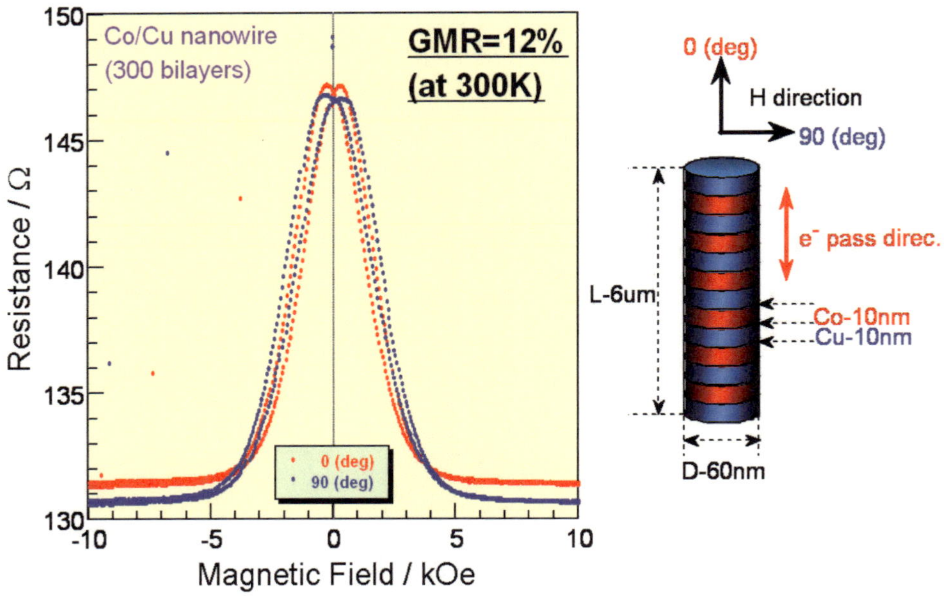

Fig. 21. Magnetoresistance of Co/Cu multilayered nanowires with length 6 μm and diameter 60 nm electrodeposited in polycarbonate nanochannels. The layer thickness of Co and Cu is 10 nm. Magnetic field was applied to in-plan direction (blue line) and perpendicular direction (red line) to the membrane film plan. The perpendicular direction to the membrane film plan corresponds to the parallel direction to the long axis of nanowires.

5. Summary

5.1 Ni, Co and Fe nanowires

Optimum deposition potential region for growing Ni, Co and Fe nanowires was determined to be the range from -0.9 to -1.2 V. Growth rate of Ni nanowires was ca. 6 nm•sec^{-1} at the cathode potential of -1.0 V. Growth rate of nanowires was increased up to around 30 nm•sec^{-1} with increasing cathode potential up to -1.2 V.
The cylindrical shape was precisely transferred from the nanochannels to the nanowires and the aspect ratio reached up to ca. 150. The each nanowire was consisted of single crystalline domain. Basis on the uni-axial magnetic anisotropy and single magnetic domain structure of the ferromagnetic metal nanowires, the magnetic hysteresis loops showed typical perpendicular magnetization behavior and the coercive force reached up to around 1 kOe.

5.2 Ni-Fe alloy nanowires

Optimum deposition potential for electrodeposition of Ni-Fe alloy is determined to be ca. -1.0 V. Fe ratio in deposit was condensed more than 10 times higher than Fe^{2+} ratio in bath. This phenomena was explained by the anomalous codeposition mechanism due to the

formation and adhesion of $Fe(OH)_2$ on cathode. According to the mechanism, $Fe(OH)_2$ on cathode would suppress and inhibit the electrodeposition of Ni.
Coercive force of Ni-22at.Fe alloy film was ca. 1 Oe, which is quite smaller than that of Ni film (ca. 110 Oe). Ni-Fe alloy films with low coercive force were obtained over the wide range of Fe content from 20 % to 60 %.
Growth rate of Ni and Ni-Fe alloy nanowires was ca. 6 nm•sec^{-1} at the cathode potential of -1.0 V. The cylindrical shape was precisely transferred from the nanochannels to the nanowires as well as the case of pure Ni and pure Fe nanowires.
Coercive force of Ni-22at.Fe alloy nanowires was ca. 100 Oe, which is almost half value of Ni nanowires (ca. 200 Oe). Magnetic hysteresis loops show unique soft magnetic behavior due to decreasing the uni-axial magnetic anisotropy resulting from the alloying effect of Fe.

5.3 Co/Cu multilayered nanowires

Optimum deposition potentials of Cu and Co are determined to be about –0.3 and –1.0 V (vs. Ag/AgCl), that is, at potentials nobler than the diffusion limit region of each metal ion. Typical deposition rates of Cu and Co were roughly 10 nm s^{-1} (at –0.3 V) and 200 nm s^{-1} (at –1.0 V). Co/Cu multilayered nanowires were electrodeposited by alternatingly changing the cathode potential from -0.3 V vs. Ag/AgCl (for Cu layer) to -1.0 V (for Co layer). Co-rich alloy electrodeposited at –1.0V consisted of 81%-Co and 19%-Cu.
GMR ratio of Co/Cu multilayered nanowires with length 6 μm, diameter 60 nm and layer-thickness 10 nm reaches up to 12 %. GMR curves of the Co/Cu multilayered nanowires showed only a small angular dependence. It was suggested that the interlayer structure will be disappeared and each layer will form alloy with increasing number of layers. Co/Cu multilayered nanowires with short length and short diameter would show larger GMR value than that of the multilayered nanowires with long length and long diameter due to the magnetic domain structure.

6. References

Kawai, S. & Ueda, R. (1975). Magnetic properties of anodic oxide coatings on aluminum containing electrodeposited Co and Co-Ni. *J. Electrochem. Soc.*, Vol.122, Issue 1, (January 1975) 32-36, ISSN 0013-4651

Kawai, S. (1975). Retentivities of anodic oxide coatings on aluminum containing electrodeposited cobalt, nickel, and cobalt-nickel. *J. Electrochem. Soc.*, Vol.122, Issue 8, (August 1975) 1026-1029, ISSN 0013-4651

Kawai, S. & Ishiguro, I. (1976). Recording characteristics of anodic oxide films on aluminum containing electrodeposited ferromagnetic metals and alloys. *J. Electrochem. Soc.*, Vol.123, Issue 7, (July 1976) 1047-1051, ISSN 0013-4651

Tsuya, N.; Saito, Y.; Nakamura, H.; Hayano, S.; Furugohri, A.; Ohta, K.; Wakui, Y. & Tokushima, T. (1986). A perpendicular magnetic recording medium by alumite. *J. Magn. Magn. Mater.*, Vol.54-57, Part 3, (February 1986) 1681-1682, ISSN 0304-8853

Tsuya, N.; Tokushima, T.; Shiraki, M.; Wakui, Y.; Saito, Y.; Nakamura, H. & Katsumata, Y. (1987). Magnetic alumite disc for perpendicular recording. *IEEE Trans. Magn.* Vol.23, Issue 1, (January 1987) 53- 55, ISSN 0018-9464

Huysmans, G. T. A.; Lodder, J. C. & Wakui, J. (1988). Magnetization curling in perpendicular iron particle arrays (alumite media). *J. Appl. Phys.*, Vol.64, Issue 4, (August 1988) 2016-2021, ISSN 0021-8979

Ishiia, Y. & Sato, M. (1989). Magnetic behavior of a film with columnar structure. *J. Magn. Magn. Mater.*, Vol.82, Issues 2-3, (December 1989) 309-312, ISSN 0304-8853

Cheng, T. -J.; Jorné, J. & Gau, J. -S. (1990). Magnetic anisotropy of electrodeposited cobalt on alumite substrate. *J. Electrochem. Soc.*, Vol.137, Issue 1, (January 1990) 93-95, ISSN 0013-4651

Zhang, L. C. & Lodder, J. C. (1990). The influence of the packing density on the magnetic behaviour of alumite media. *J. Magn. Magn. Mater.*, Vol.88, Issues 1-2, (July 1990) 236-246, ISSN 0304-8853

Martin, C. R. (1991). Template synthesis of polymeric and metal microtubules. *Adv. Mater.*, Vol.3, Issue 9, (September 1991) 457-459, ISSN 0935-9648

Aimawlawi, D.; Coombs, N. & Moskovits, M. (1991). Magnetic properties of Fe deposited into anodic aluminum oxide pores as a function of particle size. *J. Appl. Phys.*, Vol.70, Issue 8, (October 1991) 4421-4425, ISSN 0021-8979

Whitney, T. M.; Jiang, J. S.; Searson, P. C. & Chien, C. L. (1993). Fabrication and magnetic properties of arrays of metallic nanowires. *Science,* Vol.261, Issue 5126, (September 1993) 1316-1319, ISSN 0036-8075

Martin, C. R. (1994). Nanomaterials: a membrane-based synthetic approach. *Science,* Vol.266, Issue 5193, (December 1994) 1961-1966, ISSN 0036-8075

Piraux, L.; George, J. M.; Despres, J. F.; Leroy, C.; Ferain, E.; Legras, R.; Ounadjela, K. & Fert, A. (1994). Giant magnetoresistance in magnetic multilayered nanowires. *Appl. Phys. Lett.*, Vol.65, Issue 19, (November 1994) 2484-2486, ISSN 0003-6951

Blondel, A.; Meier, J. P.; Doudin, B. & Ansermet, J.-Ph. (1994). Giant magnetoresistance of nanowires of multilayers. *Appl. Phys. Lett.*, Vol 65, Issue 23, (December 1994) 3019-3021, ISSN 0003-6951

Dubois, S.; Marchal, C.; Beuken, J. M.; Piraux, L.; Duvail, J. L.; Fert, A.; George, J. M. & Maurice, J. L. (1997). Perpendicular giant magnetoresistance of NiFe/Cu multilayered nanowires. *Appl. Phys. Lett.*, Vol.70, Issue 3, (January 1997) 396-398, ISSN 0003-6951

Hulteen, J. C. & Martin, C. R. (1997). A general template-based method for the preparation of nanomaterials. *J. Mater. Chem.*, Vol.7, Issue 7, (July 1997) 1075–1087, ISSN 0959-9428

Nielsch, K.; Müller, F.; Li, A. P. & Gösele, U. (2000). Uniform nickel deposition into ordered alumina pores by pulsed electrodeposition. *Adv. Mater.*, Vol.12, Issue 8, (April 2000) 582-586, ISSN 0935-9648

Evans, P. R.; Yi, G. & Schwarzacher, W. (2000). Current perpendicular to plane giant magnetoresistance of multilayered nanowires electrodeposited in anodic aluminum oxide membranes. *Appl. Phys. Lett.*, Vol.76, Issue 4 (January 2000) 481-483, ISSN 0003-6951

Forrer, P.; Schlottig, F.; Siegenthaler, H. & Textor, M. (2000). Electrochemical preparation and surface properties of gold nanowire arrays formed by the template technique. *J. Appl. Electrochem.*, Vol.30, No.5, (May 2000) 533-541, ISSN 0021-891X

Sellmyer, D. J.; Zheng, M. & Skomski, R. (2001). Magnetism of Fe, Co and Ni nanowires in self-assembled arrays. *J. Phys.: Condens. Matter.,* Vol.13, Issue 25 (June 2001) R433-R460, ISSN 0953-8984

Sauer, G.; Brehm, G.; Schneider, S.; Nielsch, K.; Wehrspohn, R. B.; Choi, J.; Hofmeister, H. & Gösele, U. (2002). Highly ordered monocrystalline silver nanowire arrays. *J. Appl. Phys.*, Vol.91, Issue 5, (March 2002) 3243-3247, ISSN 0021-8979

Ohgai, T.; Hoffer, X.; Gravier, L.; Wegrowe, J. E. & Ansermet, J.-Ph. (2003). Bridging the gap between template synthesis and microelectronics: spin-valves and multilayers in self-organized anodized aluminum nanopores. *Nanotechnology,* Vol.14, Issue 9, (September 2003) 978-982, ISSN 0957-4484

Ohgai, T.; Hoffer, X.; Fabian, A.; Gravier, L. & Ansermet, J.-Ph. (2003). Electrochemical synthesis and magnetoresistance properties of Ni, Co and Co/Cu nanowires in nano-porous anodic oxide layer on metallic aluminum. *J. Mater. Chem.*, Vol.13, Issue 10, (October 2003) 2530-2534, ISSN 0959-9428

Enculescu, I.; Siwy, Z.; Dobrev, D.; Trautmann, C.; Toimil, Molares, M. E.; Neumann, R.; Hjort, K.; Westerberg, L. & Spohr, R. (2003). Copper nanowires electrodeposited in etched single-ion track templates. *Appl. Phys. A: Mater. Sci. Process.,* Vol.77, No.6, (November 2003) 751-755, ISSN 0947-8396

Ohgai, T.; Gravier, L.; Hoffer, X.; Lindeberg, M.; Hjort, K.; Spohr, R. & Ansermet, J.-Ph. (2003). Template synthesis and magnetoresistance property of Ni and Co single nanowires electrodeposited into nano-pores with wide range of aspect ratios. *J. Phys. D: Appl. Phys.*, Vol.36, Issue 24, (December 2003) 3109-3114, ISSN 0022-3727

Oh, J.; Tak, Y. & Lee, J. (2004). Electrodeposition of Cu_2O nanowires using nanoporous alumina template. *Electrochem. Solid-State Lett.*, Vol.7, Issue 3, (January 2004) C27-C30, ISSN 1099-0062

Wu, M. T.; Leu, I. C.; Yen, J. H. & Hon, M. H. (2004). Preparation of Ni nanodot and nanowire arrays using porous alumina on silicon as a template without a conductive interlayer. *Electrochem. Solid-State Lett.*, Vol.7, Issue 5, (March 2004) C61-C63, ISSN 1099-0062

Wang, W.; Huang, Q.; Jia, F. & Zhu, J. (2004). Electrochemically assembled p-type Bi_2Te_3 nanowire arrays. *J. Appl. Phys.,* Vol.96, Issue 1, (July 2004) 615-618, ISSN 0021-8979

Ohgai, T.; Hoffer, X.; Gravier, L. & Ansermet, J.-Ph. (2004). Electrochemical surface modification of aluminum sheets for application to nanoelectronic devices: anodization aluminum and electrodeposition of cobalt-copper. *J. Appl. Electrochem.,* Vol.34, No.10, (October 2004) 1007-1012, ISSN 0021-891X

Chu, S. Z.; Inoue, S.; Wada, K.; Kanke, Y. & Kurashima, K. (2005). Fabrication and characterization of integrated ultrahigh-density Fe-Pt alloy nanowire arrays on glass. *J. Electrochem. Soc.*, Vol.152, Issue 1, (December 2004) C42-C47, ISSN 0013-4651

Ohgai, T.; Gravier, L.; Hoffer, X. & Ansermet, J.-Ph. (2005). CdTe semiconductor nanowires and NiFe ferro-magnetic metal nanowires electrodeposited into cylindrical nano-pores on the surface of anodized aluminum. *J. Appl. Electrochem.,* Vol.35, No.5, (May 2005) 479-485, ISSN 0021-891X

Zhang, X.; Hao, Y.; Meng, G. & Zhang, L. (2005). Fabrication of highly ordered InSb nanowire arrays by electrodeposition in porous anodic alumina membranes. *J. Electrochem. Soc.,* Vol.152, Issue 10, (August 2005) C664-C668, ISSN 0013-4651

Fukunaka, Y.; Motoyama, M.; Konishi, Y. & Ishii, R. (2006). Producing shape-controlled metal nanowires and nanotubes by an electrochemical method. *Electrochem. Solid-State Lett.*, Vol.9, Issue 3, (January 2006) C62-C64, ISSN 1099-0062

Carlier, D. & Ansermet, J.-Ph. (2006). Electrochemical synthesis and magnetic properties of $CoFe_2O_4$ nanowire arrays. *J. Electrochem. Soc.*, Vol.153, Issue 5, (March 2006) C277-C281, ISSN 0013-4651

Ohgai, T.; Enculescu, I.; Zet, C.; Westerberg, L.; Hjort, K.; Spohr, R. & Neumann, R. (2006). Magneto-sensitive nickel nanowires fabricated by electrodeposition into multi- and single-ion track templates. *J. Appl. Electrochem.*, Vol.36, No.10, (October 2006) 1157-1162, ISSN 0021-891X

Wang, X. W.; Fei, G. T.; Chen, L.; Xu, X. J. & Zhang, L. D. (2007). Orientation-controllable growth of Ni nanowire arrays with different diameters. *Electrochem. Solid-State Lett.*, Vol.10, Issue 4, (February 2007) E1-E3, ISSN 1099-0062

Friedman, A. L. & Menon, L. (2007). Optimal parameters for synthesis of magnetic nanowires in porous alumina templates. *J. Electrochem. Soc.*, Vol.154, Issue 4, (February 2007) E68-E70, ISSN 0013-4651

Katkar, R. A. & Tait, G. B. (2007). The effect of stationary ultraviolet excitation on the optical properties of electrochemically self-assembled semiconductor nanowires. *J. Appl. Phys.*, Vol.101, Issue 5, (March 2007) 053508, ISSN 0021-8979

Inguanta, R.; Sunseri, C. & Piazza, S. (2007). Photoelectrochemical characterization of Cu_2O-nanowire arrays electrodeposited into anodic alumina membranes. *Electrochem. Solid-State Lett.*, Vol.10, Issue 12, (October 2007) K63-K66, ISSN 1099-0062

Riveros, G.; Gómez, H.; Schrebler, R.; Marotti, R. E. & Dalchiele, E. A. (2008). An in situ EIS study during the electrochemical growth of copper nanowires into porous polycarbonate membranes. *Electrochem. Solid-State Lett.*, Vol.11, Issue 3, (December 2007) K19-K23, ISSN 1099-0062

Kazeminezhad, I. & Schwarzacher, W. (2008). Electrodeposited Ni–Cu alloy nanowires with arbitrary composition. *Electrochem. Solid-State Lett.*, Vol.11, Issue 3, (January 2008) K24-K26, ISSN 1099-0062

Zhu, R.; Zhang, H.; Chen, Z.; Kryukov, S. & DeLong, L. (2008). Horizontally aligned single array of Co nanowires fabricated in one-dimensional nanopore array template. *Electrochem. Solid-State Lett.*, Vol.11, Issue 6, (April 2008) K57-K60, ISSN 1099-0062

Ohgai, T.; Hjort, K.; Spohr, R. & Neumann, R. (2008). Electrodeposition of cobalt based ferromagnetic metal nanowires in polycarbonate films with cylindrical nanochannels fabricated by heavy-ion-track etching. *J. Appl. Electrochem.*, Vol.38, No.5, (May 2008) 713-719, ISSN 0021-891X

Ishizaki, T.; Saito, N. & Takai, O. (2009). Surfactant-assisted fabrication of tin oxide nanowires through one-step electrochemically induced chemical deposition. *J. Electrochem. Soc.*, Vol.156, Issue 10, (August 2009) D413-D417, ISSN 0013-4651

Motoyama, M.; Dasgupta, N. P. & Prinz, F. B. (2009). Electrochemical deposition of metallic nanowires as a scanning probe tip. *J. Electrochem. Soc.*, Vol.156, Issue 10, (August 2009) D431-D438, ISSN 0013-4651

Walton, A. S.; Górzny, M.; Bramble, J. P. & Evans, S. D. (2009). Photoelectric properties of electrodeposited copper (I) oxide nanowires. *J. Electrochem. Soc.*, Vol.156, Issue 11, (September 2009) K191-K195, ISSN 0013-4651

4

Annealing Effects on the Metal and Semiconductor Nanowires Loaded Inside the Alumina Pores

Arūnas Jagminas
Institute of Chemistry
Lithuania

1. Introduction

Alumina template synthesis is a simple and cheap method for preparing arrays of nanostructural materials, such as wires, fibrils, tubes, which entails the synthesis of desired material within the cylindrically shaped pores by the respective method. To date various materials possessing anisotropic, tunneling, and giant magnetoresistance, superconductivity, semiconductivity, unusual non-linear optics, surface enhanced Raman scattering, and enhanced catalytic properties have been loaded inside the alumina pores and applied in practice. Currently the advancement of nm-scaled materials technology has prompted investigations devoted to their stability during storage in air, exploitation, irradiation, and heat-treatments under various environments. Furthermore, the heat treatment of some materials such as iron oxyhydroxides deposited inside the alumina pores seems still probably the most appropriate route to fabricate prospective magnetic nanowire arrays. The knowledge of semiconducting nanospecies thermal behavior is also crucial in their successful use to destroy selectively the tumor cells.

This paper presents our recent progress in transformation, design, characterization and potential applications of nanowired materials encased within the alumina pores upon heat treatment in air. Unexpected behavior of some metal nanowires loaded inside the alumina pores, such as iron and tin, upon heat treatment in air is presented here for the first time. As will be shown below, the heating of such arrays at proper conditions results in the formation of spinel-type $FeAl_2O_4$ films and SnO/SnO_2 barbed-shaped nanowire arrays. Also, the transformations of some semiconducting nanowires, particular FeO(OH), Cu_3Se_2 and $Cu_{2-x}Se$, upon heating are also presented herein. Finally, new fields of prospective applications of annealed nanowire arrays owing to gained magnetic and non-linear optical properties are discussed.

The paper is organized as follows: In the next subsection, the formation, structure and composition of most popular porous alumina templates are briefly presented. In subsection 2.2, a short overview of annealing effects, which have been reported to date for nanowire arrays loaded inside the alumina pores, is given. Our results obtained by heat treatment of nanowires of metallic iron, tin and semiconducting iron oxyhydroxide, FeO(OH), encapsulated within the alumina pores, are presented in section 3, while section 4 is devoted to annealing effects of copper selenide nanowire arrays. Finally, section 5 presents a short summary and conclusions.

2.1 Porous alumina

Porous oxide films (alumina) prepared *via* anodization of highly pure and smooth aluminum surface in an aqueous solution of sulfuric, oxalic, and o-phosphoric acid at proper concentration, temperature, and voltage are typical templates for fabrication of arrays of various nanowires in a quite uniform diameter and spacing with well-defined product dimensions at a packing density of 10^9 - 10^{11} species/cm^2. Recently, highly-ordered structure alumina with a broad range of pore diameters as large as 300 nm (Quin et al. 2005) and as small as 5 nm can be fabricated in an unlimited size area. The pore diameter, cell size, and barrier layer thickness positioned at the bottom of alumina pores (Fig. 1) are all linearly dependent on the anodizing voltage with some exceptions (Jagminas et al. 2001), while the depth of pores is simply controlled by anodizing time (Metzger et al. 2000). According to the final applications the thickness of alumina templates is usually limited to 20-30 μm, but much thicker templates (Römer & Steinem 2004), as well as very thin (Kokonou et al. 2007), are sometimes required.

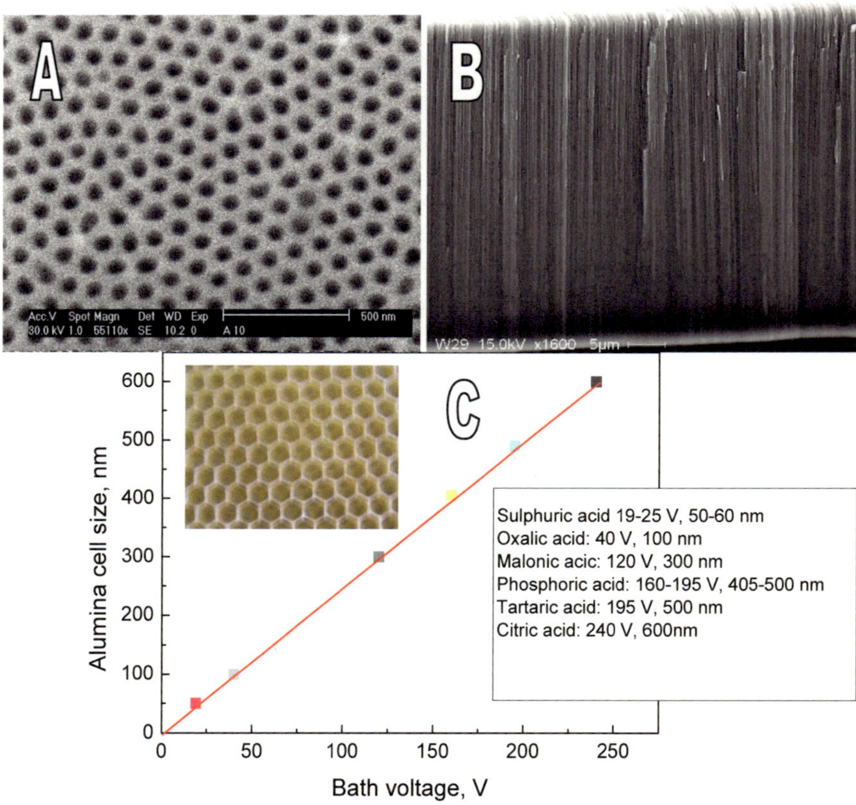

Fig. 1. A typical top-side (A) and cross-sectional (B) SEM view of porous alumina templates. (C) demonstrates variations of alumina cell size on the anodizing voltage for various electrolytes.

The chemical composition of alumina films differs depending on the composition of anodizing solution as well as anodizing conditions due to incorporation of acid anions and water molecules into the outer part of alumina cells. For example, the sulfuric acid alumina films contain high amounts of anionic species (12-14 wt.% sulfate), while the phosphate and oxalate contents in corresponding alumina films are respectively 6-8 and 2-4 wt.% (Thompson 1997). Note that incorporated anion species produce a negative surface charge of the pore walls influencing the pore filling process by one or another material. Therefore, the hydrophobic/hydrophilic pore wall properties play a significant role for nanowire (nanotube) growth in the precursor solution. We suggest that the wetability of pore walls could also influence the nanowire behavior during their storage in the matrix and subsequent treatments, such as annealing (see further).

2.2 Annealing effects in nanowires

Highly-ordered porous aluminas possess quite uniform and parallel nanochannel structure (Fig.1). Therefore, the nanowires loaded within the alumina pores usually possess almost the same diameter as the diameter of pores. The exception here demonstrates nanowires deposited by the pulse current where various diameters of Bi nanowires were obtained *via* variation of the pulse duration (Li et al. 2005). However, in the case of complete filling of the alumina pores, the pore walls that separate the nanowires prevent nanowire diameter variation during high temperature measurements because the thermal expansion coefficient of alumina films is significantly lower ($\sim 8.8 \times 10^{-6}$ K^{-1}) than that of the metals: Cu ($\sim 2.1 \times 10^{-5}$ K^{-1}), Ag ($\sim 1.7 \times 10^{-5}$ K^{-1}), Ni ($\sim 1.7 \times 10^{-5}$ K^{-1}), Bi ($\sim 1.4 \times 10^{-5}$ K^{-1}). Thus, at higher heating temperatures the alumina cell walls might produce a radial pressure on the nanowire. As reported (Cai et al. 2008), the radial pressure can be relaxed by the axial expansion if the deposited nanowires are single crystalline demonstrating the complex thermal behavior during annealing in vacuum. In contrast, in the case of polycrystalline nanowires, there are enough of liquid-like spaces between nanocrystallites (Gleiter 2000) to release the stress from the lattice expansion and, thus, the thermal expansion coefficients of such nanowires are almost the same as that of the corresponding bulk.

Wang et al. (2004) claimed that copper nanowires in alumina templates have a thermal expansion coefficient close to zero and this is attributed to the diffusion of vacancy defects from the body to the nanowire surface. The thermal expansion coefficient of Ni nanowires grown within the alumina pores ($Ø_{pore}$ ~60 nm) at -2.0 V by direct current deposition was also studied *in situ* by specimen heating up to 600 °C in vacuum (Cai et al. 2008). These nanowires after deposition were found to be in the polycrystalline state with nanocrystallites up to 40 nm in size. The thermal expansion coefficient of Ni nanocrystallites in the polycrystalline Ni nanowires was estimated to be 1.7×10^{-5} K^{-1} as in the bulk Ni, enabling to conclude that there is, as well as in bulk Ni, enough space between nanocrystallites to release the stress from the lattice expansion during heat treatment.

The thermal expansion of single crystalline Ag nanowires with the axis direction along the [220] crystallographic direction deposited within the alumina pores ($Ø_{pore}$ ~55 nm) was studied by determining the axial lattice parameter and the state of vacancies in nanowires for the first and repeated annealing in vacuum (Xu et al. 2006). In this study it was shown that the plots of variation of axial lattice parameter for as-grown and annealed silver nanowires can be divided into two regions. When the temperature is bellow 650 °C for as-grown and 350 °C for annealed specimen, the axial lattice parameter of silver nanowires

keeps nearly constant, but has a slight increase above the indicated temperatures. Thus, the authors concluded that the thermal expansion coefficient of the single crystalline nanowires is determined by two factors: the axial expansion resulted from the radial contraction, which was induced by the surface pressure, and the contraction caused by vacancies. It has also been postulated that after annealing in vacuum the contribution of vacancies to the axial expansion in Ag nanowires decrease due to their congregation into larger voids.
The annealing in vacuum of alumina template, encased with metal alloy nanowire arrays, such as Ag/Cu assembled by nanoparticles of copper and silver, results in the formation of solid solution Ag(Cu) nanowires (Xu et al. 2008). Furthermore, the surface plasmon resonance of Ag nanowires in an average diameter of about 25 nm centered at 410 nm (3.02 eV) red shifted after annealing to ~495 nm (2.50 eV) when the molar ratio of copper to silver increased up to ~40 %.

3. Annealing in air

The heating of alumina templates with nanowires, such as Fe, Cu, FeO(OH) in air can lead to the formation of thermodynamically more stable nanowired metal oxides sometimes demonstrating unique magnetic or optical properties. Also, the annealing procedure has been successfully used for clustering of magnetite, Fe_3O_4, nanoparticles, encased in the alumina template pores by vacuum suction, into nanowires achieving a dramatic effect on their magnetic properties: the coercitivity of these arrays at cryogenic temperature increased from 350 to 8000 Oe (Choi et al. 2007).

3.1 Fe nanowire arrays

During recent two decades several works have been reported regarding the electrochemical synthesis of iron nanowires within the alumina pores and their size-dependent magnetic properties (AlMawlawi et al. 1991; Sun et al. 2001; Qin et al. 2005) and corrosion behavior (Jagminas et al. 2008). However, there has been no study trying to fabricate Fe_xO_y nanowire arrays by thermal annealing of embedded iron nanowires within the alumina pores or clarify their transformations during annealing. Thus, in this study we investigated these phenomena as a function of annealing conditions and the size of Fe nanowires by the SEM, XRD and Mössbauer spectroscopies in more detail. In contrast to earlier studies where for electrochemical deposition of iron an aqueous solution composed mainly of iron (II) sulphate and boric acid has been applied, we used a solution of iron (II) sulphate, magnesium sulphate, $MgSO_4$, and citric acid. We have found that the growth of iron nanowires in this solution under the alternating current mode at a constant average current strength proceeds faster (Fig. 2a) and uniformly (Fig. 2b). The prolonged deposition leads to complete filling of alumina pores and the powdered iron layer formation onto the template surface. X-ray diffraction measurements show only two diffraction peaks for as-deposited material, e.g. a strong peak at 2Θ 44.67° and a week one at 2Θ close to 65°. In agreement with the data of the polycrystalline bulk Fe (PDF: No 006-0696), these peaks indicate the formation of metallic α-Fe nanowires with a preferrential growth in direction [110] along the crystallographic axes. The decrease in the diameter of alumina pores, $\varnothing_{pore}$, results in the broadening of diffraction peaks, as expected, without any obvious changes in the growth direction of Fe nanowires (Fig. 3).

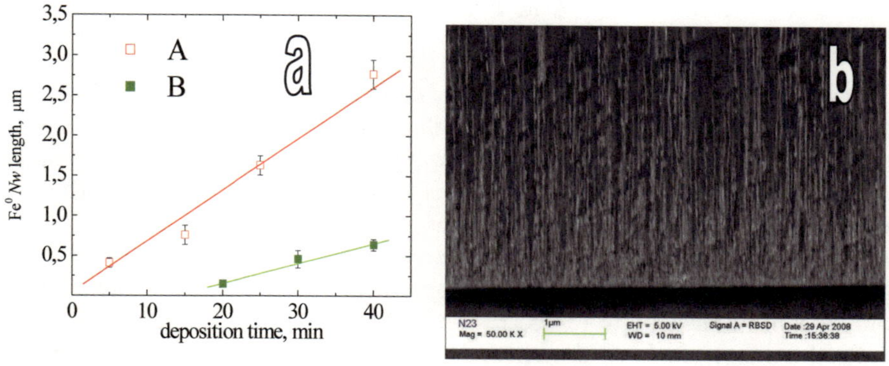

Fig. 2. (a):Variation of average height of Fe nanowires in alumina template pores ($\varnothing_{pore}$ ~15 nm) during alternating current (50 Hz) deposition at a constant current density of 0. 25 A/dm^2 from the solution (in M): 0.05 $FeSO_4$ + 0.025 $MgSO_4$ + 0.05 citric acid (A) and 0.1 $FeSO_4$+ 0.7 H_3BO_3 + 0.0085 ascorbic acid (B). (b): A typical cross-sectional FESEM view of alumina template ($\varnothing_{pore}$ ~28 nm) encased with Fe nanowires by deposition from (A) solution.

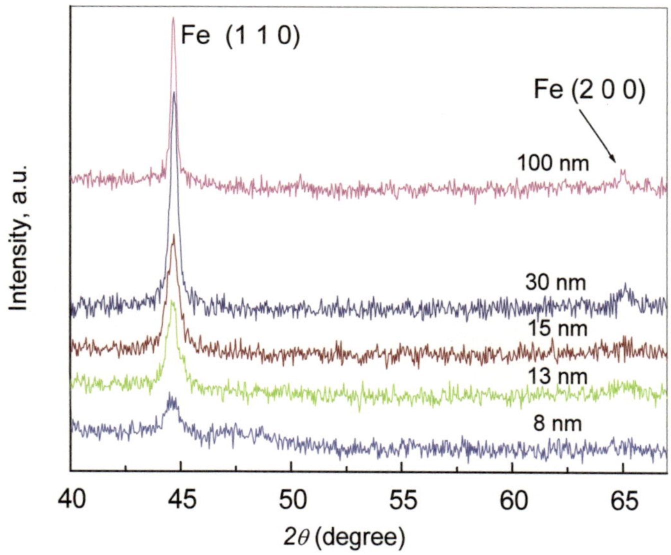

Fig. 3. XRD patterns of alumina films varying the diameter of pores following iron deposition by alternating current in an aqueous solution of 0.1 $FeSO_4$, 0.025 $MgSO_4$ and 0.1 M citric acid (pH 2.95) at 0.3 A·dm^{-2} and 17 °C for 20-30 min.

The Mössbauer spectra recorded in the conversion electron registration mode from the top and the back-side of alumina matrix, encased with Fe nanowires, after several days storage of not sealed specimen under ambient conditions are shown in Figure 4. As can be seen, these spectra are quite different. A back-side alumina Mössbauer spectrum (Fig. 4b) can be simply fitted using the sextet assignable to ferromagnetic iron, α-Fe. According to (Kuzmann et al. 2003), the ratio of intensities of lines in the Mössbauer spectra sextet is given as:

$$3:\frac{4\sin^2\theta}{1+\cos^2\theta}:1:1:\frac{4\sin^2\theta}{1+\cos^2\theta}:3 \qquad (1)$$

where θ is the angle between the direction of γ rays and magnetic moments in an iron atom. We have found that for all cases the relative intensities of the second and fifth lines in MS of as-grown Fe nanowires are close to zero, implying that the magnetic moments of the Fe nanowires are along their growth direction. On the contrary, the top-side Mössbauer spectrum recorded for not sealed specimen after several days storage in air demonstrates additionally a typical doublet with the Mössbauer parameters: the isomer shift respectively to α-Fe δ = 0.36 mm·s^{-1} and the quadrupole splitting Δ = 0.74 mm·s^{-1} assignable to Fe(III). The doublet could arise due to the corrosion and formation of $Fe(OH)_3$ or $Fe(OH)_3 \cdot nH_2O$ phases. Furthermore, we have determined that the sealing of alumina matrix just after deposition of iron in boiling water protects the Fe nanowires from corrosion during storage of samples under ambient conditions.

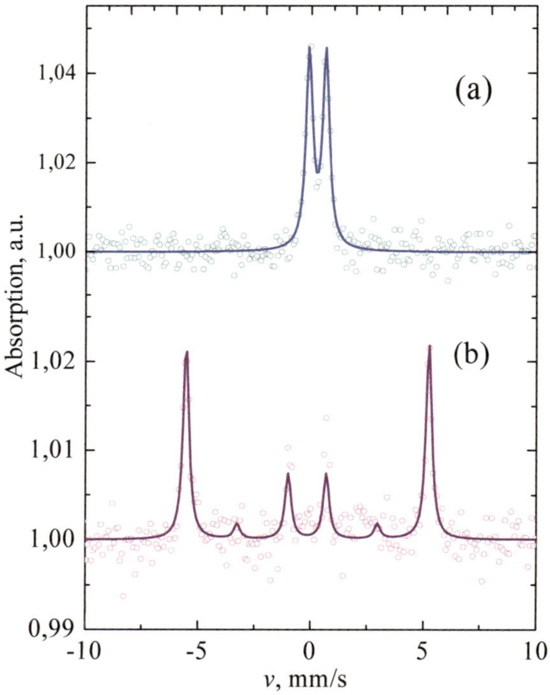

Fig. 4. Mössbauer spectra of alumina template encased with a Fe nanowire array ($\varnothing_{pore}$ 15 nm, h_{Fe} ~ 7 μm), taken under the conversion electrons mode from the top (a) and back (b) side of the matrix, indicate that the acmes of iron nanowires corroded forming $Fe(OH)_3$ or $Fe(OH)_3$ nH_2O if alumina after depositions remained unsealed.

Figure 5 shows the XRD spectra of alumina templates with the diameter $\varnothing_{pore}$ of 15 nm filled with Fe nanowires after annealing. Note that in this experiment the initial weight of analyte and the content of deposited iron were close. As can be seen, the increase in annealing temperature results in the lowering of the content of metallic iron remained after processing. In the case of annealing of the specimens at 450 °C for 1.5 hour (curve 3), the peaks of metallic iron in the XRD spectrum of the resulting product disappeared.

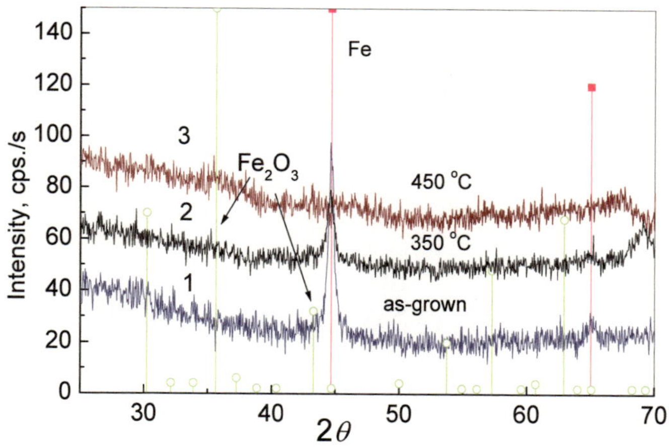

Fig. 5. The XRD patterns of alumina template with encapsulated Fe nanowires by alternating current deposition ($Ø_{pore}$15 nm) following template sealing in boiling water for 30 min (1) and annealing at 300 (2) and 450 °C (3) for 1.5 hour.

The effects of thermal annealing in air on the composition of as-grown Fe^0 nanowires in sealed alumina pores of a different diameter $Ø_{pore}$ are shown in Figures 6 and 7. All Mössbauer spectra were analyzed by the least squares fitting using the required subspectra and determining the hyperfine parameters, i.e. the isomer shift, δ, the quadrupole splitting, Δ, and the hyperfine field, B. It has been assumed that a relative area of the subspectrum is proportional to the amount of Fe in different compounds. Figure 6a shows a typical Mössbauer spectrum of alumina specimens with a pore diameter of 30 nm after deposition of Fe nanowires and alumina sealing. This spectrum, recorded in the transmission geometry, is of ferromagnetic iron α-Fe fitted using a sextet which has appropriate hyperfine parameters. For Fe nanowires, the parameters were found to be the same as for bulk α-Fe, i.e. the isomer shift $\delta = 0 \pm 0.01$ mm s^{-1}, the quadrupole splitting $\Delta = 0 \pm 0.01$ mm s^{-1}, the hyperfine field $B = 33.1 \pm 0.1$ T and the line width $\Gamma = 0.28 \pm 0.02$ mm s^{-1}. No considerable variation in the hyperfine parameters for different Fe nanowire diameters in the 8 to 30 nm range was obtained.

The annealing of specimens results in the complication of Mössbauer spectra due to transformations of the Fe nanowires composition. These transformations were found to be dependent on the annealing temperature, T_{ann}, and the diameter of pores, $Ø_{pore}$ (Jagminas et al. 2009).

In the case of annealing the Fe^0 nanowires encased in alumina pores with the diameter $Ø_{pore}$ of 30 nm at $300 \leq T_{ann} < 600$ °C, for example at 400 °C (Fig. 6b), the Mössbauer spectra were fitted with three components, i.e. a sextet demonstrating the presence of ferromagnetic α-Fe and additionally two doublets. These doublets had characteristic Mössbauer spectra hyperfine parameters which can be summarized as follows for all samples: for the first one, the isomer shift $\delta = (0.9 - 1.1) \pm 0.1$ mm s^{-1}, the quadrupole splitting $\Delta = (1.9 - 2.3) \pm 0.1$ mm s^{-1} and a line width $\Gamma = (0.3 - 0.7) \pm 0.1$ mm s^{-1} and, for the second one, $\delta = (0.3 - 0.5) \pm 0.1$ mm s^{-1}, $\Delta = (0.8 - 1.1) \pm 0.1$ mm s^{-1} and $\Gamma = (0.6\text{-}0.8) \pm 0.1$ mm s^{-1}, respectively. On the basis of Mössbauer spectra parameters the first doublet should be attributed to Fe(II) paramagnetic compound, and can be most probably assigned to the spinel-type $FeAl_2O_4$ compound (Botta et al., 2003). Moreover, from the large line width of this doublet peaks it

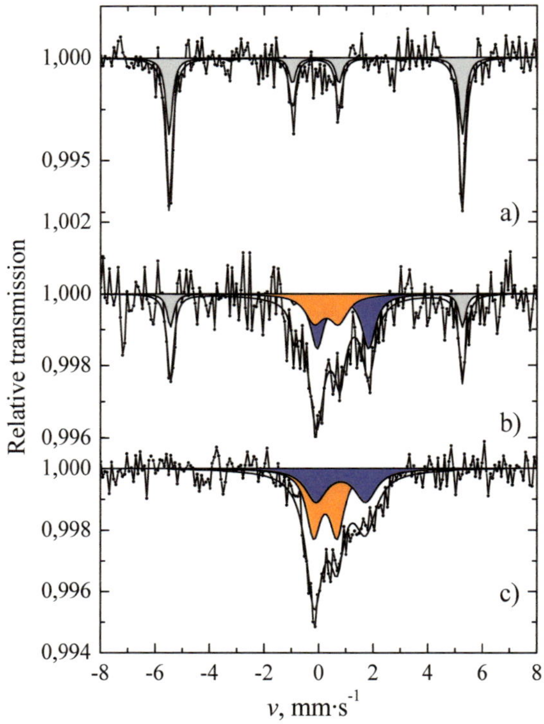

Fig. 6. The Mössbauer spectra recorded in the transmission mode for the Fe nanowires array synthesized within the alumina pores ($Ø_{pore}$30 nm) by alternating current deposition and template sealing in boiling water (a) and annealing at 400 (b) and 600 °C (c) for 1.5 hour.

can be concluded that thermally formed $FeAl_2O_4$ should be nonstoichiometric. We assigned a second doublet, typical of Fe(III) paramagnetic compounds, to a-$(Fe_xAl_{1-x})_2O_3$ which is obtained when iron incorporates into alumina. As for this component only a paramagnetic doublet is observed, and there are no traces of sextets of iron oxides, x should be below 0.1 (Kuzmann et al. 2003).

The annealing of the Fe nanowire array, encased in the alumina matrix, at $T_{ann} \geq 600$ °C results in some fading of the black color of alumina. With a further increase in T_{ann} these changes intensified. A typical Mössbauer spectrum of the specimen with the 30 nm diameter after annealing at 600 °C is shown in Figure 6c. In this case, the fitting results revealed a complete transformation of ferromagnetic iron into two components, e.g. $FeAl_2O_4$ and $(Fe_xAl_{1-x})_2O_3$ with a similar content.

Similar variations in the composition of Fe nanowires upon annealing were determined for iron nanowires encapsulated within the alumina pores with the 15 nm diameter. However, in this case the ferromagnetic component disappeared at about 450 °C as a result of full transformations into paramagnetic Fe(II) and Fe(III) components with a somewhat higher amount of Fe(III). It is noteworthy that with a further increase in T_{ann} the ratio between Fe(III) and Fe(II) components changes significantly. For example, after annealing at 600 °C the resultant products contain ~ 80 % of Fe(III) and only about 20 % of Fe(II).

The compositional investigations of Fe nanowires encased within the alumina template pores with the 13 nm diameter, produced by alternating current deposition under the

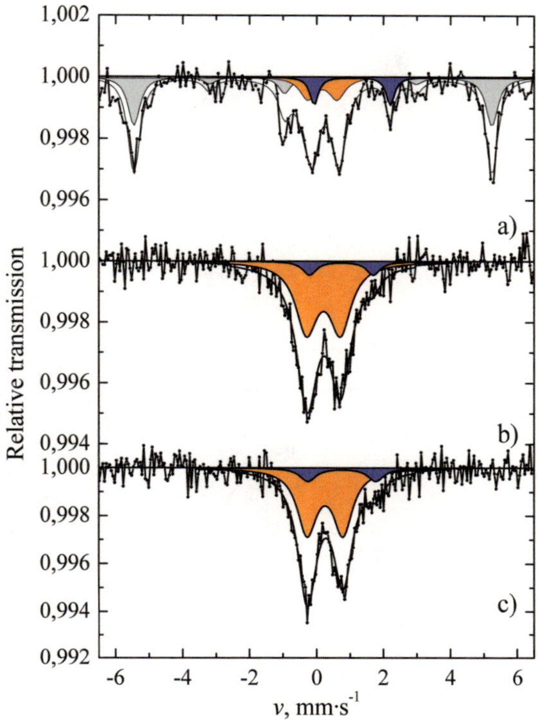

Fig. 7. The Mössbauer spectra recorded in the transmission mode for the Fe nanowires array in the alumina template with an average pore diameter of 8 nm after sealing (a) and annealing at 400 (b) and 510 °C (c) for 1.5 hours. To increase sensitivity of measurements, the deposition solution additionally contained 0.5 g·dm^{-3} of ^{57}Fe.

conditions of this study, revealed two additional features (see Fig. 8c). Firstly, some part of deposited iron within the alumina pores even before annealing was found to be in the form of paramagnetic Fe(III) implying that the formation of $(Fe_xAl_{1-x})_2O_3$ is possible as a result of the alumina template sealing in boiling water. Secondly, in the case of the 13 nm diameter template, a ferromagnetic component disappeared in the vicinity of 400 °C being transformed fully into paramagnetic Fe(III) and Fe(II) components in proportions similar to a case of 15 nm diameter template annealed at 600 °C.

The Mössbauer spectrum of alumina template with an average diameter of 8 nm filled with iron nanowires even after the sealing (Fig. 7a) was found to be more complicated and composed of three components similar to those following the annealing of specimens with $Ø_{pore} \geq 15$ nm. The quantities of paramagnetic Fe(III) and Fe(II) components were found to be larger making up together more than 40 % of the total amount of deposited iron (Fig. 8d). Moreover, annealing of these specimens results in more rapid transitions of ferromagnetic Fe(0) to paramagnetic Fe(III) and Fe(II) compounds at obviously lower temperatures with prevailing Fe(III) formation.

The results obtained indicate that the diameter of primary iron nanowires is a key parameter which determines the transformations when annealing at the same temperature. These transformations became especially obvious even at ~100 °C if the nanowires diameter is in

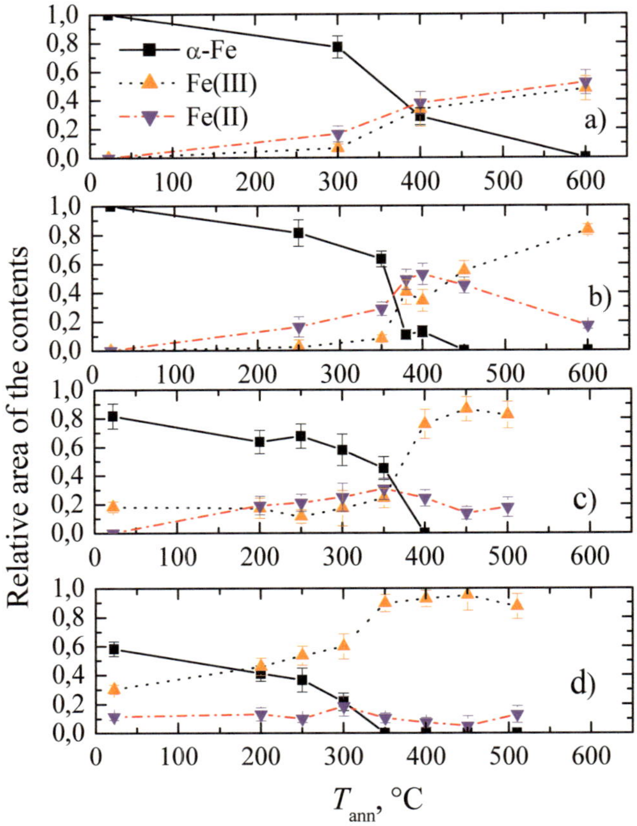

Fig. 8. Dependences of the amount of iron phases formed within the alumina template pores *via* alternating current deposition and annealing on the annealing temperature and the average pore diameter (in nm): (a) 30; (b) 15; (c) 13 and (d) 8.

the order of the Debye length. We suggest that this effect is linked to the increase in a free surface energy impact of thin nanowires modifying the temperature of thermal diffusion of material molecules into alumina bulk.

3.2 Sn nanowire arrays

Due to chemical stability and a wide band gap at room temperature [3.6-4.0 eV (Ji et al., 2004)] SnO_2 thin films are promising materials for solar cells (Fererre et al. 1997), short wavelength optoelectronic (Song et al., 2004), and blue photoluminescence (Presley et al. 2004) devices. Besides, tin-based oxides are promising candidates to replace the carbon-based anode materials because of their large capacity for lithium insertion by the formation of $Li_{4.4}Sn$ alloy (Wang et al. 2005). During the recent decade SnO_2 nanotubes (Qin et al. 2008), nanoparticles and nanorods (Firooz et al. 2008), and nanobelts (Fujihara et al. 2004) have been fabricated by different methods such as sol-gel process (Gu et al. 2003), hydrothermal synthesis (Firooz et al. 2008; Fujihara et al. 2004), chemical vapour deposition (CVD) (Kong et al. 2003; Sundqvist et al. 2006), metalorganic chemical vapour deposition (MOCVD) (Feng et al. 2008), and etching of tin foils in alkaline solution (Peng et al. 2008). Despite these proposals there is a great need to develop reliable and cheap synthesis methods for fabrication of densely-packed, uniform-shaped tin oxide nanowire arrays for practical applications.

Here we report the way for fabrication arrays of the core (SnO)/shell (SnO_2) barbed-shaped nanowires *via* electrochemical formation of densely-packed tin nanowires inside the alumina pores and their subsequent heat treatment in air at a proper protocol. The morphology of Sn nanowire arrays fabricated within alumina pores was investigated before and after heat treatments using field emission scanning electron (FESEM) and transmission electron (TEM) microscopy in alumina cross-sections followed by a complete template dissolution. Mössbauer spectra were recorded in a constant acceleration mode using a spectrometer Wissenschaftliche Elektronik GMBH and an $^{119}CaSnO_3$ source. The velocity scale was calibrated relative to $CaSnO_3$. All experimental spectra were fitted to Lorentzian lines using a least-squares fitting algorithm.

Good uniformity and completeness of alumina pores filling by tin nanowires in the solution containing $SnSO_4$, tartaric acid, $C_4O_6H_6$, hydrazine sulphate, and H_2SO_4 by alternating current deposition can be evidenced from the FESEM images of alumina templates in their cross-sections (Fig. 9A) and after the following template etching (Fig. 9B). We found that annealing of these specimens results in the change of alumina black colour towards puce and gold tints, as well as the decrease in Sn^0 diffraction peaks and the appearance and growth of new ones with increase in annealing temperature, T_{ann}. The first diffraction peaks of SnO_2 appeared for samples annealed at 400 °C. Note that all peaks attributable to metallic tin disappeared only for $T_{ann} \geq 500$ °C, e.g. at a temperature much higher than the melting point of bulk tin (231.9 °C). In this case, the diffraction lines at $2\Theta = 26.58, 33.87, 51.76°$ can be assigned to tetragonal rutile-type crystalline phase of tin(IV) oxide, SnO_2 (PDF file no. 41-1445).

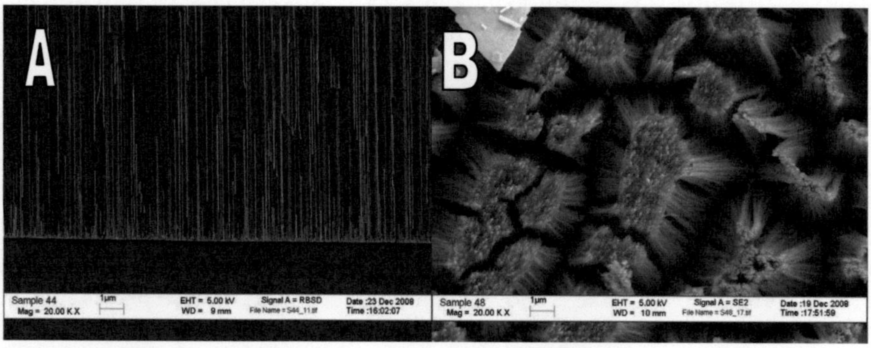

Fig. 9. A typical cross-sectional (A) and in-plane (B) FESEM view of alumina templates encased with tin nanowires by alternating current deposition from the solution (in M): 0.05 $SnSO_4$, 0.03 tartaric acid, 0.015 hydrazine sulfate and sulfuric acid (pH = 1.15) at a constant average current density of 2.75 mA cm^{-2} for 30 min. Pore diameters are 15 (A) and 28 nm (B). The alumina in (B) is partially dissolved.

In the case of a 2 hour long annealing at 500 °C of the templates with pore diameters of 15 nm, the crystallite size of SnO_2 particles calculated from the XRD line broadening using TOPAS software, based on the Rietveld structure refinement Scherer equation, is about 9.7 ± 1.5 nm. It is worth noticing that the XRD pattern of specimens treated at 450 °C, except the diffraction lines attributable to SnO_2, contained additional lines whose positions depend on the annealing time, τ_{ann}. These lines could be assigned to the crystalline phase of Sn(II) oxide, SnO (PDF file no. 06-0395) (Fig. 10). Nevertheless, the peak positions of these lines were found to be slightly shifted towards lower values of diffraction angles, indicating the distortion of crystalline lattice, since a = 3.802 and c = 4.900 Å instead of a = 3.802 and c = 4.836 Å.

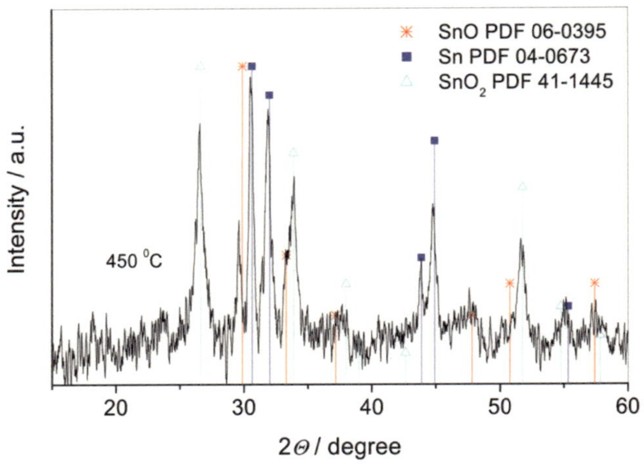

Fig. 10. X-ray diffraction pattern of alumina template encased with tin nanowires ($Ø_{Sn}$ ~15 nm) by deposition from the solution as in Fig. 9 at the average current density (50 Hz) of 2.75 mA cm^{-2} for 30 min after annealing at 450 °C for 3 hours.

To qualitatively and quantitatively further assess the effect of annealing on the transformation of Sn^0 nanowires, encased within the alumina pores by electrodeposition, Mössbauer spectroscopy was performed and a typical Mössbauer spectrum of the annealed products is presented in Figure 11a. In the case of as-deposited nanowired material, a singlet line in Mössbauer spectrum with the isomer shift, δ, equal to 2.49 ± 0.02 mm/s, attributable to β-Sn, can be only observed.

Our investigations revealed that after alumina template sealing in boiling water and annealing the Mössbauer spectrum of Sn nanowires loaded inside alumina pores changes drastically. In the case of annealing the specimen at $T_{ann} \geq 450$ °C, the resulting Mössbauer spectrum is composed of two broadening doublets characteristic of SnO and SnO_2 (Fig. 11a). The quadrupole splitting (Δ) and isomer shift (δ) values for Sn^{2+} and Sn^{4+} states in annealed specimens are presented in Table 1. However, the magnitude of these parameters coincides with the well known ones only for the crystalline SnO_2 (Collins, et al. 1979). For SnO, in all annealed specimens both Δ and δ values obtained are larger than those for stable tetragonal (black) SnO powder, as can be seen in Fig. 11 and Table 1. We suppose that differences in SnO parameters are due to the formation of nanocrystalline nonstoichiometric form of tetragonal SnO with possible contribution of amorphous SnO as an intermediate oxide phase through the sequence of reactions: $Sn^0_s \Rightarrow Sn^0_{liq} \Rightarrow SnO_{amorph}(SnO_{cryst}) \Rightarrow SnO_2$.

Notably, an increase in T_{ann} up to 550 °C results in a decrease in the SnO content. For example, in the case of annealing at 450 °C, the content of remained SnO was 1.75 times larger than that at 500 °C.

Typical transmission electron microscopy images on SnO/SnO_2 blend are depicted in Fig. 12. First, we see that Sn^0 rod-shaped nanowires transform into barbed-type nanowire products upon heat treatment. We also see that the barbed nanowires are 2 or 3 times

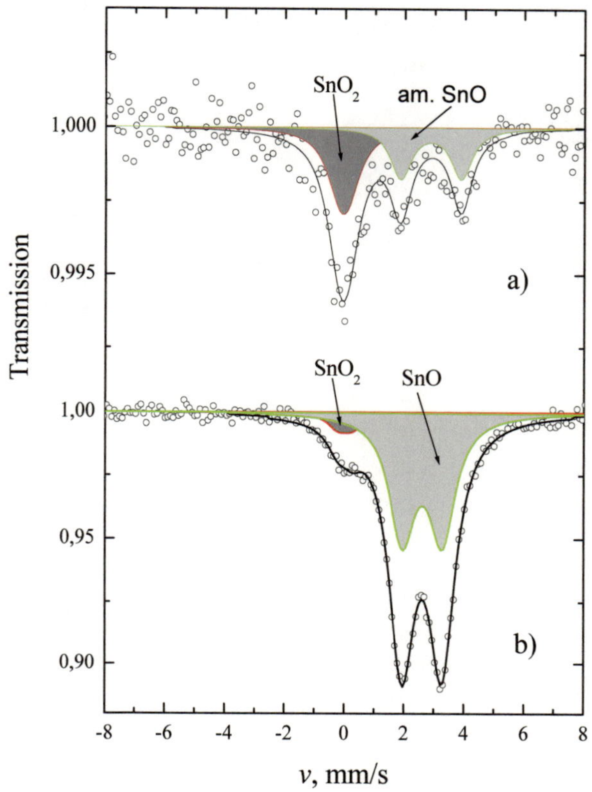

Fig. 11. Room temperature ^{119}Sn Mössbauer effect spectrum of Sn^0 nanowires array fabricated in the pores of alumina template by electrodeposition as in Fig. 10 after 2 hours annealing at 450 °C (a) and the same spectrum for commercial SnO powder (b).

Sample	β-Sn		SnO			SnO_2		
	δ, mm s-1	Γ, mm s-1	δ, mm s-1	Δ, mm s-1	Γ, mm s-1	δ, mm s-1	Δ, mm s-1	Γ, mm s-1
As-deposited.	2.49 ±0.02	0.94 ±0.05	–	–	–	–	–	–
Annealed 450 °C, 2 h.	–	–	2.88 ±0.04	2.02 ±0.07	0.86 ±0.04	-0.05 ±0.03	0.3 ±0.2	1.0 ±0.2
Annealed at 500 °C 2h	–	–	2.87 ±0.05	2.08 ±0.09	0.86 fixed	0.02 ±0.01	0.62 ±0.02	0.88 ±0.05
Annealed at 500 °C, 2h, (160 K MS)	–	–	3.18 ±0.05	2.38 ±0.09	0.95 ±0.17	0.09 ±0.02	0.69 ±0.03	0.87 ±0.06
Tetr. SnO sample	–	–	2.60 ±0.01	1.34 ±0.01	1.06 ±0.01	0.02 ±0.04	0.5 ±0.1	0.83 ±0.15

Table 1. The parameters of Mössbauer spectra for Sn^0 nanowire arrays annealed at indicated temperatures when the template pore diameter is 15 nm.

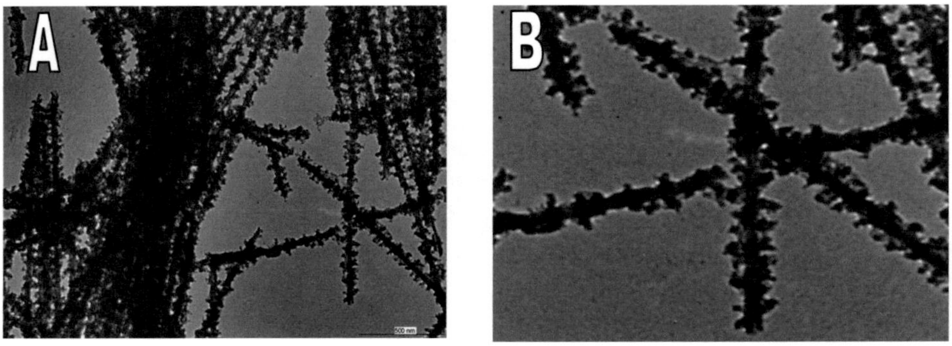

Fig. 12. Typical TEM images of SnO/SnO_2 barber-shaped nanowires released from the support. $Ø_{pore}$ = 28 nm. Sn^0 nanowire array in a sealed alumina template was heat treated at 500 °C in air for 2 hours.

thicker at the thorn sites. To the best of our knowledge, such a previously unthinkable architecture of nanowires has never been seen in nanotechnology. Especially interesting is the formation of numerous mechanically quite stable thorns. It is assumed that these thorns of up to 25 nm height (for 28 nm pore template) are composed of crystalline SnO_2 as more energetically favourable. We suggest that these thorns grow up through the all-out thermal and pressure induced diffusion of molten tin inside the walls of alumina nanochannels and that oxygen-bandings result in strong compression and distortion of the matrix cells, thereby destroying the matrix due to increase in size and shape of a novel nanowired product.

3.3 *a*- and γ-Fe_2O_3 composite nanowire arrays *via* electrodeposition and thermal decomposition of lepidocrocite, γ-FeOOH

Highly-ordered, uniform and densely-packed arrays of iron oxide nm-scaled architectures, such as nanowires and nanotubes, receive considerable attention due to their prospective use in solar cells, gas sensing and super-high-density recording devices. Magnetite, Fe_3O_4, maghemite, γ-Fe_2O_3, and hematite, α-Fe_2O_3, are the crystallographic materials of iron oxides differing in structure and properties. Magnetite and maghemite are ferromagnetics while bulk hematite is anti ferromagnetic material with the Neal temperature 950 K and Morin transition temperature 263 K (Cornell & Schwertmann 1996). Note that composites from mixed iron oxides due to magnetic exchange coupling between magnetic particles frequently demonstrate a shift of hystereses and coercitivity (Christensen et al., 1982) currently paving a way to their relevant applications in catalysis, high density recording contours, and biomedical diagnostics (Blanc-Beguin et al. 2009).

Bulk ferromagnetic maghemite is a form of iron oxide that can be obtained by thermal decomposition of lepidocrocite, γ-FeOOH (Özdemir & Dunlop 1993). γ-Fe_2O_3 is a transition form, and upon further heating undergoes a sequence of phase transformations up to hematite, α-Fe_2O_3 (Gehring & Hofmeister 1994). On the whole, the temperature at which dehydroxilation reaction begins depends on the preparation and heating conditions of bulk lepidocrocite crystals (Morales 1998). Thus, the activation energy of this reaction varies from 13.1 to 26.7 kcal mol^{-1}, depending on the crystallinity of γ-FeOOH.

The thermal decomposition of iron oxyhydroxide polymorphs, such as goethite, *a*-FeOOH, akaganeite, *β*-FeOOH, lepidocrocite, and amorphous FeOOH · nH_2O has always been an interesting subject seeking to form Fe_3O_4 and *a*-, *γ*-, or ε-Fe_2O_3 materials and their composites.

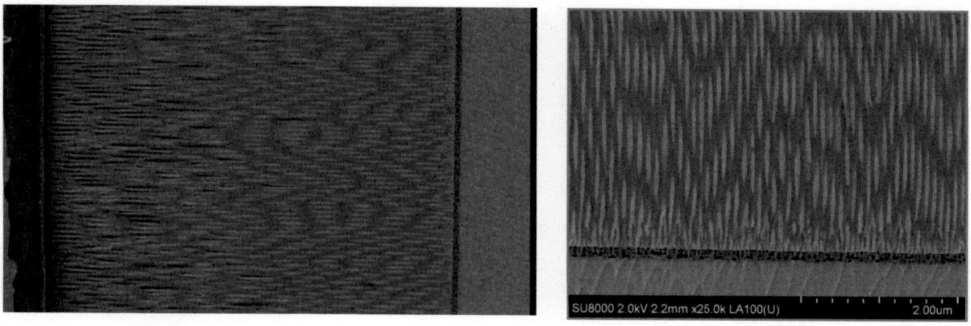

(a) (b)

Fig. 13. Cross-sectional FESEM view of alumina templates loaded with lepidocrocite nanowires by alternating current (50 Hz) deposition from 0.022 M $Fe(NH_4)_3(C_2O_4)_3$ and 0.011 M $^{57}Fe(C_2O_4)_3$ kept at pH 5. 75. $\varnothing_{pore}$: (a) 45 nm, (b) 150 nm.

However, to date only akaganeite nanowires have been successfully encased within the alumina pores by electrodeposition (Gao et al. 2002; Zhang et al. 2007) transforming them later to hematite nanowires by heat treatment in air at $T_{ann} \geq 500$ °C. Recently we have developed a route for uniform synthesis of amorphous lepidocrocite nanowires within the alumina pores using an alternating current mode (50 Hz) and an aqueous solution of iron(III) ammonium oxalate friendly to alumina matrices (Jagminas et al. 2009b). A nice filling of alumina pores with lepidocrocite nanowires in this solution by deposition through the barrier layer at a constant alternating current density of 0.35 A dm^{-2} was evidenced here both for 45 and 150 nm $\varnothing_{pore}$ templates looking at the parts (a) and (b) of Fig. 13, which show the FESEM images taken from the template cross-sections. This is also evidenced by uniform colouring of alumina matrices in deep orange tints. Again, the growth of lepidocrocite nanowires was determined here using the Mössbauer spectra at cryogenic temperatures (Fig.14) and $^{57}Fe(C_2O_4)_3$ precursor.

The set of Mössbauer spectra for γ-FeOOH nanowires encased in the alumina template with $\varnothing_{pore} \sim 45$ nm and heat-treated at temperatures from 150 to 650 °C for 3 hours ($\partial T/\partial t = 10$ °/min) is shown in Fig. 15. There the group of spectra recorded at room temperature is labeled by (a) while (b) shows a similar spectra recorded at a low temperature. The first clear feature from Fig. 15 is that lepidocrocite nanowires remain quite stable up to 400 °C although the color tint of alumina templates after annealing at $T_{ann} \geq 150$ °C changes evidently. In line of these observations it has been found that quadrupole splitting of Mössbauer spectra doublet, attributable to γ-FeOOH, increases with annealing temperature likely due to elimination of the absorbed and of the molecular water as in the case of akaganeite (β-FeOOH) calcination (Gao et al. 2002).

Noteworthy that the sextets, attributable to Fe_2O_3 phases appear only for specimens heat-treated at $T_{ann} \geq 500$ °C. In fact, the heat treatment at 500 °C results in the largest Mössbauer spectra changes of the product deposited in alumina pores; with a further temperature increase these changes become less pronounced. Thus, in the next setup we analyzed the composition of the specimen calcinated at 500 °C in more detail using both room and low temperature Mössbauer spectra. The results obtained are presented in Fig. 16. To fit these spectra, we used up to two sextets and one-two doublets. The sextets were ascribed to the hematite with a hyperfine field of ~50.1 T at room temperature and to Fe_2O_3 with hyperfine field of ~47.9 T. Again, doublets were ascribed to the species in a superparamagnetic state

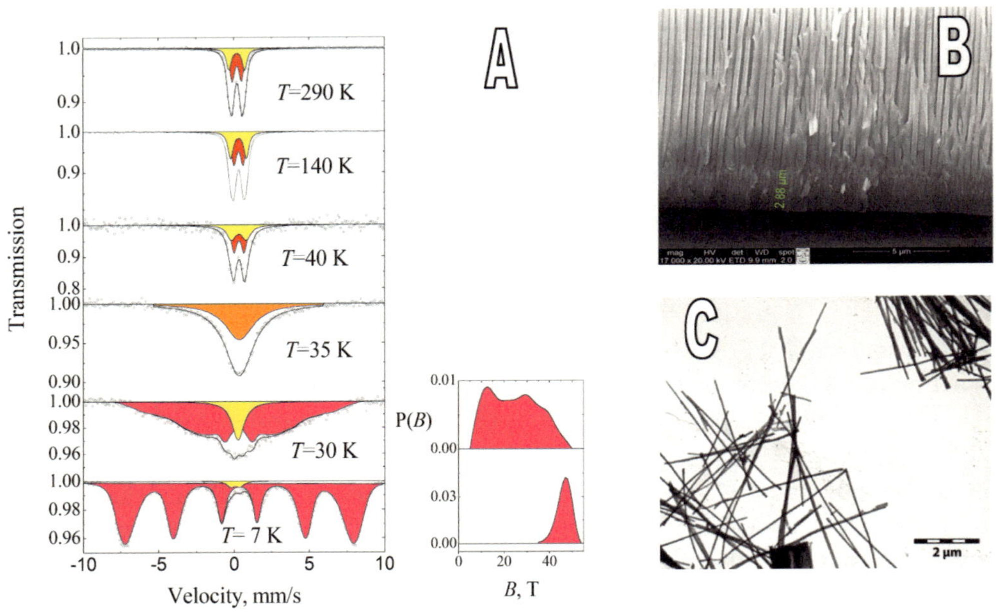

Fig. 14. The transmission Mössbauer spectra taken at indicated temperatures show the magnetic transition below 35 K for as-deposited FeOOH nanowires inside the alumina pores (B,C) implying the formation of quite pure γ-FeOOH phase. The right side in (A) shows the hyperfine fields distribution for description of the spectra.

and/or to paramagnetic irons. The refined spectral parameters are listed in Table 2. From the analysis of these data the presence of hematite phase in the annealed specimen is clear. Furthermore, the low temperature measurement in Fig. 15 proved that a part of the doublets correspond to superparamagnetic particles of $\gamma\text{-}Fe_2O_3$ whereas the rest can be assigned to the paramagnetic phase of maghemite, $\gamma\text{-}Fe_2O_3$. Note that formation of superparamagnetic species demonstrates the increase of the central part in the low temperature Mössbauer spectra. This increase can be due to the existence of small species of hematite as well as due to maghemite. Two subspectra in the low temperature spectrum split part are characteristic by different quadrupols. Without any doubt one of them (Δ = 0.37 mm/s), typical at a temperature lower than Morin temperature, T_M, is related by the presence of hematite. The second one with Δ = 0.04 mm/s, as an intermediate value close to zero, could be attributed either to $\gamma\text{-}Fe_2O_3$ or to very small particles of hematite in which the Morin transformation effect, similar to that as in the case of (Schroeer & Nininger 1967), might take place.

By the analogy with the thermal decomposition of lepidocrocite in a colloidal state (Gendler et al., 2005), it is thought that the transformation reaction $\gamma\text{-FeOOH} \Rightarrow \gamma\text{-}Fe_2O_3 \Rightarrow a\text{-}Fe_2O_3$ in nanowired amorphous lepidocrocite, $\gamma\text{-FeOOH} \cdot nH_2O$, proceeds *via* a sequence of several steps. The first one, associated with desorption of physisorbed molecular water, takes place even at T_{ann} around 150 °C, because the color and optical properties of alumina templates loaded with lepidocrocite nanowires evidently change under these conditions. This observation is in line with the indication (Gehring & Hofmeister, 1994) that, on the

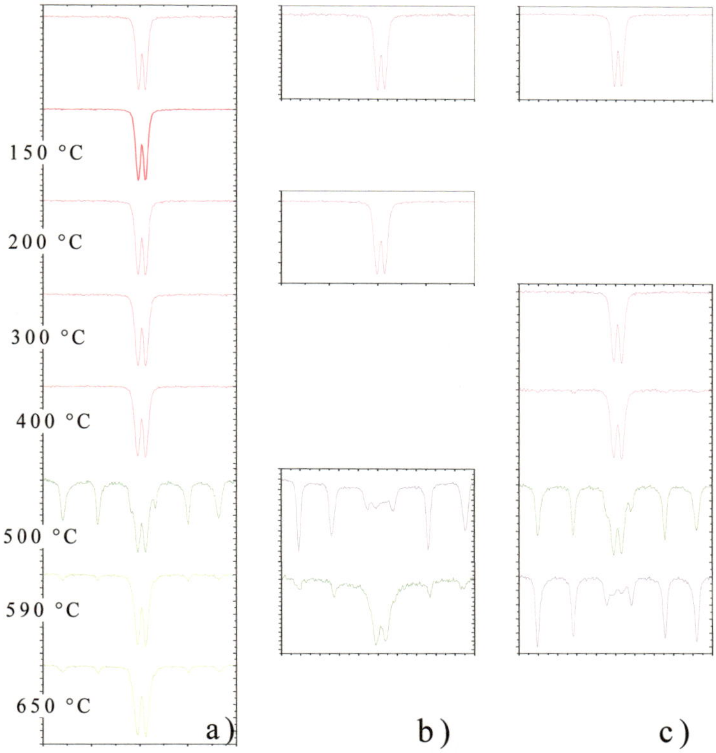

Fig. 15. The transmission Mössbauer spectra taken at room (a, c) and 140 K (b) temperatures for lepidocrocite, γ-FeOOH, nanowire arrays heat treated at indicated temperatures for 3 hours. $Ø_{pore}$: (a, b) 45 nm, (c) 150 nm.

molecular level, the removal of OH groups in the lepidocrocite begins at 142-155 °C. Based on the results obtained here by Mössbauer transmission spectroscopy and Raman scattering (not shown here) it seems likely that the second step, corresponding to the dehydroxylation of iron oxyhydroxide nanowires to *γ*- and *α*-Fe_2O_3, completely proceeds in the vicinity of 500 °C although from literature data (Bernal et al. 1957) the rhombic structure of synthetic lepidocrocite reconstructs to cubic maghemite in the 230-250 °C region which at 375 °C is transported completely to hematite with a hexagonal structure, e.g. at much lower temperatures.

4. Annealing effects of copper selenide nanowire arrays

The formation of copper selenide nanowire arrays, in particular those of $Cu_{2-x}Se$ and Cu_3Se_2 within alumina pores has been demonstrated by us in an aqueous solution of $CuSO_4$, $MgSO_4$ and H_2SeO_3 at proper concentrations and pH by alternating current deposition (Jagminas et al. 2006). However, only an almost pure phase of $Cu_{1,75}Se$ showing both a direct energy bandgap, $E_{g,dir}$, of about 2.3 eV and a defect related optical absorption band close to 1.1 eV, has been deposited using sulfuric acid alumina templates with an average pore diameter, $Ø_{pore}$, of 15 nm. One should note that such values are close to those reported for $Cu_{1.85}Se$ thin films where $E_{g,dir}$= 2.2 eV and E_{def}= 1.4 eV (Herman & Fabrick 1983). Very recently, the optimal composition for the deposition of a quite pure phase Cu_3Se_2 nanowires inside the

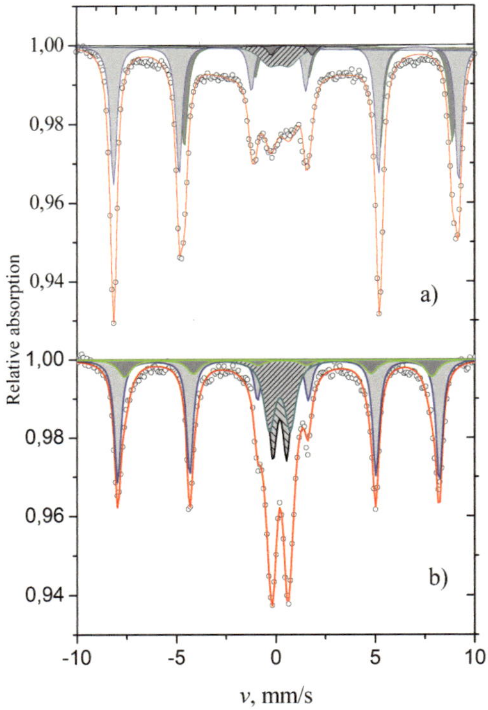

Fig. 16. The transmission Mössbauer spectra recorded at 140 K (a) and room (b) temperatures for lepidocrocite nanowire arrays loaded inside the alumina matrix ($Ø_{pore}$ 45 nm) after heat treatment at 500 °C in air for 2 hours.

		S, %	δ, mm/s	Δ, mm/s	B, T	A23
Room temperature	1	37	0.37±0.01	-0.22±0.01	50.1±0.1	2.83±0.06
	2	19	0.36±0.01	-0.21±0.02	47.9±0.2	2.45±0.18
T=140 K	1	39	0.47±0.01	0.37±0.01	53.7±0.1	2.81±0.06
	2	37	0.49±0.01	0.04±0.01	52.8±0.1	3.13±0.07

Table 2. Calculated parameters of Mössbauer spectra sextets (Fig. 16) for 45 nm diameter lepidocrocite nanowire array heat treated at 500 °C for 2 hours.

alumina pores has also been reported (Jagminas et al. 2009c). Furthermore, it has been demonstrated that nm-sized crystals and nanowires of $Cu_{1.75}Se$ and Cu_3Se_2 encased in the transparent alumina template demonstrate nonlinear absorption features at 1.06 μm (Statkutė et al. 2008; Juška, et al. 2009a). We have found that Cu_3Se_2/alumina greenish tints, characteristic of these films, upon its annealing in ambient atmosphere even at a moderate temperature of 150 °C became yellow. Upon further increase in annealing temperature, these changes became more obvious.

Figure 17 presents the XRD profiles of the alumina templates filled by Cu_3Se_2 nanowires before and after annealing for three hours at various temperatures. An as-deposited material pattern shows twelve well-defined peaks with relative intensities completely consistent with those of crystalline tetragonal Cu_3Se_2 [a = 4.27 Å and c = 6.40 Å (PDF 47-1745), umangite]. The thermal treatment in air of a Cu_3Se_2 nanowire-shaped material results in decomposition

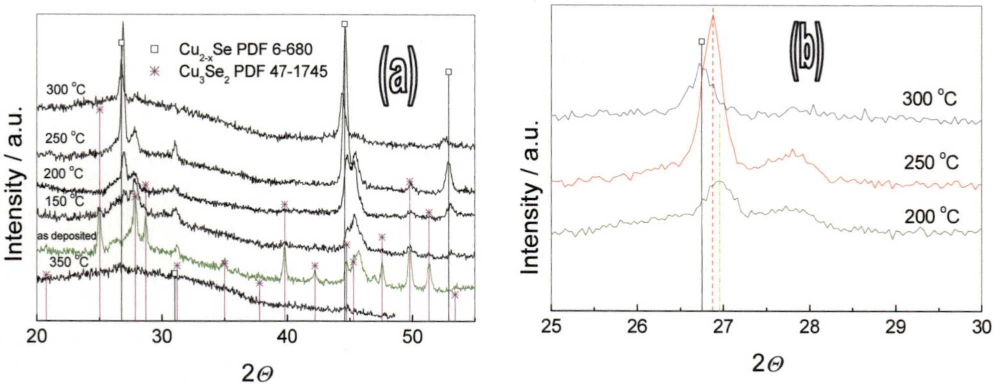

Fig. 17. (a) XRD patterns of alumina matrices encased with copper selenide by alternating current deposition in the solution (in M): 0.03 $CuSO_4$, 0.015 H_2SeO_3, 0.03 $Al_2(SO_4)_3$, 0.005 triethanolamine, and H_2SO_4 (pH = 1.15) at 0.5 A dm^{-2} for 45 min before and after annealing in air for one hour. (b) XRD pattern fragments within a 25 to 30° 2Θ range for the same as in (a) Cu_3Se_2/alumina specimen annealed at 200, 250 and 300 °C.

of copper selenide depending on T_{ann}. At $T_{ann} \approx 200$ °C a part of Cu_3Se_2 transforms into $Cu_{2-x}Se$. With further increase in the annealing temperature the amount of $Cu_{2-x}Se$ component increases becoming dominant at ~ 300 °C. The most significant fact is the stoichiometry of $Cu_{2-x}Se$ phase found to be also dependent on T_{ann}, as judged from the diffraction peak position in the XRD pattern fragment within the range of 2Θ from 26.5 to 27.5° (Fig. 17b). We have found that annealing of nanowire-shaped Cu_3Se_2 at 250 °C results in formation of $Cu_{1,85}Se$, while annealing at lower, e.g. 200 °C, or higher, e.g. 300 °C, temperatures yields $Cu_{2-x}Se$ with $x > 0.15$ and $x < 0.15$, respectively. This effect possibly originates from the diffusion of selenium into the alumina bulk upon heating.

Transmission spectra of alumina templates encased with almost pure Cu_3Se_2 nanowires before and after annealing at various temperatures were recorded for wavelengths ranged from 190 to 3125 nm (Fig. 18a). It is obvious that the shape of absorption spectrum for as-deposited Cu_3Se_2 nanowires array (curve 1) differs from the ones annealed (curves 2-5). The main feature of the first is a wide absorption band in the visible and near infrared starting just below the edge of a fundamental band gap and covering almost the whole dip of the spectra. Analysis performed on this part of absorption spectra (see Figure 18b) revealed two Loretzians $\{w/(4(h\nu - h\nu_0)^2 + w^2)\}$ superimposed, each of them with peaks at energy $h\nu_0 = 1.14$ and 1.65 eV with FWHM – $w = 0.58$ and 1.48 eV, respectively. These results suggest that along with Cu_3Se_2 another selenide, like $Cu_{2-x}Se$ in the amounts undetectable for XRD, takes part in the absorption.

Band gap energy (E_g) values of deposited and annealed products were calculated using: $a(h\nu) = A(h\nu - E_g)^{n/2}$, where α is the absorption coefficient, A is a constant, E_g is the band gap value, $h\nu$ is the photon energy and n is dependent on the nature of transition: 1 or 3 for direct and 4 or 6 for indirect transitions. This estimation gives that direct band gap values ($E_{g,dir}$) for as-grown and annealed specimens increase with annealing temperature from ~ 2.16 eV (before annealing) to 2.44 eV (T_{ann} = 300 °C).

Optical nonlinearities of as-grown and thermally treated specimens were studied in the experimental set-up based on the mode-locked Nd:glass laser (the wavelength of 1053 nm,

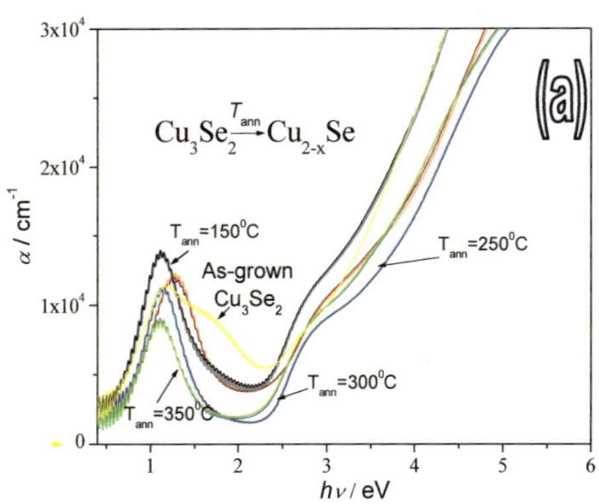

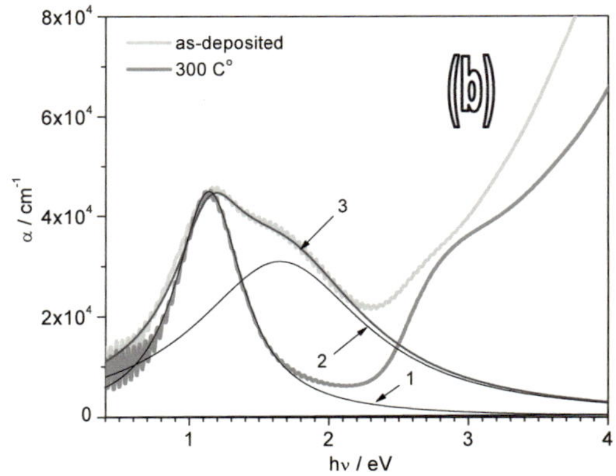

Fig. 18. (a) Variations of the absorption coefficient with photon energy ($h\nu$) for nanowired Cu_3Se_2 arrays in the alumina template pores before (1) and after (2-5) annealing for 90 min. The average diameter, spacing and height of species were 45 nm, 108 nm and approx. 0.5 µm, respectively. (b) Decompositional analysis of absorption spectra for as-deposited Cu_3Se_2 nanowired products. The calculated curves are Lorentzian's [$\sim(w/(4(h\nu-h\nu_0)^2+w^2)$, where $w \approx 0.58$ eV (FWHM), centered at 1.14 eV] for (1) and [$\sim(w/(4(h\nu-h\nu_0)^2+w^2)$, where $w \approx 1.48$ eV (FWHM), centered at 1.65 eV] (2), respectively. Curve (3) represents superposition of curves 1 and 2.

pulse duration of 2 ps, and the pulse repetition rate of 0.66 µJ). A Gaussian laser beam was focused with a 15 cm focal length lens to beam waist diameter of ~ 100 µm. A total power of the beam transmitted through the sample was measured in the field as a function of the sample position along the beam axis Z. This measurement corresponded to the standard open-aperture Z-scan procedure (Sheik-Banae et al. 1990). For the dynamic characterization of the optical nonlinearities, the laser beam was divided into two cross-polarized beams by a thin-film polarizer (Pačebutas et al. 2006). Using half-wave/polarizer pairs, beam intensities were changed independently; one of the beams was delayed in an optical delay line. The intensity of the probe beam was 30 times lower than that of the pump beam.

The results of a single-beam Z-scan measurement are presented in Figure 19. As seen from this Figure, the increase in intensity of the beam transmitted through copper selenide nanowires array at sample positions corresponding to the highest beam intensities can be evidenced for all the samples investigated. As the laser wavelength is within the defect absorption band of the samples, it could be assumed that this effect is caused by the bleaching of the defect-to-band transitions of the electrons. The density of defects participating in this transition can be estimated from the onset energy of the bleaching effect as ~ 10^{20} cm^{-3}. The increase in the transmittance is especially pronounced for the specimens annealed at 150 and 200 °C (see curve 2). It correlates with the observed narrowing of the absorption peak in the range from 1 to 1.3 eV, and supports the idea of more homogeneous defect distribution obtained after annealing of the samples.

The dynamics of photoinduced transmission change in as-grown and annealed Cu_3Se_2 nanowire arrays studied by means of the pump-probe technique is presented in Figure 20. Relatively long photoexcited carrier lifetimes for all thermally treated specimens were obtained, indicating that this material system can be prospective for the manufacture of various optoelectronic devices, e.g. photovoltaic solar cells.

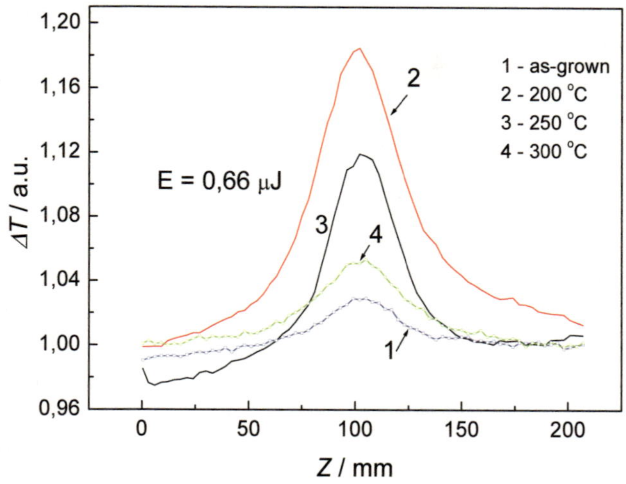

Fig. 19. Open-aperture Z-scan experimental data for Cu_3Se_2 nanowire arrays encased in alumina template by alternating current deposition before (1) and after (2-4) annealing at 200 (2), 250 (3) and 300 °C (4) for an hour. (pulse energy - 0.66 µJ, pulse duration – 2 ps).

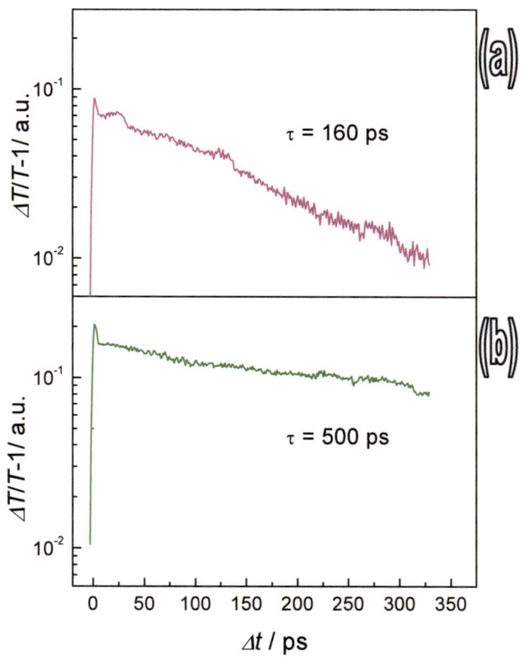

Fig. 20. Photoinduced transmission kinetics measured for Cu_3Se_2 nanowire arrays encased in alumina template by alternating current deposition in the solution as in Fig. 17 at 0.5 A dm^{-2} for 6 min after thermal treatment at 200 (a) and 250 °C (b). Pulse characteristics are the same as in Fig. 19.

A traditional femtosecond two-beam pump-probe spectroscopy technique was employed to study the kinetics of ultrafast non equilibrium charge carrier relaxation in copper selenide nanowire arrays entrapped in alumina templates depending on the nanowires diameter, composition and excitation energy (Juška et al. 2009 a,b). The experimental results demonstrate that the transient absorption kinetics is almost independent of the excitation and probe wavelength and show a biexponential charge carrier recombination with the excitation intensity dependent decay rates. The initial ultrafast relaxation, which gets slower at higher excitation intensities, is followed by the slower decay component emerging at high intensities. These relaxation peculiarities were described by a theoretical model of two concurrent relaxation channels involving deep and shallow impurity levels (Fig. 21). Occupation of the valence band and shallow trapping levels leads to the absorption bleaching, which competes with the induced absorption due to transitions from these energy levels to higher energy states. The combined result of these competing optical effects is the bleaching of the impurity absorption band and the induced absorption at higher energies (Fig. 22).

We have found that the transient absorption properties are almost independent of the diameter of copper selenide nanowires within the range of 8 to 25 nm implying that low dimension effects such as spatial confinement and surface processes are of minor importance to relaxation processes. Recently we have also found that steady-state absorption spectra of copper selenide differing in the length of nanowires clearly differ. However, nonlinear optical effects in these arrays follow a very similar tendency, e.g.

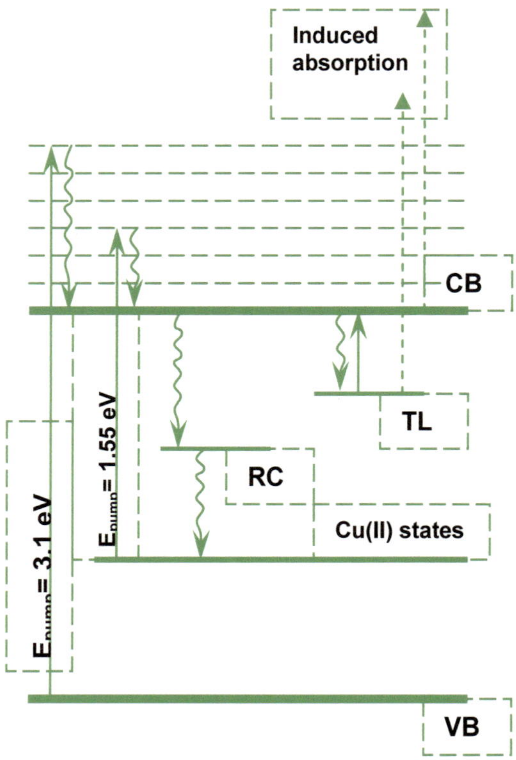

Fig. 21. Energetic diagram of $Cu_{2-x}Se$ nanowires. Solid arrows indicate steady state absorption and excited state relaxation transitions. Dashed lines demonstrate induced absorption transitions.

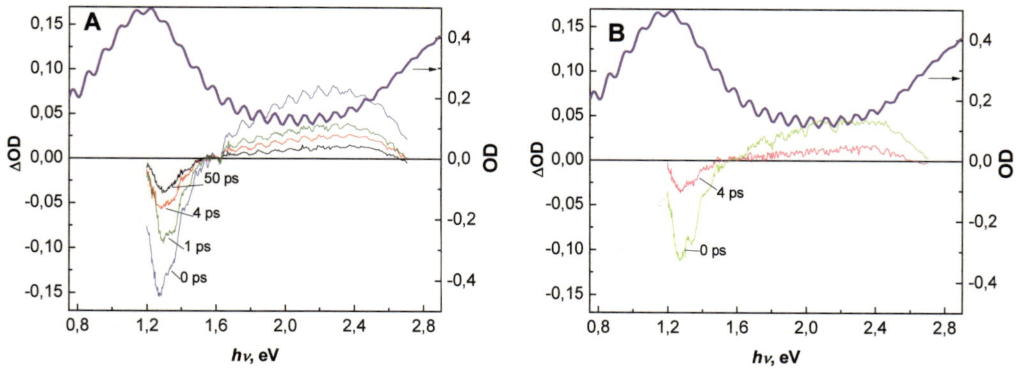

Fig. 22. Transient absorption spectra of the alumina matrix encased with $Cu_{2-x}Se$ nanowire (Ø 13 nm) recorded at different delay times under 1.55 eV photon energy excitation with pulse intensity of 23 mJ/cm^2 (A) and under 3.1 eV excitation with pulse intensity of 13 mJ/cm^2 (B). Thick lines show the optical density of the sample. Spectra modulations are due to interference effects.

induced absorption dominates in the visible light spectra region while the bleaching band appears in IR. In contrast, the composition of copper selenide nanowires, changing upon annealing (see Fig. 18), is crucial for absorption peculiarities. In the case of nearly pure Cu_3Se_2 nanowire arrays, the initial relaxation is more expressed and leads to the formation of additional middle-gap states resulting in the rise of absorption bleaching in the visible light spectrum range. Furthermore, a higher concentration of defect states in Cu_3Se_2 and $Cu_{2-x}Se$ mixture nanowires leads to an ultrafast and nearly complete recovery of temporal optical features.

The ability to adjust nonlinear optical features in copper selenide nanowires is very attractive for applications. As-formed and not annealed copper selenide nanostructures might be of a particular interest for nonlinear optical elements that require a fast optical response and a nearly complete recovery of bleaching. Annealed samples with a lower density of defects and local states can be used for devices that require longer lifetimes of nonequilibrium carriers.

5. Conclusion

A brief overview of the effect of heat treatments of some metallic (Fe, Sn) and semiconducting [FeO(OH), Cu_3Se_2, $Cu_{2-x}Se$] nanowire arrays loaded inside the alumina template pores is presented herein for our recent research. Original results have been obtained for annealing of iron nanowire arrays eventually forming $FeAl_2O_4$ films. We suspect that annealing of some other metal nanowires, such as nickel and cobalt, encapsulated inside the alumina template pores and capable of forming spinel-type structures, can also result in the formation of $MeAl_2O_4$ films. These materials, demonstrating the new properties, are the focus of ongoing research in the field of spintronics. In contrast, the heating of nanowired metals, like tin, which can not form spinel-type materials results in the formation of metal oxide nanowire arrays. In this manner it is possible to fabricate nanowire arrays of novel barbed-shaped design due to diffusion of metal atoms into the

alumina cell walls robbing the oxygen from Al_2O_3 and destroying the matrix. These nanowires stand out for their extremely high surface area and core/shell morphology. Therefore, they can found interesting applications in catalysis.
With formation a new highly-ordered systems of nanowired arrays of Cu_2Se_3 and $Cu_{2-x}Se$ in alumina matrices we are observing new non-linear optical behaviour of these semiconducting nanowire arrays that we are just beginning to interpret in terms of bleaching of the defect-to-band transitions of electrons in nanowires. Other semiconducting materials, such as iron oxyhydroxide, γ-FeO(OH) (lepidocrocite) are being deposited inside the alumina pores and explored upon the heating. It should be noted that by this way, we are able to form nanowired arrays of maghemite/hematite heterostructures, which have prospective applications in magnetic storage devices.
I hope that the idea of heat treatment of nanowired materials, encapsulated in alumina matrices, creates a simple and cheap route to manipulate their composition and properties, gaining a novel prospective functionalities and applications. This approach, although being still in a very early stages of development, is expected be of fundamental interest for future nanotechnologies.
Financial support of Lithuanian Science and Study Foundation through Research Project no. C07035 is acknowledged. I am very grateful for the indispensable help of my colleagues from Institute of Physics: Dalis Baltrūnas, Vidmantas Gulbinas, Kestutis Mažeika and Jonas Reklaitis; Arūnas Krotkus from Institute of Semiconductor Physics, Rolandas Tomašiūnas from Vilniaus University, Zenonas Jusys, Remigijus Juškėnas and Marytė Kurtinaitienė from Institute of Chemistry and Gulio Paolo Veronese from Instituto per la Microelettronica ed I Microsistemi del CNR, Bologna.

6. References

D. AlMawlawi, N. Coombs & M.Moskovits. (1991). Magnetic properties of Fe deposited into anodic aluminium oxide pores as a function of particle size. *J. Appl. Phys.* Vol. 70(8), pp. 4421-4425, ISSN: 0021-8979

J. D. Bernal, D. R. Dasgupta & A. L. Mackay. (1957). Oriented transformations in iron oxides and hydroxides. *Nature*, Vol. 180, pp. 645-647, ISSN:0028-0836

F. Blanc-Beguin, S. Nabily, J. Gieraltowski, A. Turzo, S. Querellou, P.Y. Salaun. (2009). Cytotoxicity and GMI bio-sensor detection of maghemite nanoparticles internalized into cells. *J. Mag. Mag. Mater.* Vol. 321(3) pp. 192-197, ISSN:0304-8853

P. M. Botta, R. C. Mercader, F. F. Aglietti & J. M. Porto Lopez. (2003). Synthesis of Fe-$FeAl_2O_4$-Al_2O_3 by high-energy ball milling of Al-Fe_3O_4 mixtures. *Scripta Materialia*, Vol. 48, pp. 1093-1098, ISSN: 1359-6462

Q. Cai, J. Zhang, X. Chen, Z. Chen, W. Wang, G. Mo, Z. Wu, L. Zhang & W. Pan. (2008). Structural study on Ni nanowires in an anodic alumina membrane by using *in situ* heating extended x-ray absorption fine structure and x-ray diffraction techniques. *J. Phys.: Condens. Matter.*, Vol. 20, p. 115205, ISSN:0953-8984

K.-H. Choi, S.H. Lee, Y.-R. Kim, R. Malkinski, A. Vovk, Y. Barnakov, J.-H. Park, Y.-K. Jung & J.-S. Jung. (2007). Magnetic behaviour of Fe_3O_4 nanostructure fabricated by template method. *J. Mag. Mag, Mater.*, Vol. 310, pp. e861- e863, ISSN: 0304-8853

A. N. Christensen, M. S. Lenmann & P. Convert. (1982). Deuteration of crystalline hydroxides. Hydrogen bonds of gamma AlOO(H,D) and gamma FeOO(H,D). *Acta Chem. Scand.*, Ser. A36, pp. 303-308 , ISSN:0904-213X

G. S. Collins, T. Kachnowski, N. Benczer-Koller & M. Pasternak. (1979). Application of the Mössbauer effect to the characterization of an amorphous tin-oxide system. *Phys. Rev. B*, Vol. 19(3), pp. 1369-1374, APS journal

P. Colomban, S. Cherifi & G. Despert. (2008). Raman identification of corrosion products on automotive galvanized steel sheets, *J. Raman Spectrosc.*, Vol. 39 (7), pp. 881-886, Willey InterScience, DOI:10.1002/jrs.1927

R. M. Cornell & U. Schwertmann. (1996). *The Iron Oxides: Structure, Properties, Reactions, Occurrence and Uses*; VCH, Berlin

X. Feng, J. Ma, F. Yang, F. Ji, F. Zong, C. Luan & H. Ma. (2008). Structural and photoluminescence properties of single crystalline SnO_2: In films deposited on *a*-Al_2O_3 (0001) by MOCVD. *J. Cryst. Growth*, Vol. 310(16), pp. 3718-3721, ISSN: 0022-0248

S. Ferrere, A. Zaban & B.A. Gregg. (1997). Dye sensitization of nanocrystalline tin oxide by perylene derivatives. *J. Phys. Chem. B*. Vol. 101(23), pp. 4490-4493, ACS Publication, ISSN:1520-6106

A. A. Firooz, A. R. Mahjoub & A. A. Khodadadi. (2008). Preparation of SnO_2 nanoparticles and nanorods by using a hydrothermal method at low temperature. *Mater. Lett.*, Vol. 62(12-13), pp. 1789-1792, ISSN: 0167-577X

S. Fujihara, T. Maeda, H. Ohgi, E. Hosono, H. Imai & S.-H. Kim. (2004). Hydrothermal routes to prepare nanocrystalline mesoporous SnO_2 having high thermal stability. *Langmuir*, Vol. 20, pp. 6476- 6481, ISSN:0743-7463

C. X. Gao, Q. F. Liu & D. S. Xue. (2002). Fabrication and characterization of amorphous β-FeOOH nanowire arrays. *J. Mater. Sci. Lett.*, Vol. 21, No. 22, pp. 1781-1783, ISSN: 0261-8028

A. U. Gehring & A. M. Hofmeister. (1994). The transformation of lepidocrocite during heating; a magnetic and spectroscopic study.*Clays Clay Miner*. Vol. 42, pp. 409-415, ISSN: 1552-8367

H. Gleiter. (2000). Nanostructured materials: basic concepts and microstructure. *Acta Mater.*, Vol. 48, pp. 1-29, ISSN:1359-6454

F. Gu, S.F. Wang, C.F. Song, M.K. Lü, Y.X. Qi, G.J. Zhou, D. Xu & D.R. Yuan. (2003). Synthesis and luminescence properties of SnO_2 nanoparticles. *Chem. Phys. Lett.*, Vol. 372(3-4), pp. 451-454, ISSN:0009-2614

A. M. Hermann & L. Fabick. (1983). Research on polycrystalline thin-film photovoltaic devices. *J. Cryst. Growth* , Vol. 61(3), pp. 658-664, ISSN:0022-0248

A. Jagminas, D. Bigelienė, I. Mikulskas, R. Tomašiūnas. (2001). Growth peculiarities of aluminium oxide at high voltages in diluted phosphoric acid. *J. Cryst. Growth*, Vol. 233, pp. 591-598, Elsevier, ISSN:0022-0248

A. Jagminas, R. Juškėnas, I. Gailiūtė, G. Statkutė & R. Tomašiūnas. (2006). Electrochemical synthesis and optical characterization of copper selenide nanowire arrays within the alumina pores. *J. Cryst. Growth*, Vol. 294(2), pp. 343-348, ISSN: 0022-0248

A. Jagminas, K. Mažeika J. Reklaitis, M. Kurtinaitienė & D.Baltrūnas. (2008). Template synthesis, characterization and transformations of iron nanowires while aging. *Mat. Chem. Phys.* Vol. 109, pp. 82-86, ISSN: 0254-0584

A. Jagminas, K. Mažeika J. Reklaitis, V. Pakštas & D.Baltrūnas. (2009a). Annealing effects on the transformations of Fe nanowires encapsulated in the alumina template pores. *Mat. Chem. Phys.*, Vol. 115, pp. 217-222, ISSN: 0254-0584

A. Jagminas, K. Mažeika, E. Juška, J. Reklaitis, D. Baltrūnas, R. Schmidt. M(2009b). Fabrication and characterization of lepidocrocite (γ-FeOOH) nanowire arrays. Appl. Surf. Sci. (subm.)

A. Jagminas, R. Tomašiūnas, A. Krotkus, R. Juškėnas & G. Aleksejenko. (2009c). Fabrication and phase variation in annealed Cu_3Se_2 nanowire arrays. *Appl. Surf. Sci.*, Vol. 255(17), pp. 7739-7742, Elsevier, ISSN:0169-4332

O. Jessensky, F. Müller & U. Gösele. (1998). Self-organized formation of hexagonal pore arrays in anodic alumina. *Appl.Phys. Lett.*, Vol. 72, pp. 1173-1175, ISSN: 0003-6951

Z. Ji, Z. He, Y. Song, K. Liu & Y. Xiang. (2004). *Thin Solid Films*, Vol. 460, pp. 324-326, ISSN: 0040-6090

G. Juška, A. Jagminas & V. Gulbinas. (2009a). Excitation relaxation in copper selenide nanowires. *Phy. Status Solidi B*, Vol 246 (5), pp. 1082-1087, Wiley Inter Sci., ISSN:

G. Juška, V. Gulbinas & A. Jagminas. (2009b). Transient absorption of copper selenide nanowires of different stoichiometric structure. Lithuan. J. of Phys. (subm.)

M. Kokonou, K.P. Giannakopoulos & A. G. Nassiopoulou. (2007). Few nanometer thick anodic porous alumina films on silicon with high density of vertical pores. *Thin Solid Films*, Vol. 515, pp. 3602-3606, Elsevier, ISSN:0040-6090

X. Kong, D.Yu & Y. Li. (2003). Synthesis of SnO_2 nanoribbons by direct oxidation of tin powders. *Chem. Lett.*, Vol. 32, pp. 100-101, The Chemical Soc. of Japan, ISSN:0366-7022

E. Kuzmann, S. Nagy and A. Vertes. (2003). Critical review of analytical applications of Mössbauer spectroscopy illustrated by minerological and geological examples. *Pure Appl. Chem.*, Vol. 75(6), pp. 801-858, IUPAC publications, ISSN:0033-4545

L. Li, Y. Zhang, Y. W. Yang, X. H. Huang, G. H. Li & L. D. Zhang. (2005). Diameter-dependent thermal expansion properties of Bi nanowire arrays. *Appl. Phys. Lett.*, Vol. 87, p. 031912, ISSN: 0003-6951

H. Masuda & K.Fukuda. (1995). Ordered metal nanohole arrays made by a two-step replication of honeycomb structures of anodic alumina. *Science*, Vol. 268, pp. 1466-1468, ISSN:1466-1468

R. M. Metzger, V. V. Konovalov, M. S, T. Xu, G. Zangari, B. Xu, M. Benakli & W. D. Doyle. (2000). Magnetic nanowires in hexagonally ordered pores of alumina. *IEEE Trans. Magn.*, Vol. 36(1), pp. 30-35, ISSN:0018-9464

M. P. Morales, M. J. Munoz-Aguado, J. L. Garcia-Palacios, F. J. Lazaro & C. J. Serna. (1998). Coercitivity enhancement in γ-Fe_2O_3 particles dispersed at low volume fraction. *J. Magn. Magn. Mater.*, Vol.183, pp. 232-240, ISSN:0304-8853

Ö. Özdemir & D. J. Dunlop. (1993). Chemical remanent magnetization during gamma-FeOOH phase transformations. *J. Geophys. Res.* Vol. 98, No *B3*, pp. 4191-4198, ISSN:0148-0227

V. Pačebutas, G. Aleksejenko, A. Krotkus, J. W. Ager, W. Walukiewicz, H. Liu & W. J. Schaff. (2006). Optical bleaching effect in InN epitaxial layers. *Appl. Phys. Lett.*, Vol. 88, p. 191109, ISSN:0003-6951

X. Peng, G. Wu, P. Holt-Hindle & A. Chen. (2008). Growth and characterization of free-standing singe crystalline tin and tin oxide nanobelts. *Mater. Lett.*, Vol. 62, pp. 1969-1972, ISSN:0167-577X

R. E. Presley, C. L. Munsee, C.-H. Park, H. Hong, J. F. Wager & D.A. Keszler. (2004). Tin oxide transparent thin-film transistors. *J. Phys. D: Appl. Phys.*, Vol. 37(20), pp. 2810-2813, Institute of Physics Publishing, ISSN:0022-3727

D. Qin, P. Yan, G. Li, J. Xing & Y. An. (2008). Self-construction of SnO_2 cubes based on aggration of nanorods. *Mater. Lett.*, Vol. 62(16), pp. 2411-2414, ISSN:0167-577X

J. Qin, J. Nogues, M. Mikhaylova, A. Roig, J. S. Munoz & M. Muhammed. (2005). Differences in the magnetic properties of Co, Fe, and Ni 250-300 nm wide nanowires electrodeposited in amorphous anodized alumina templates. *Chem. Mater.* Vol. 17(7), pp. 1829-1834, ACS Journal, ISSN:1520-5002

W. Römer & C. Steinem. (2004). Impedance Analysis and Single-Channel Recordings on Nano-Black Lipid Membranes Based on Porous Alumina. *Biophysical Journal,* Vol. 86, pp. 955-965, ISSN:0006-3495

D. Schroeer & R. C. Nininger. (1967). Morin transition in a-Fe_2O_3 microcrystals. *Phys. Rev. Lett.*, Vol. 19, pp. 632-654, ISSN:1079-7114

M. Sheik-Bahae, A. A. Said, T. H. Wei, D. J. Hagan & E. W. Van Stryland. (1990). Sensitive measurement of optical nonlinearities using a single beam. *IEEE J. Quantum Electron*, Vol. 26(4), pp. 760-769, ISSN: 0018-9197

J.-O. Song & T.-Y. Seong (2004). Highly transparent Ag/SnO_2 ohmic contact to *p*-type GaN for ultraviolet light-emitting diodes. *Appl. Phys. Lett.*, Vol. 85, p. 6374, American Institute of Physics, DOI:10.1063/1.1834990

G. Statkutė, A. Jagminas & R. Tomašiūnas. (2008). Photo-induced transmittance in copper-selenide nanowires. *Opt. Mater.*, Vol. 30, pp. 743-745, Elsevier, ISSN:0925-3467

J. Sundqvist, J. Lu, M. Ottosson & A. Harsta. (2006). Growth of SnO_2 thin films by atomic layer deposition and chemical vapour deposition: A comparative study. *Thin Solid Films*, Vol. 514, pp. 63-68, ISSN: 0040-6090

G. E. Thompson. (1997). Porous anodic alumina: fabrication, characterization and applications. *Thin Solid Films*, Vol. 297, pp. 192-201, ISSN:0040-6090

Y. Wang, J. Yang, C. Ye, X. Fang & L. Zhang. (2004). Thermal expansion of Cu nanowire arrays. *Nanotechnology*, Vol. 15, pp. 1437-1440, ISSN:0957-4484

Y. Wang, J. Y. Lee & H. C. Zeng. (2005). Polycrystalline SnO_2 nanotubes prepared via infiltration casting of nanocrystallites and their electrochemical application. *Chem. Mater.*, Vol. 17(15), pp. 3899-3903, ACS Publications, DOI:101021/cm050724f

X. J. Xu, G. T. Fei, W. H. Yu, L. Chen & L. D. Zhang. (2006). *In situ* x-ray diffraction study of the thermal expansion of the ordered arrays of silver nanowires embedded in anodic alumina membranes. *Appl. Phys. Lett.*, Vol. 88, p. 211902, ISSN: 0003-6951

X. Xu, Z. Wang, T. Zhang, X. Zeng, W. Xu, J. Zhang, J. Yan, J. Zhang, L. Zhang. (2008). Manipulation of optical properties of Ag/Cu alloy nanowire arrays embedded in anodic alumina membranes. *Appl. Surf. Sci.*, Vol. 254, pp. 3845-3848, ISSN:0169-5332

Li-Y. Zhang, J. Fen & De-S. Hue. (2007). An investigation of thermal decomposition of β-FeOOH nanowire arrays assembled in AAO templates. *Mater. Lett.*, Vol. 61(6), pp. 1363-1367 ISSN:0167-577X

5

Fabrication and Applications of Metal Nanowire Arrays Electrodeposited in Ordered Porous Templates

Giray Kartopu[1] and Orhan Yalçın[2]
[1]*University of Nottingham, Nottingham NG7 2RD,*
[2]*Niğde University, 51240 Niğde,*
[1]*UK*
[2]*Turkey*

1. Introduction

At present, there is a huge anticipation that the realm of nanotechnology will soon be realized, and the life will become easier and more enjoyable - thanks to numerous new products and apparatus that will be operating on 'nano-facts' and nano-sized objects. Material properties and functionalities of the bulk tend to differ when one or more of its dimensions are reduced down to between 100 to 1 nm, the so called 'nanosize regime'. Thin films (two dimensional, or 2D), nanowires/nanotubes (one dimensional, or 1D), and nanoparticles/quantum dots (zero dimensional, or 0D) constitute the basic classes of nanomaterials. Among these, particularly the 1D-materials are highly desirable, as their geometric shape and high surface area impart high functionally. For example, metal nanowires (MNWs) are likely to become an integral part of future nanodevices, at least as the elements interconnecting the functional components such as 'nano'-transistors.

Additional to the provision of electrical connection to nano-circuits, the MNWs are expected to be utilized as the functional components in various applications ranging from high density perpendicular data storage to nano-sensors, from high-sensitivity nano-electrodes to meta-materials, and so on. In this chapter, we shall focus on the potential applications of MNWs that are synthesized via the template fabrication method.

Template synthesis of MNWs arrays comprises the electrochemical reduction of the ions of one/more desirable metals inside the nano-pore channels of an insulating membrane fabricated via self-assembly. Owing to its cost effectiveness, versatility and high throughput the template fabrication, being a bottom up synthesis method, offers significant promise for the production of versatile, tailor-made MNWs.

For most end applications, it is desirable to have MNWs with high aspect ratio (ratio of wire length to diameter) and/or specific surface area, production reproducibility as well as uniformity in wire size and shape in conjunction with high spatial ordering. Thus, the self-organized porous anodic aluminium oxide (AAO) membrane, displaying most of the desirable template properties, has become the centre of focus for MNWs synthesis.

In what follows we shall first examine the fabrication and properties of the AAO template. Next, the electrochemical template synthesis of both single-component and multilayered/ superlattice MNWs will be described. Finally, major applications of these MNWs will be presented.

2. Preparation and properties of anodic aluminum oxide templates

2.1 History and background

Formation of porous as well as barrier-like alumina films by aluminium anodization is a well known procedure which has been studied since 1950s [Keller et al., 1953]. Today it has found its way to aviation industry [Avcorp Inc., 2009], and even to our everyday life, e.g. through kitchen utensils, where an industrial scale hard-anodization is performed to give the aluminium a protective finish rendering it thermally and mechanically resistant and non sticky.

Anodization of aluminium usually takes place in a suitable acidic solution comprising of chromic/phosphoric/sulphuric/oxalic/malonic acid in water at a fixed current density (galvanostatic) or, more commonly, at a fixed bias (potentiostatic). In the case of porous films, the pore and cell dimensions as well as pore regularity are strong functions of the applied potential. For example, the established growth rate for the cell size, or the pore (center-to-center) distance, is 2.5 nm/V [O'Sullivan & Wood, 1970]. Steady-state porous film formation is maintained by the equilibrium between Al oxidation and oxide-dissolution at the pore base. Since oxygen is incorporated in the anodic film (in the form of Al_2O_3), the volume (thickness) of the formed film is higher than that of the consumed Al. Optimal conditions for the self assembly of highly ordered pore arrangements were discovered only after 1995 [Masuda & Fukuda, 1995; Li et al., 1998]. Today, such ordered porous films are widely employed as mass fabrication tools, i.e. 'templates', for a myriad of 1D and composite nanomaterials in the name of fundamental and/or technological interests [Martin, (1994); Wade & Wegrowe, (2005)]. The template synthetic method is so versatile [Hurst et al., 2006; Das, B. & Singaraju, P. (2005); Liang et al., (2002)] that its applications are as vast as the imagination!

2.2 Ordered porous aluminum oxide films by double-anodization

Pore formation in aluminium oxide films initiates at random locations on the aluminium surface, and as the anodization progress a self organisation process takes place, which dictates termination/merging of 'bad' pores, resulting in an ordered 2D hexagonal arrangement (Fig. 1a). This principle forms the basis idea in the formation of highly ordered AAO templates via the 'double-anodization' method [Masuda & Fukuda, 1995]. In this method, the initially formed porous oxide film, with rather less organised top layers, is removed and the anodization is repeated under similar conditions. As the pores reaching at the aluminium surface have tube-end-like shapes, removal of the first oxide film leaves the aluminium surface with highly ordered semi-spherical etch pits (Figs. 1b, 2a), which thereupon act as seeds for the ordered pore formation via the second anodization, resulting in non-intersecting, uniform pore channels running straight through the AAO film (Figs. 1c, 2b).

For template synthesis applications, double-anodized AAO templates are highly preferable over single-anodized or commercially available AAO membranes (e.g. Whatman Anodisc™ filters) as the latter lacks uniformity in pore size and shape and has little or no long-range pore ordering. Further, often the quoted (nominal) average pore size in commercial specimens shows strong deviations throughout the film.

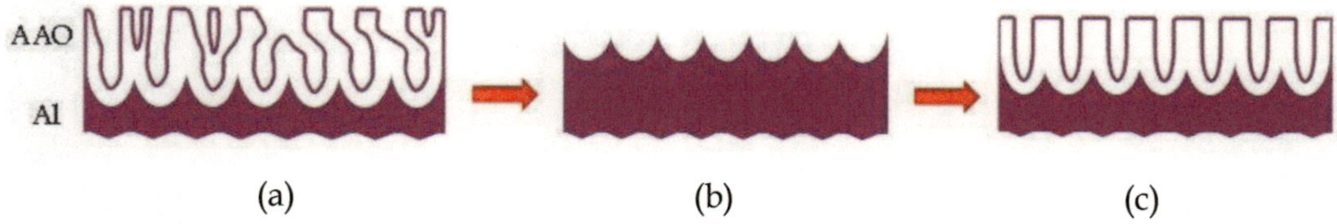

Fig. 1. Schematic representation of ordered AAO formation via double-anodization: (a) first anodization, (b) removal of the first (disordered) oxide, and (c) second anodization.

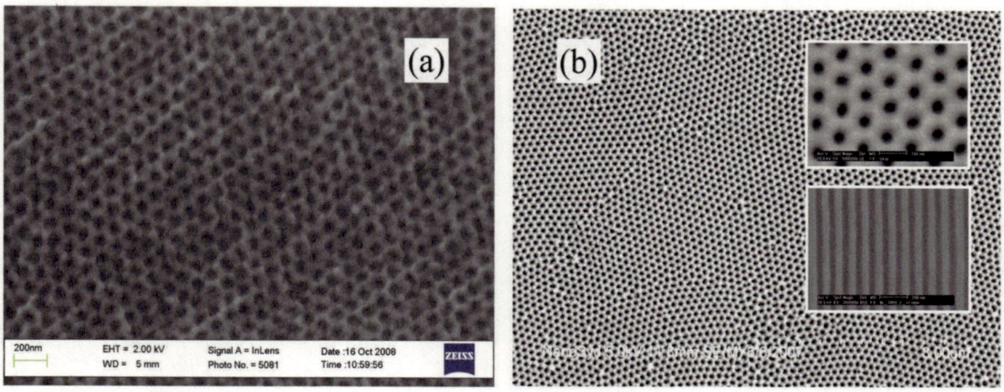

Fig. 2. SEM image of (a) the Al surface upon removal of the first anodic oxide film, and (b) the second anodic oxide film (insets: close-up view of the pore channels). Anodization was performed in 0.3M oxalic acid, at 40V and 15 °C (24h first anodization).

Commercial templates are typically prepared and lifted off from the supporting aluminium substrate using a single anodization procedure employing a step-wise voltage reduction method [Furneaux & Rigby, 1989]. In this process, the anodization voltage is gradually reduced near the end to about 0V, causing pore splitting, i.e. by formation of cracks in the alumina 'barrier layer' at the pore bottoms, followed by the detachment of the entire AAO film (owing to its diminished bonding to the Al substrate). Consequently, pores are much smaller and disordered at membrane surfaces compared to the film interior, albeit it is typically the surface porosity that is quoted in product descriptions as these membranes are usually produced for filtration applications. Figs. 3a and 3b compares tilted cross-sections of an AAO Anodisc™ membrane having nominal pore size of 200 nm with an AAO film double-anodized in 0.3M oxalic acid at 40V (to enhance imaging pores in AAO film were enlarged using an isotropic chemical etching following the second anodization). The contrast in the pore uniformity between these templates is further highlighted in Figs. 3c and 3d where mechanical polishing is applied subsequent to metal-pore filling by electroplating (see Section 3 for details on electrochemical template synthesis).
Detailed structural characterizations performed by transmission electron microscopy (TEM) analysis on AAO films anodized under different conditions reveal that porous alumina necessitates a porosity of about 10% to be self ordered [Nielsch et al., 2002]. There are three well-established anodizing regimes that yield in AAO films of this type (Table 1) [Li et al., 1998]. As the pH of solution is lowered the dissolution rate at the pore base increases, and

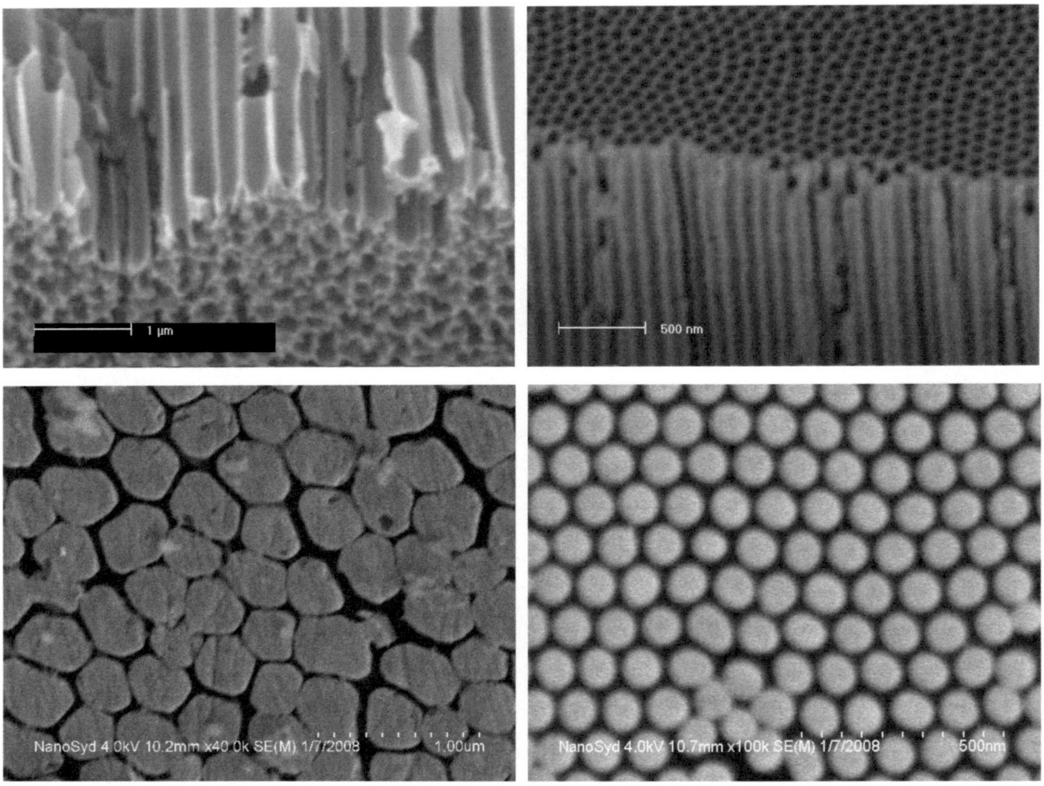

Fig. 3. Cross sectional view of (a) Anodisc™ AAO (quoted pore size 200 nm) and (b) doubleanodized AAO (0.3M oxalic acid, 40V). (c, d) Topographic view of the same templates upon pore filling and subsequent mechanical polishing (bright features correspond to metal nanowires; dark regions correspond to AAO pore walls).

Electrolyte	Bias (V)	Interpore distance (nm)	Pore density (per cm^2)
H_2SO_4 (sulphuric acid)	25	63	~10^{11}
$C_2H_2O_4$ (oxalic acid)	40	105	~10^{10}
H_3PO_4 (phosphoric acid)	195	500	~10^9

Table 1. The three well-established anodization conditions that lead to self-ordered porous AAO films.

thus, lower potentials are favored for the self-sustainable growth of ordered AAO films. Therefore, different self ordering voltages exist for different acid types. As pointed out earlier, the interpore distance, or the cell size, (L) is defined by the applied potential with a proportionally constant of 2.5 nm/V [O'Sullivan & Wood, 1970]. In accord, commonly observed values for self ordered AAO films, if obtained in the above regimes, are $L \approx 63$, 105 and 500 nm, respectively (Table 1). Outside the optimal range of the anodization potential the porosity can be remarkably different. For example, reducing the potential from 195 to 160 V in phosphoric acid increases the film porosity to about 40% [Nielsch et al., 2002]. Increasing the potential from 40 to 120-150V in oxalic acid (under extreme cooling of the Al/electrolyte

interface) decreases the porosity to about 3% [Lee et al., 2006]. The electrolyte temperature and concentration, on the other hand, have only minor effects on template parameters and are usually set to 0-20 °C and 0.1-0.4 M, respectively. For example, increasing the concentration of sulphuric acid from 1 to 10 wt %, i.e. from about 0.1 to 1 M, results in a mere 20% reduction in the cell parameter [Nielsch et al., 2002]. Volume expansion factor ($R = V_{Al2O3}/V_{Al}$) of ordered AAO having 10% porosity is determined to be about 1.2 [Jessensky et al., 1998].
In fact, additional to double-anodization strategy outlined above there are other factors used to improve the template parameters, e.g. the ordered pore domain size. Firstly, Al has to be annealed for 1-4 h in a non-oxidizing atmosphere, such as that of Ar or N_2, close to the melting temperature of Al (typically between 400-500 °C). Secondly, the Al surface has to be de-greased by an ultrasonic treatment in acetone, and then, electro-polished to a mirror finish in a perchloric acid/ethanol electrolyte, typically of 1:5 vol. ratio, at a bias of 15-25V for 15-60s (in this process, too, Al is connected as the anode). Finally, the initially non-circular pore channels can be rounded up, as well as intentionally widened, using post-anodization chemical etch, e.g. in dilute phosphoric acid at 30-35 °C for 5-15 min.

2.3 Highly ordered porous aluminum oxide films by aluminum pre-texturing

Size of the ordered pore domains in double-anodized AAO extends to a maximum of about 5 μm; large numbers of pore defects are still observed at the domain boundaries. In order to obtain ideally ordered films surface pre-texturing of aluminum has to be realized. Several methods have been tested for this purpose, and formation of even mono-domain AAO films has been demonstrated. As it can be predicted, despite its costliness and additional complexities, lithography aided Al pre-texturing does away the need to perform double/multiple anodizations since high pore ordering is readily obtained through a single anodization process. The following are the major routes taken to achieve ideal pore arrangements in anodic alumina:

- Imprint lithography (pattern transfer via embossing with a SiC mold [Masuda et al., 1997], SiN mold [Choi et al., 2003], commercial grating [Mikulskas et al., 2001], etc.),
- Focused ion-beam lithography (direct writing on Al [Peng et al., 2005; Robinson et al., 2007] or via photoresist patterning [Liu et al., 2003]),
- Interference/holographic lithography (laser interference [Sun & Kim, 2002; Krishnan & Thompson, 2007]),
- Pattern transfer via nano-sphere lithography (self assembly of colloidal particles followed by embossing [Fournier-Bidoz et al., 2004] or reactive-ion etching [Kim et al., 2007]), and
- Pattern transfer via (Al) deposition on nano-structured substrates [Nishio et al., 2008].

Each of these approaches has its own advantages and disadvantages. For example, the molds, i.e. nano-patterned masters, used for embossing of Al have to be custom built for each application as they must contain nanoscale features that are compatible with the anodization conditions that lead to self assembly (see Table 1). Despite its cost efficiency and simplicity nano-sphere lithography cannot yield in mono-domains that extend over large scales since the self organisation of colloidal particles is inherently limited in perfection. Nevertheless, depending on the nature of the aimed end-product or the application, a particular method may stand out due to considerations related to the device geometry, functionality and/or financial issues. Fig. 4 presents two examples of ideally ordered AAO films obtained through Al pre-texturing.

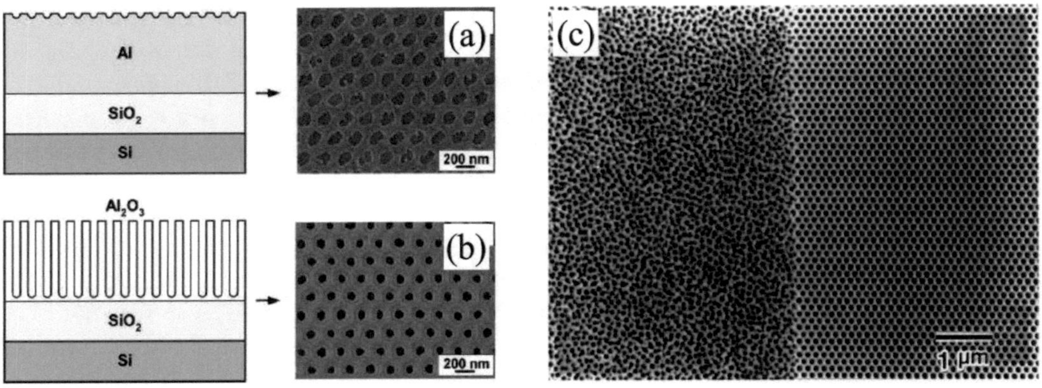

Fig. 4. Fabrication of ideally ordered porous alumina films using Al pre-texturing processes: Laser interference lithography (a, b) [Krishnan & Thompson, 2007. Copyright Wiley-VCH Verlag GmbH & Co. KGaA. Reproduced with permission.] and imprint lithography (c) (Reprinted with permission from [Masuda et al., 1997]. Copyright [1997], American Institute of Physics.). (a) Surface of an Al thin film (deposited on oxidized Si substrate) is patterned using interference lithography and wet-etching techniques, (b) pre-patterned Al film is anodized (at 86V in 5% H_3PO_4 solution), and (c) single-anodized (at 60V in 0.3M $C_2H_2O_4$ solution) Al foil surface without (left) and with (right) pre-texturing via imprint lithography.

3. Electrochemical synthesis of metal nanowires in anodic alumina templates

Electrochemical synthesis into anodic alumina pore channels can be realized once a conductive path is formed through the pore channels between the electrolyte (containing metal ions) and the deposition electrode at the bottom of the pores. This proves to be particularly difficult for anodic alumina films grown on Al foils as these films inherently have a thick (several tens of nm), insulating oxide (alumina) barrier layer that separates the pores from the Al metal.

3.1 Deposition into free-standing membranes

In order to circumvent the above problem there are two classical approaches one can follow. In the first case, first a sufficiently thick alumina film is formed on Al foil and then released to obtain a self-standing membrane with through pores using sequential chemical etching of the nonoxidized Al (Fig. 5a) and the alumina barrier layer (Fig. 5b). Physical vapour deposition of a thin metal film on one of the membrane surfaces (Fig. 5c) completes the process of obtaining the desired sample geometry to begin electro-deposition process. (This is also the geometry attained when using other free-standing membranes, such as commercial polycarbonate filters [Schönenberger et al., 1997; Whitney et al, 2003], i.e. prior to electro-deposition). Electrodeposition of MNWs can then be performed using either two- or three-electrodes conventional direct-current (DC) plating (Fig. 5d). The MNWs grow from the bottom electrode in a bottom-up fashion. Thus, the length of wires can be controlled at will simply via changing the deposition time, or more correctly, the charge density. The process is governed by the Faraday equation, which takes into account the template parameters (pore size, pore density, porosity, etc.) as well as filling efficiency [Schonenberger et al., 1997].

Fig. 5. Fabrication of MNWs via free-standing AAO membranes: a) dissolution of non-oxidized Al, b) removal of alumina barrier layer, c) metallization, and d) DC electroplating. Al can be selectively dissolved in an aqueous solution of $HgCl_2$, or $CuCl_2$:HCl, or $SnCl_4$ at the room temperature or under cooling. Barrier layer can be removed (and the pores rounded) by chemical etching in dilute (~5 wt. %) H_3PO_4 at 30-35 °C.

Using the free-standing AAO membranes, it is possible to substantially widen the pores by chemical etching, which in turn provides more flexibility in the tuning of nanowires' diameter and their associated physical properties. These membranes are resistant to heat treatments up to ~400 °C, after which they start to crystallize and, more importantly, become distorted via curling up. Furthermore, since it is difficult to obtain thick, self-standing AAO films with small pores this method is not suitable for the synthesis of MNWs having small diameters, i.e. <20 nm.

3.2 Deposition into on-substrate anodic alumina/Al templates

In the second case, the AAO layer is kept on the Al substrate while the barrier layer is thinned/cracked (Fig. 6a), using in-situ voltage reduction method and/or ex-situ (post-anodization) chemical etching, so as to facilitate its electrical conductivity. Nevertheless, as the barrier layer cannot be eliminated completely (to preserve the structural integrity of the AAO/Al pair), electrical conduction remains limited and hence it is necessary to use alternating-current (AC), rather than DC, deposition (Fig. 6b) employing a high voltage (~10-25V) and a high frequency (several hundred Hz) [Yin et al., 2001; Metzger et al., 2000].

Fig. 6. Fabrication of MNWs via on-substrate AAO/Al template: a) barrier layer thinning, and b) AC electro-deposition.

As an advantage of this method, both thin and thick on-substrate AAO/Al templates can be used to prepare MNWs, thus permitting access to the sub-20 nm (pore/wire diameter) range. The process is relatively simpler compared to that utilizes free-standing membranes. However, the synthesized MNWs are often porous and polycrystalline, and are not well bonded to the Al electrode, which hinder the electrical characterization and electrical properties of the as-synthesized MNWs. Additionally, due to the existence of the thin 'in-between' alumina layer, it is not feasible to obtain self-standing MNW arrays on the (Al) substrate by the removal of the AAO template. Another disadvantage of this method is that the pores cannot be widened to relatively large values, as otherwise delamination of the

AAO film from the Al surface takes place [Nielsch et al., 2000]. Further, superlattice-type multilayer MNWs (see Section 3.4) cannot be prepared using the AC deposition method. The filling efficiency, i.e. percentage of filled pores, is largely dependent on the spatial uniformity of the method followed to thin down/crack the barrier layer. Finally, ex-situ heating experiments on the MNWs can be performed, in principle, only up to the melting temperature of Al (650 °C).

3.3 Deposition into supported thin film anodic alumina/metal/substrate templates

As for AAO films obtained through anodization of thin (~0.5-1 μm) Al films deposited on conductive or metal-coated substrates, often realized is the unique, bridge-like shape and thinness of the barrier layer [Crouse et al., 2005; Foong et al., 2009]. It is thus relatively easy to open the pore bottoms to the (conducting) substrate via a brief chemical etching (Fig. 7a). The MNWs can then be grown in the conventional bottom-up manner using DC electroplating (Fig. 7b).

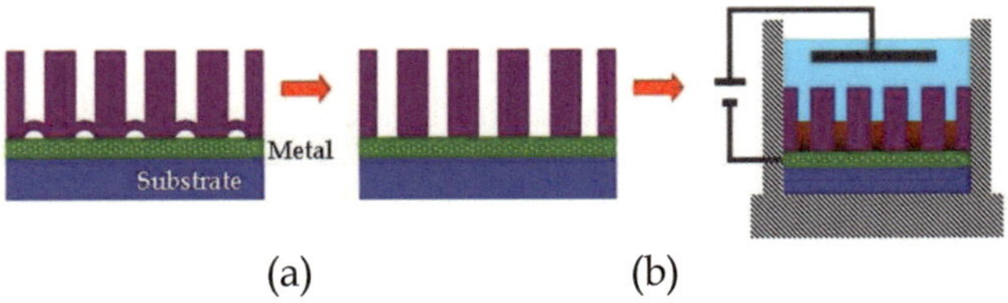

Fig. 7. Fabrication of MNWs via supported AAO film: a) removal of bridge-like barrier layer, and b) DC electro-deposition.

Major benefits this method are that it allows ex-situ annealing temperatures up to 1000 °C [Mallet et al., 2005] and that robust free-standing MNW arrays can be obtained by etching of the thin film template [Matefi-Tempfli et al., 2009]. One of its drawbacks is that high vacuum deposition is required for the growth of Al films as well as several metallic interfacial layers, which typically entails an inert (Au or Pt) film (bottom electrode for electrodeposition) and a buffer layer(s), such as Ti, Nb, etc., used for adhesion and prevention of substrate oxidation via the anodization process. Moreover, the degree of pore ordering remains relatively poor, even if double-anodization is applied, because of the limited thickness of the starting Al film. Thus, for cases where pore/wire ordering is of crucial importance the Al pre-texturing method appears to be the most viable option.

3.4 Deposition of multilayer metal nanowires

Simple multilayered MNWs can be produced by changing the electrolyte and deposition bias for each component at a time. More complicated multilayer designs, such as the A/B/A/B/... type superlattice [Piraux et al., 1994], on the other hand, can be synthesized from a single bath by alternating the deposition voltage, albeit one of the layers always turning slightly impure (due to doping with other metals in the bath). In this process, the ionic concentration of the metal that can be reduced at lower (over)potentials is kept relatively very low (~1:100). At low potential pulse only these ions are deposited in layer A, whereas at higher potential pulse more than one type of ions is reduced in layer B. However, as there is a marked difference in ionic concentrations layer B is considered as nearly pure in the more concentric metal. Galvanostatic pulses can also be used to form multilayer MNWs. Fig. 8 illustrates close-up view of a Fe/Au (magnetic/optical) multilayer nanowire obtained via galvanostatic pulses in an Anodisc™ AAO membrane with ~200 nm pores. Such MNWs combine useful properties of the respective components, offering enhanced functionalities with respect to those of the individual layers.

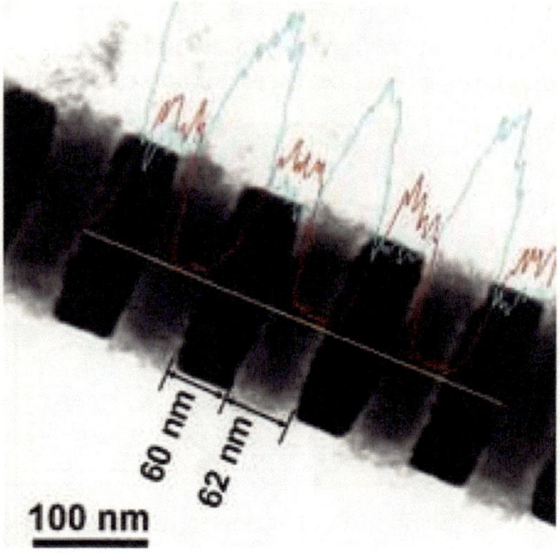

Fig. 8. Transmission electron micrograph and energy-dispersive X-ray elemental line scanning result (blue/light Fe, red/dark Au) of a superlattice nanowire obtained using 10 mA cm^{-2} (for Fe) and 0.5 mA cm^{-2} (for Au) pulses from an electrolyte containing $FeSO_4$, $KAu(CN)_2$, and H_3BO_3 with pH~3.5 [Lee et al., 2007. Copyright Wiley-VCH Verlag GmbH & Co. KGaA. Reproduced with permission.].

4. Applications of metal nanowires

4.1 Perpendicular magnetic data storage media

Recent roadmaps and reviews on hard disc drives indicate that a transition from longitudinal to perpendicular magnetic recording is taking place in these years, which may lead the storage density to increase to the 1 Tbit/in^2 levels [Kaitsu et al., 2006; Richter et al, 2007]. To realize perpendicular recording, it is necessary to have thin films containing high density of single-domain hard magnetic grains, or nano-particles, with out-of-plane anisotropy. In other words, magnetization easy axis of the individual recording elements (bits) must be aligned in the out-of-plane direction. Such anisotropy can be imposed, for example, by controlling either or both of the magneto-crystalline (via crystal orientation) and shape anisotropies of the magnetic particles. In this context, high aspect ratio nanoparticles or appropriately textured grains are desirable.
Due to the shape anisotropy effect, it is much easier to magnetize a 1D ferromagnet to its saturation using a longitudinal field. As depicted in Fig. 9a, for ferromagnetic nanowires below a critical diameter the single-domain state can be attained if the applied magnetic field is aligned parallel to the wire axis. On the other hand, multi-domain state results in if the field is applied in transverse direction (Fig. 9b). Compact nanowire arrays are, therefore, among the candidates for perpendicular storage applications. However, as for any magnetic system, the overall (effective) anisotropy of the nanowire array, and hence its suitability for these applications would be determined by many factors via the equation

$$H_{eff} = H_{shp} + H_{crys} + H_{elas} + H_x \quad (1)$$

where H_{eff} is the effective anisotropy field, H_{shp} the shape anisotropy field, H_{crys} the magnetocrystalline anisotropy field, H_{elas} the magneto-elastic (or strain related) anisotropy field, and H_x represents other fields present, e.g. the external field, magnetostatic (inter-particle) interaction fields, etc. Nanowire arrays embedded in ordered nano-templates thus represent a test-bed for the observation and investigation of these individual anisotropies.

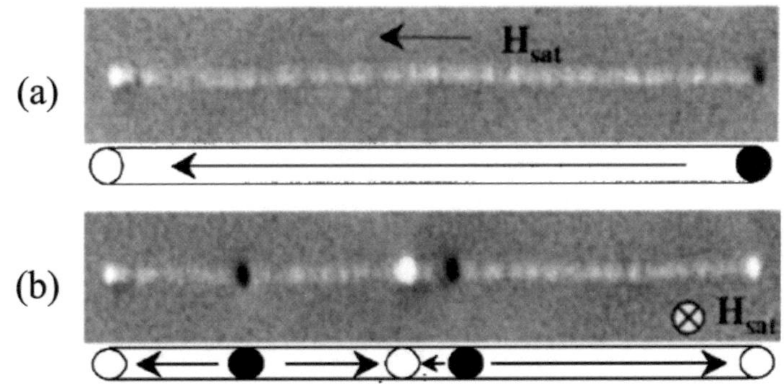

Fig. 9. Domain-imaging of a single 35 nm Co nanowire by magnetic force microscopy after saturation in a magnetic field applied (a) parallel and (b) perpendicular to wire axis (Reprinted with permission from [Ebels et al., 2000]. Copyright (2000) by the American Physical Society.). (For Co wires with aspect ratio of 10, the critical diameter is 140 nm.)

Ordered and high pore density AAO templates offer an opportunity to study the role of magnetostatic interactions [Clime et al., 2006; Fodor et al., 2008] as well as other anisotropies [Kumar et al., 2006] in determining the magnetic behaviour of densely packed ferromagnetic nanowire array films. Recently, size effects were investigated in uniform Co wires embedded in double-anodized oxalic acid AAO templates [Kartopu et al., 2008a]. Fig. 10 shows that the templates and arrays are well ordered. Magnetization data (Fig. 11) indicate that the axial hysteresis loops are sheared, that is quite large fields are required to saturate these arrays, due to strong magnetostatic (wire-wire) interactions. Effective anisotropy is determined by the interplay between the shape anisotropy and magnetostatic fields. As the wire diameter is increased magnetization easy-axis rotates from out-of-plane to in-plane direction. Further, as opposed to dipolar approximation, which suggests a continuous increase in interaction field with particle volume (length) [Grimsditch, 1998; Rivas et al., 2002], interestingly the H_{eff} or magnetostatic fields first tend to increase with wire length, i.e. the aspect ratio (τ) and then saturates. This behaviour is satisfactorily predicted by calculating the interaction energy for a pair of cylindrical ferromagnetic wires with high aspect ratio ($\tau > 2$) [Clime et al., 2006; Beleggia, 2004].

4.2 Giant-magnetoresistive multilayer nanowires

One of the early applications of template synthesized nanowires involved the use of magnetic/non-magnetic superlattice MNWs in realization of the giant magnetoresistance (GMR) effect in the current-perpendicular-to-plane (CPP) mode [Piraux et al., 1994]. The GMR effect, discovered in vacuum deposited (Fe/Cr) multilayer thin films by the late 1980s [Baibisch et al., 1988; Binasch et al., 1988] and commercialized in the late 1990s [Gochowski et al., 2003], has brought the 2007 Nobel Prize in Physics to its discoverers (Grünberg & Fert) [The Nobel Foundation, 2007]. High sensitivity of the GMR magnetic sensors ('readout

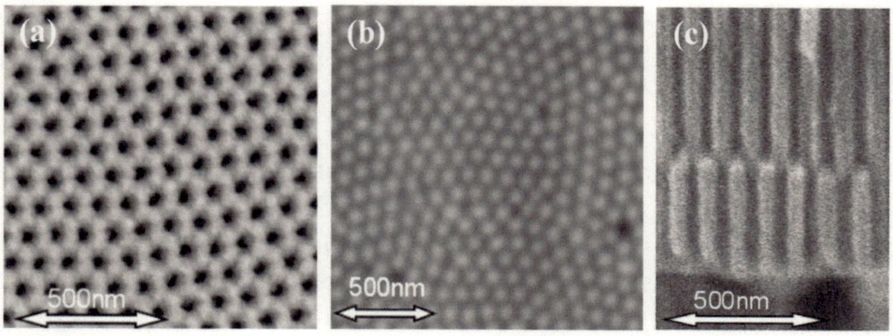

Fig. 10. Typical SEM image of (a) a double-anodized oxalic acid AAO film (~40 nm pore size), (b) topographic and (c) cross-sectional view of Co wire arrays deposited inside AAO pore channels (Reprinted with permission from [Kartopu et al., 2008a]. Copyright [2008], American Institute of Physics.).

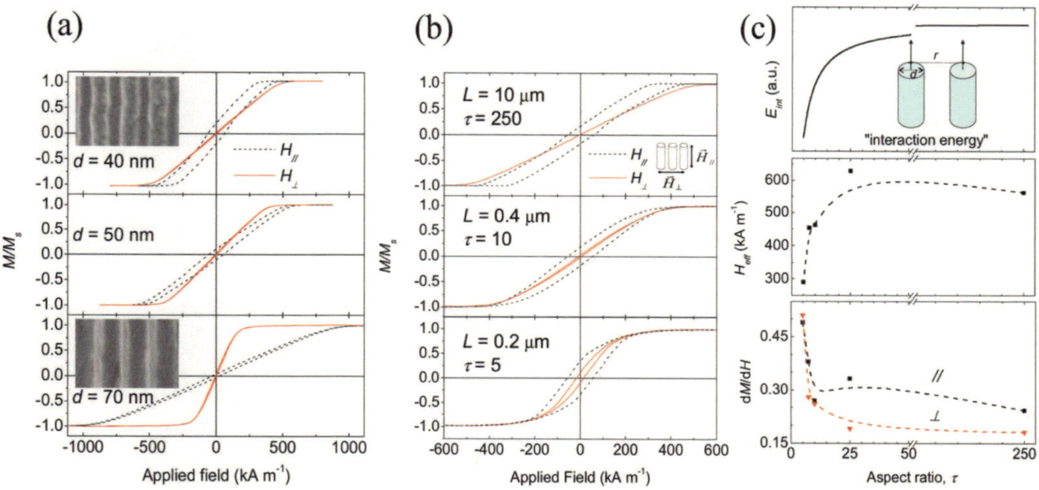

Fig. 11. Magnetization (M-H) curves as functions of (a) wire diameter (40-70 nm), and (b) wire length (0.2-10 μm) for Co nanowires in AAO templates collected parallel (dashed) and perpendicular (line) to wire axis. In (a) wire lengths are min. 4 μm; in (b), the wire diameter is fixed to about 40 nm. Insets in (a) show cross-sectional SEM images of the AAO templates used (dark lines correspond to pores). (c) Slope at zero field (*dM*/*dH*) and the effective anisotropy field (H_{eff}) calculated using the *M-H* curves in (b). Also shown (top panel) is the calculated magnetostatic interaction energy (E_{int}) for two axially magnetized intreracting wires with inverse distance of $\rho(r/d) = 0.4$ and $\tau > 2$ (Reprinted with permission from [Kartopu et al., 2008a]. Copyright [2008], American Institute of Physics.).

heads') enabled the enhancement of hard disc storage density. When the spins in the magnetic layers are aligned (e.g. upon application of an external field), resistance of the structure is reduced with respect to the initial 'random' spin state. In all-metal based thin film structures the GMR effect can practically be observed only when the flow of current is in the film plane (CIP mode) since the film resistances, and so the resulting GMR signal (fractional change in film resistance under applied field), are too low in the CPP configuration due to the insufficient film thickness (~1 μm). This is a drawback considered larger GMR values are expected in the CPP geometry [Pratt et al, 1991; Valet & Fert, 1993].

However, this is not the case for high aspect ratio MNWs in which the current naturally flows through the wire. Further, due to the fact that the relevant scaling lengths are the electron mean free path (λ_{mfp}) for the CIP geometry and the spin diffusion length (λ_{sdl}) for the CPP geometry and that $\lambda_{sdl} >> \lambda_{mfp}$ the 1D wire-like structure tolerates thicker layers. This, in turn, is beneficial to maintain a layer thickness uniformity and reproducibility from the production point of view.

Fig. 12 presents the first example of GMR-MNWs [Piraux et al., 1994]. The wires consisted of alternating Cu/Co layers pulsed deposited from a sulphate bath containing 10^{-3} M Cu ions and 0.5 M Co ions into the 10 μm long pore channels of a polycarbonate filter. At -0.2 V pulses almost exclusively Cu is deposited, while at -0.9 V pulses Co is deposited along with about 10% Cu. A thin Cu film evaporated onto the template surface ensured contacting of many wires to the bottom electrode (also Cu) deposited prior to electrodeposition process. Ferromagnetic characteristic of the Co layers was verified by magnetization measurements, and the maximum room temperature GMR measured about 15% (Fig. 12b).

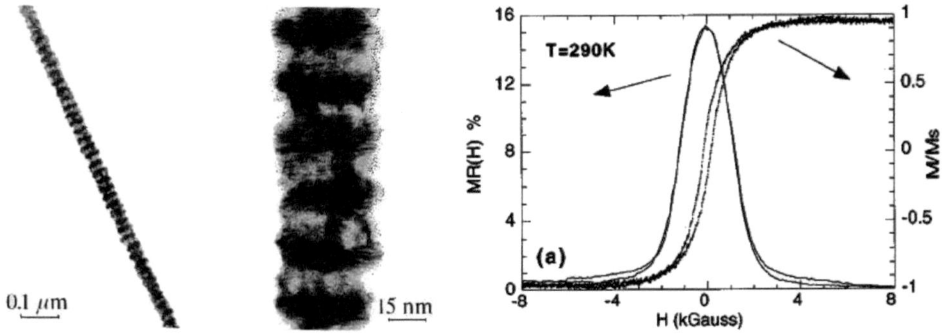

(a) (b) (c)

Fig. 12. (a, b) TEM images of [Co (10 nm)/Cu (10 nm)] multilayer nanowires (Co dark bands, Cu light bands), and (c) M-H curve and percentage magnetoresistance for a [Co (7 nm)/Cu (3 nm)]*1000 sample at room temperature (Reprinted with permission from [L. Piraux et al., 1994]. Copyright [1994], American Institute of Physics.).

In subsequent years, more studies on multilayer GMR-MNWs were conducted also utilizing Anodics™ [Evans et al., 2000; Tang et al., 2007] and double-anodized AAO [Ohgai et al., 2003; Tan et al., 2008; Huang et al., 2009] templates. It is claimed that the alumina template based nanowires provided larger values of GMR with respect to those grown in polycarbonate membranes [Evans et al., 2000; Ohgai et al., 2003]. Nevertheless, compared in general to the multilayer thin film systems the typical GMR values obtained from MNWs remain superior to those of electrodeposited thin films, and rather close to those of vacuum deposited thin films [Bakonyi & Peter., 2009]. Temperature dependent investigations on the effects of layers thickness and periodicity allowed the calculation of important physical values, such as λ_{sdl} and interface and bulk scattering parameters [Piraux et al., 1998], for the used metals. For example, λ_{sdl} of Co at room temperature is calculated to be ~38 nm, and a crossover from interface-dominated to bulk-dominated GMR occurs around Co layer thickness of 8 nm in Co/Cu multilayer MNWs. In another study, it is verified that the GMR of nanowires increased with the number of magnetic/non-magnetic bi-layers [Ohgai et al., 2003]. It is also observed that the nanowire arrays having isotropic properties, i.e. without definite easy axis of magnetization, display the highest magnetoresistance [Tan et al., 2008; Huang et al., 2009].

4.3 High resolution scanning probe microscopy tips

The spatial resolution of scanning probe microscopies depends heavily on the geometry and sharpness of the tip. Atomic force microscopy (AFM) enabled by carbon nanotube (CNT) tips is well known (see review article [Wilson & Macpherson, 2009]) and recently became commercialized [Nanosensors, 2008]. Metal based tips, on the other hand, are appealing for conducting AFM (also called electrical force microscopy, EFM), scanning surface potential microscopy (SSPM), and magnetic force microscopy (MFM) as well as AFM combined with near-field scanning optical microscopy (NSOM) or tip-enhanced Raman spectroscopy (TERS) [Renishaw plc, 2009; AIST-NT, 2009]. The CNT-tips can be obtained, for example, by chemical vapour deposition (CVD) from a catalyst particle at the end of a regular micro-fabricated AFM tip [Yenilmez et al., 2002], or by 'picking-up' from a substrate by scanning with a silicon tip [Hafner et al., 2001]. Metallic tips can be formed by electrochemical etching of a metal wire/tip [Gingery & Buhlmann, 2007], physically coating a CNT-tip [Deng et al., 2004], or by self assembly [Garcia-Martin et al., 2004]. Fig. 13 depicts an example of self assembly-prepared Co nanowire tip and its potential in MFM imaging.

(a) (b)

Fig. 13. (a) A self-assembled 30 nm Co nanowire MFM tip, and (b) magnetic domains on NiFe micro-patterns (sides 400 and 600 nm) revealed using a tip as that in (a). Spatial resolution is about the wire diameter. [Garcia-Martin et al., 2004. IOP Publishing. Reproduced with permission.]

For a high aspect ratio Co wire the shape anisotropy field, $H_{shp} \approx 2\pi M_s$, amounts to 700 kA/m (~8800 Oe), where M_s is the saturation magnetization. Thus, even hard magnetic specimens can be studied using such nanowire-tips, without inflicting a change in the direction of tip's magnetic field, that is for sample (stray) fields of up to H_{shp}. Morever, it should be possible to improve the spatial resolution of the MFM imaging by further reducing the wire diameter.

4.4 Substrates for surface-enhanced Raman scattering

Surface-enhanced Raman scattering (SERS), a phenomenon known since 1970s [Fleischmann et al., 1974], relies on the electromagnetic enhancement (EM) on noble metal nanostructures for enhancing the Raman signal of surface-adsorbed chemical molecules. While ordinary enhancements can be up to 10^6, in the late 1990s, even the Raman fingerprint of a single molecule was observed from certain nanoparticle aggregates, which also utilized other resonance effects [Nie & Emory, 1997]. In these experiments, Raman enhancements on the order of 10^{13}-10^{14} were calculated. There is a consensus now that the metal nanoparticles must be as close as possible, but not touching, to each other in order to provide such hot spots. Compact, ordered and controllable pore structure of porous AAO templates combined with the ease and versatility of metal nanowire/nanodot production by electrodeposition makes template fabricated metal nanostructures attractive for SERS studies. For these experiments, as illustrated in Fig. 14a, the templated metal nanostructures must be exposed to the electromagnetic excitation (typically a visible laser beam), which would also increase the area of metal surface for the adsorption of analyte molecules. Field enhancement for SERS in nanoparticle arrays has been studied by many researchers [Garcia-Vidal & Pendry, 1996; Kahl & Voges, 2000; Genov et al., 2004]. Particularly relevant here, the analytical result from [Genov et al., 2004] is plotted in Fig. 14b for an ordered vertically-aligned Ag nanowire array embedded in air at the wavelength of a common laser line (He-Ne, 633 nm):

$$G \cong \frac{\pi(m+1)^{7/2}}{2((4-\pi)m+4)\kappa^{7/2}} \times \sqrt{\frac{4t^2+9}{(t^2+1)^{3/2}} - \frac{t(4t^4+15t^2+15)}{(t^2+1)^3}} \tag{2}$$

where $m = |E_0|/E_1$, $t = (m/\gamma - 1)$, $E_m = E_0(1 - i\kappa)$ dielectric function of the metal, κ the loss factor, E_1 the dielectric function of the medium, and γ is the inverse interparticle distance (wire diameter/width of interparticle gap).

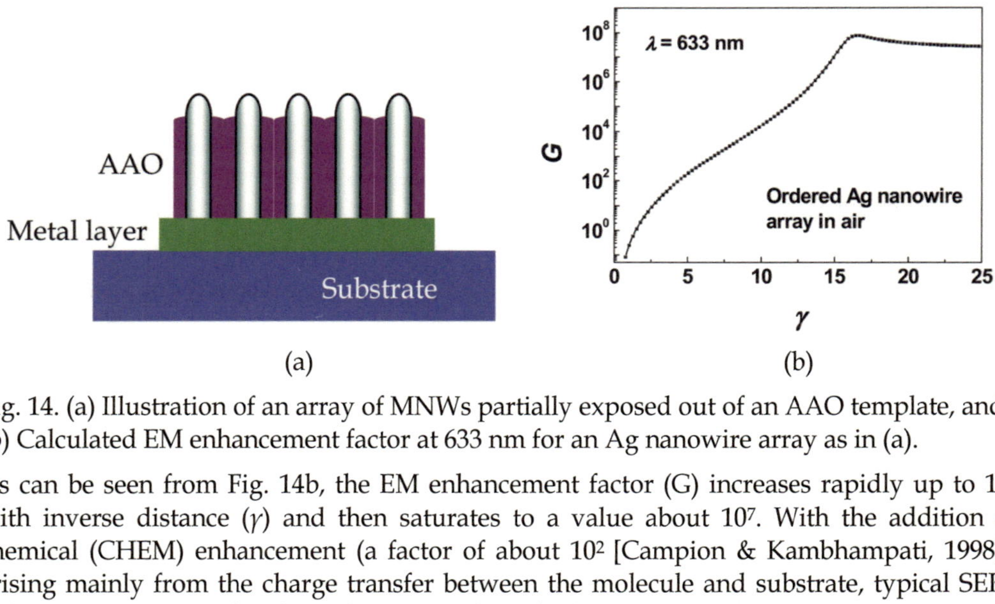

(a) (b)

Fig. 14. (a) Illustration of an array of MNWs partially exposed out of an AAO template, and (b) Calculated EM enhancement factor at 633 nm for an Ag nanowire array as in (a).

As can be seen from Fig. 14b, the EM enhancement factor (G) increases rapidly up to 10^8 with inverse distance (γ) and then saturates to a value about 10^7. With the addition of chemical (CHEM) enhancement (a factor of about 10^2 [Campion & Kambhampati, 1998]), arising mainly from the charge transfer between the molecule and substrate, typical SERS enhancement power of such a substrate can be as high as 10^{10}.

As a thumb of rule, the distribution in the length of electrodeposited nanowires in porous template channels increases with the wire length. Therefore, in order to have uniform length MNWs for SERS studies either the deposition time must be shortened, yielding nanodots rather than nanowires [Kartopu et al., 2006; Wang et al, 2006], or the initially long MNWs must be levelled by an appropriate post treatment, such as mechanical polishing [Kartopu et al., 2008b] and ion milling [Sauer et al., 2005]. In the latter case, a chemical etching may be applied to increase the exposed area of nanowires (Fig. 15a and b). An in situ chemical etching (in dilute H_3PO_4) is used to probe the temporal change in the SERS spectrum of analyte molecules adsorbed on MNWs/Ag nanowires (Fig. 15c) [Sauer et al., 2005]. The signal intensity first increases to a maximum and then decreases rapidly before it completely vanishes. In the first regime, surface coverage of molecules increases with time, i.e. the area of newly-exposed (lateral) wire surfaces, while in the second regime the order of array coarsens considerably, and eventually leads to collapsing of the wires. The molecules attached to the lateral wire surfaces may experience stronger enhancements due to (increasing) interparticle interactions, i.e. multipolar fields. If this assumption is valid, it would also explain why a maximum is observed in Raman enhancement as well as the disappearance of the signal (when the wires start to make contact/collapse).

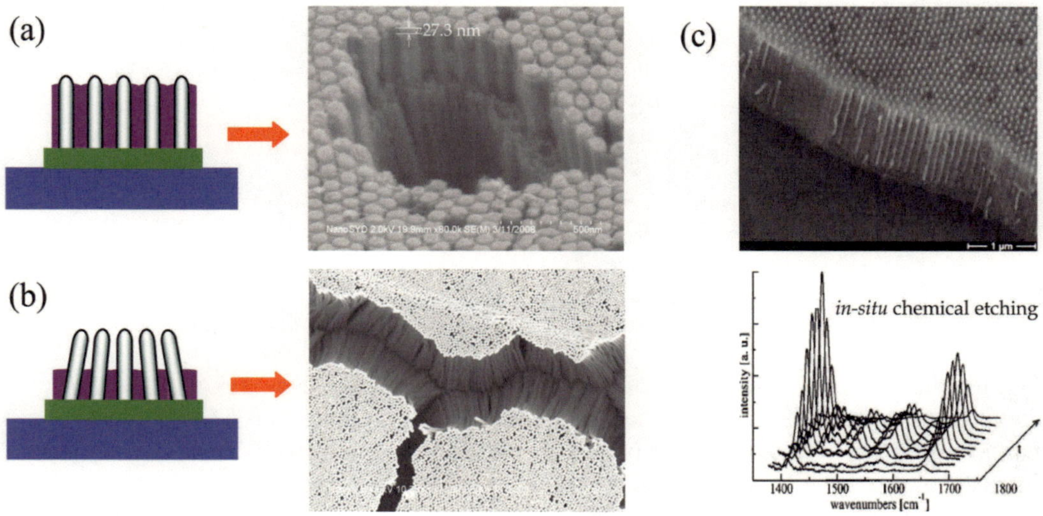

Fig. 15. Mechanically polished Ag nanowires/AAO composite after (a) 'mild-etching', and (b) 'over-etching' of the (AAO) pore walls in diluted H_3PO_4. (c) Data from [Sauer et al., 2005] (Reprinted with permission. Copyright [2005], American Institute of Physics.) showing (upper panel) a levelled Ag nanowire array in AAO and (lower panel) the dynamic change in the SERS spectrum of oxazine 170 molecules (adsorbed on Ag surfaces) excited at 633 nm with the in situ etching time. [A close-up view sample in (b) was also given as Fig. 3d].

4.5 Metamaterials

Another interesting use of MNWs fabricated in AAO templates could be in the field of negative index metamaterials, which are currently considered as the 'superlens' for high resolution imaging application. Metamaterials are artificially designed nano-composites possessing remarkable optical properties that do not exist in nature. They are capable of changing the propagation of light and can result in its negative refraction (Fig. 16a). Several theoretical models predict that metal-dielectric nano-composites based on a thin slab of AAO (dielectric) infiltrated with high aspect ratio Ag (or Au) nanowires can support a negative (effective) permittivity in the visible and infrared [Ao & He, 2005; Menon et al., 2008], meaning certain polarization of light could undergo negative refraction through this medium. Experimental evidence on this phenomenon has recently been obtained (Fig. 16) [Yao et al., 2008].

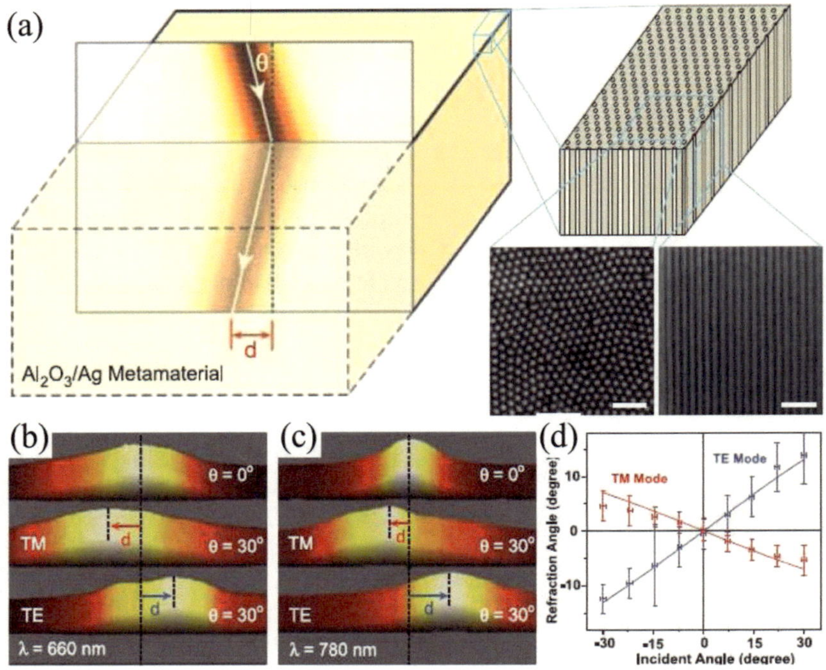

Fig. 16. The first experimental observation of negative refraction from the AAO/MNWs composite system. (a) Schematic of negative refraction through the Al_2O_3/Ag nanocomposite and close-up views of an ordered AAO template (scale bars: 500 nm). (b, c) Intensity and lateral displacement of the transmitted light (with respect to the centre of contact of the incident light) at two different wavelengths measured at the exit surface of AAO/Ag composite films of 4.5 μm (b) and 11 μm (c) thickness [horizontal size of (b) and (c) are 5 and 12 μm, respectively]. (d) Dependence of refraction angles on incidence angle for TM and TE polarizations at 780 nm. The group refractive indices of the metamaterial are calculated to be -4.0 and 2.2 for TM and TE polarizations, respectively [Yao et al., 2008] (Copyright 2008 by the American Association for the Advancement of Science).

It is commonly reported that transmission/absorption spectra of Ag or Au nanowires synthesized in AAO pores display two distinct bands depending on the excitation conditions. One of these resonances occur near the position of bulk plasmon resonance characteristic of the metal (~380 nm for Ag and ~500 nm for Au) for both *S* (or TE) and *P* (or TM)-polarizations at all angles, while the other resonance band is observed in the red (for Ag nanowires) or infrared (for Au nanowires) wavelengths for only *P* (TM)-polarized light for incidence angles grater than about 10 degrees [Evans et al., 2008; Menon et al., 2008]. Consequently, while the former band is due to transverse resonance of the nanowires, the latter band occurring at higher wavelengths can be attributed to longitudinal plasmon resonances associated to the long axis of the nanowires. The longitudinal mode is shown by appropriate modelling to be a collective property of the nanowire array, rather than that of the individual wire, as otherwise it would be expected at much higher wavelengths (in infrared) [Evans et al., 2008]. From these theoretical and experimental results it appears that the longitudinal resonance mode of nanowires has to be excited (with TM polarized light) for the AAO/MNWs composites to show the negative refraction phenomenon. Further studies should elicit further particulars of this interesting, low cost metamaterial system. Among potential application areas are waveguiding, imaging, optical communication, and biosensing [Yao et al., 2008; Kabashin et al., 2009].

4.6 Nano-optics and molecular electronics

Multi-layered or -segmented MNWs are expected to be used as novel characterization tools in fields of nano-optics and molecular electronics due to the fact that wire dimensions and composition can be precisely tuned and that the wires can be manipulated by simple solution-based processing methods [Martin & Baker, 2005; Hurst et al., 2006]. A recently developed technique, coined 'on-wire lithography (OWL)' [Qin et al., 2005], which results in notched nanowires with unprecedentedly small gaps is likely to provide huge impetus in these regards. In OWL, first an arbitrarily designed multi-layered MNW array is obtained in the pore channels of a nano-template as described in Sec. 3.4. The wires are then liberated by complete dissolution of the template and placed on a flat substrate. In order to prevent collapsing of the nanowire architecture in the final step, the top semi-surface of the wires is coated with a thin film that is resistant to chemical etching. Wires are then loosened off the substrate (by ultrasonic agitation), and then a selective chemical etch is applied to remove only certain metal segments. The result is notched, discontinuous nanowires with the backing of a thin film (Fig. 17).

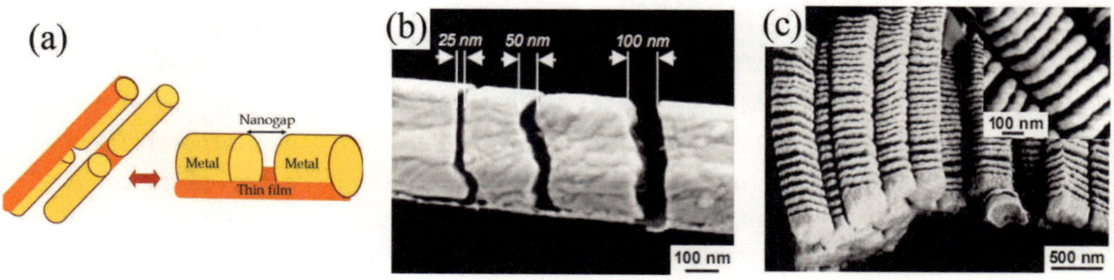

Fig. 17. OWL-produced notched nanowires: (a) Schematic illustration, and (b, c) micrographic views. Data reproduced from [Qin et al., 2005] (Copyrigt 2005 by the American Association for the Advancement of Science).

Nano-gaps down to 5 nm have been demonstrated, whereby dip-pen lithography-aided positioning of polymer molecules within 13 nm gapped-wires (contacted by standard microlithography techniques) provided evidence on the existence and electronic behavior, i.e. conductivity, of the molecules [Qin et al., 2005]. In another case, Au nano-discs with controlled sizes and gaps were obtained by OWL and employed in novel experiments designed for systematically studying hot spots produced due to interparticle couplings [Qin et al., 2006]. In these optical experiments, the electromagnetic enhancement (EM) was studied using SERS effect (see also Sec. 4.4), and it is verified that SERS enhancement factor was at least two orders of magnitude higher for any pairs/multiples of discs with a nanogap compared to an isolated disc. Further, optimum gap and disc sizes were determined at a given excitation wavelength. These experiments demonstrate the potential of multilayer MNWs in the fields of molecular electronics and nano-optics, considering that extremely small nanogaps can be produced controllably and reproducibly using a simple method without the need of expensive lithography techniques.

4.7 Biological tags

Perhaps the most straightforward application of multi-segmented MNWs is in the field of biological labelling. Wires formed of metals that support strong plasmon resonances in the visible, i.e. Ag, Au, and Cu, can be routinely prepared having unique combinations of layer thicknesses and periodicities. These metals scatter visible light so strongly that even individual segments that are much smaller than the wavelength can be viewed using a standard optical microscope system with white illumination source (Fig. 18) [Nicewarner-Pena et al., 2001; Mock et al, 2002]. If spotted on a certain biomolecule, an arbitrarily-designed (or '-coded') multi-segment wire can act as a fingerprint for the rapid and simple identification of the biomolecule (among others). To exemplify the capabilities of this encoding system, a nanowire containing 13 segments and made of two metals (e.g. Ag/Au) is calculated to have 4160 permutations [Nicewarner-Pena et al., 2001].

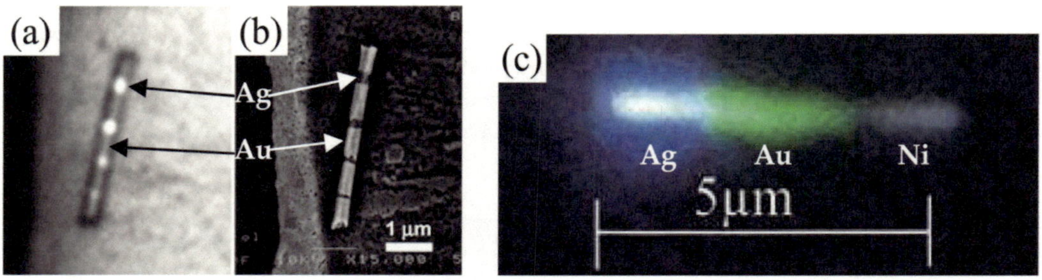

Fig. 18. (a) Optical and (b) SEM images of an Ag/Au multi-segmented nanowire with ~550 nm Au and 60, 110, 170, and 240 nm (bottom to top) Ag segments [Nicewarner- Peña et al., 2001] (Copyright 2001 by the American Association for the Advancement of Science). Ag segments look brighter in (a) and darker in (b). (c) Optical image of a three-metal nanowire (Reprinted with permission from [Mock et al, 2002]. Copyright [2002], American Chemical Society.).

4.8 Field-emission electron emitters

One-dimensional nanostructures (nanowire, nanotube) with high aspect ratio and low tip diameter are recognized as ideal field-emission (FE) electron sources, which are potentially useful as emitters in flat panel displays. Particularly carbon nanotubes (CNTs) have been the subject of considerable study [Heer et al., 1995; Collins & Zettl, 1997]. More recently, FE properties of 'nanowhiskers' of metal oxides [Luo et al., 2004] and metals [Kim et al., 2008] have also attracted interest. CNTs are usually produced by chemical vapour deposition (CVD) and hence requires high growth temperatures (>500 °C) which limits the choice of substrate. In addition, nanowiskers lack structural alignment and makes it difficult to control the density and aspect ratio of the emitters which are important parameters for tuning the FE threshold and current density. Therefore, use of template-synthesized CNTs [Quy et al., 2006] and room-temperature electodeposited MNWs [Vila et al., 2004] in ordered templates becomes attractive for FE applications. Fig. 19a depicts the common triode-type experimental set up for detecting nanotube/nanowire field emission. Typically a high voltage on the order of a few kV is applied to the cathode (sample) for it to emit brightly. The current-voltage (I-V) characteristics are usually analysed with the Fowler-Nordheim (F-N) relationship:

$$J = A\left(\frac{\beta^2 E^2}{\phi}\right)\exp\left(-\frac{B\phi^{3/2}}{\beta E}\right) \tag{3}$$

where A and B are constants, β is the field enhancement factor, ϕ work function of the metal, and the electric field is given by $E = V/d$ with V the applied voltage and d the grid-cathode distance (spacer thickness, see Fig. 19a). The enhanced local field at the wire/tube tip is given by $E_t = \beta E$. Desirable properties for a FE device are the low turn-on voltage, high current density, and good emission homogeneity. Fig. 19b presents typical results for the grid and anode currents obtained from a Co nanowire array template-synthesized on a metallised silicon substrate [Vila et al., 2004]. In these experiments, the density of the wires was kept as low as $10^7/cm^2$ to avoid screening/interference effects between the wires [Collins & Zettl, 1997]. Ratio of anode to cathode currents is constant while the calculated enhancement factor was almost the same as the aspect ratio of the Co wires. The turn-on voltage (12 V/μm) is comparable to that of CNTs (1-10 V/μm) [Davydov et al., 1999]. Deviation from the F-N relationship (solid line) at higher fields is attributed to an undesired serial resistance (R = 4 MΩ) in the experimental set-up.

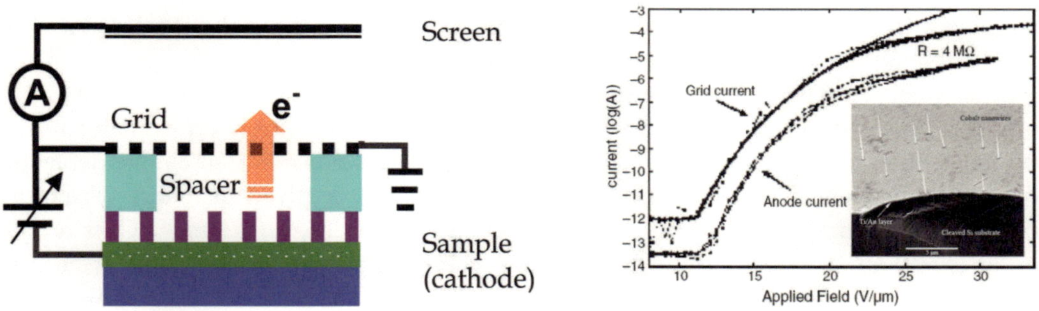

(a) (b)

Fig. 19. (a) Schematic of nanotube/nanowire FE experimental set-up (triode configuration). (b) FE characteristics of Co nanowires template-synthesized on metal-coated silicon substrate determined using a set-up similar to that in (a). Inset shows a micrograph of the Co nanowire array. (Data adapted from [Vila et al., 2004]. (Copyright [2004], American Chemical Society. Reproduced with permission.).

A unique advantage of field-emitting nanostructures' synthesis with thin film AAO templates (see also Sec. 3.3) is that established microelectronic processing technologies can also be applied to obtain FE display pixels and novel devices [Quy et al., 2006; Mátéfi-Tempfli et al., 2008]. Interest in recent theoretical studies focuses on optimum array parameters for maximum enhancement factor (β) and current density [Kim et al., 2009]. It is noted that β and current density, which are strong functions of the aspect ratio and spacing of emitters, applied field, and so on, actually do not require very dense arrays [Kim et al., 2009].

4.9 Gas sensors

Much of the work devoted to chemical sensors based on 1D nanostructures are aimed at developing high sensitivity gas sensors operating at low temperatures. Metal and metal oxide nanowires obtained on the step edges of highly-oriented pyrolytic graphite (HOPG) substrates have proven to be suitable for such sensing applications [Murray et al., 2004; Favier et al., 2001]. For example, palladium (Pd) nanowires were shown to exhibit excellent performance in the detection of high concentrations (above 2%) of hydrogen (H_2) [Favier et al., 2001]. This was explained to be due to the rather porous/discontinuous character of the Pd wires. When subjected to H_2 the Pd wires were observed, by *in situ* AFM imaging, to swell considerably (due to formation of PdH_x on Pd surfaces), thereby closing the gaps between nano-grains and allowing the wire to conduct electricity. Therefore, these peculiar wires acted effectively as hydrogen 'switches'. Despite the success of such HOGP-grown nanowires, other 'continuous' 1D architectures are desirable for sensing of low and ultra-low concentrations of H_2 and other gases (CO, NO_2, volatile organic compounds, etc.). With this aim in mind, several studies were conducted to synthesize and exploit MNWs in AAO templates [Kartopu et al., 2008c; Kim et al., 2006; Kolmakov et al., 2003]. Pd nanowires were DC electrodeposited in free-standing (Fig. 20a) as well as on-substrate thin film AAO templates (Fig. 20b). High sensitivity of such structures was verified using the set-up depicted in Fig. 20c, where the self-standing Pd array is contacted through the metallic substrate as the bottom contact and a sharp tip on the top. Fig. 20d shows that low concentrations (~0.2%) of H_2 can be detected in a reversible manner.

In another study, first metallic tin (Sn) nanowires were obtained by AC deposition in AAO templates, then released from the template and converted by thermal annealing to SnO_2 nanowires (30 nm radius), which were subsequently used in sensing experiments in the form of single-nanowire devices prepared by micro-contacting [Kolmakov et al., 2003]. It is observed that the initially conducting SnO_2 wire became dramatically an insulator in the presence of sufficient oxygen (ca. 10% in nitrogen), which indicates that charge-depletion (due to adsorption of oxidizing species in atmosphere) has deprived the entire wire of its carriers. The characteristic electron exchange length (Debye length, λ_D) for SnO_2 is on the order of ~40 nm (at 500 K). Accordingly, whilst charge-depletion remains to be a 'surface' phenomenon for the bulk, it was possible to switch low-diameter SnO_2 wires between the conducting and insulating states. Despite the promising results with phase-change MNWs, process of contact formation onto the sensing element (single wire or wire array) needs to be improved to increase sensitivity.

Recently, a simple yet effective approach has been introduced which yields metal-oxide-metal (MOM) nanowires (Fig. 21a) [Tresback et al., 2005]. In this process, first multi-segment MNWs are synthesized in AAO template, released from the template and then a segment of the wires is selectively oxidised by annealing at a temperature low enough to keep other segments metallic. Results pertaining to a single-wire (Au-NiO-Au) gas sensor (prepared by micro-contacting) are presented in Fig. 21b and c. Compared to an 'all-oxide' NiO single-wire device, the MOM sensor performs reasonably well for detecting low concentrations (<1%) of CO, despite the sensor response and recovery times are rather too long. Optimization would be needed for applications requiring compact sensors with low power consumption and high sensitivity. Such nanosensors, if combined in-parallel with others, are believed to lead to realization of the 'electronic nose' concept [Sysoev et al., 2006].

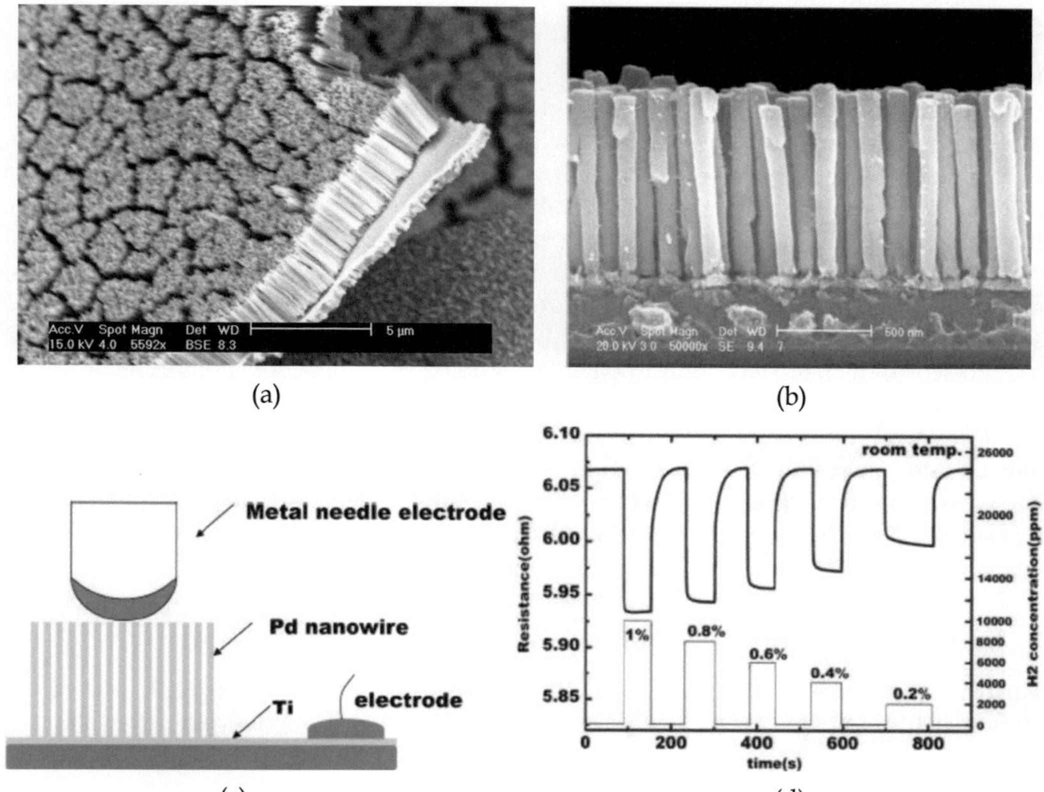

Fig. 20. Self-supporting Pd nanowires on metallic substrates prepared using free-standing (a) and supported thin film AAO (b) templates. (c) Set-up used for gas sensing experiments. (d) H_2-sensing performance of the array in (b) at room temperature. Data in (a) is reproduced with permission from [Kartopu et al., 2008c] (Copyright [2008], Elsevier). Data in (b-d) is reproduced with permission from [Kim et al., 2006] (© [2006], IEEE).

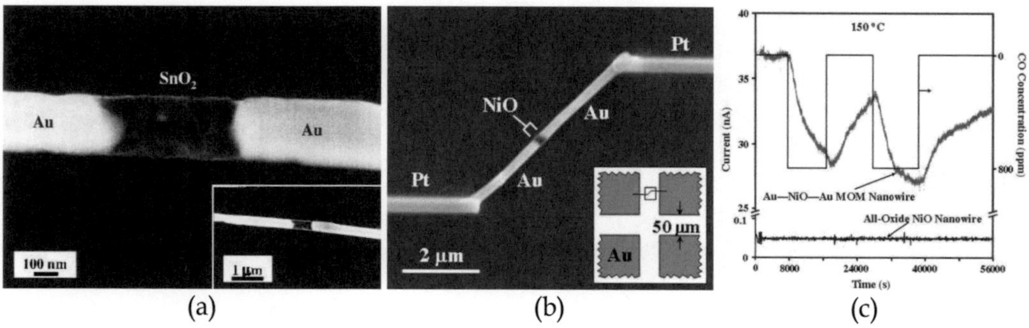

Fig. 21. (a) A template-synthesized Au-SnO_2-Au MOM nanowire at two different magnifications (Reproduced with permission from [Tresback et al., 2005] (© [2005], MRS). (b) Micrograph and (c) CO gas-sensing performance of a Au-NiO-Au MOM nanowire contacted using Pt-wires grown by focused ion beam (FIB) synthesis. In (c), also shown is the response of an all-oxide NiO nanowire synthesized and contacted in similar fashion (Reproduced with permission from [Tresback et al., 2008] (© [2008], MRS).

5. Conclusion

Fabrication, properties, and prominent applications of electrodeposited metal nanowires in porous anodic alumina templates have been presented. Both single component and multilayer nanowires can be prepared with ease, high reproducibility, and versatility. Compatibility of the produced structures with further processing methods, such as simple solution-based processing methods and established microelectronic processing technologies, makes these systems very attractive for a myriad of applications, ranging from chemical sensors to magnetic sensors, perpendicular data storage to field-emission electron sources, and from metamaterials to molecular electronics. Due to the limited spacing, and overlaps with previous reviews on the topic, several other interesting applications/properties, such as high-sensitivity nano-electrode arrays [Martin, 1994] and pH sensors [Antohe et al., 2009], superconductivity [Kline et al, 2006], thermoelectric power generation [Sommerlatte et al., 2007], etc. have not been presented. This chapter merely aimed at introduction of the basics and updated information on this young and dynamic research field, hoping that it will encourage both researchers and industry to make new initiations, and yield further nano-technological concepts and fruitful applications in various disciplines of science and engineering.

6. Acknowledgements

Financial support of TÜBİTAK (project 107T635) is gratefully acknowledged. One of the authors (GK) thanks to EU and Deutsche Forschung Gemeinschaft for several fellowships.

7. References

AIST-NT, (2009). Nanofinder R 30 - SmartSPM™ 1000: Confocal Raman / AFM combined system. http://www.aist-nt.com/wp-content/uploads/2009/04/reflection-ters-with- aist-afm.pdf

Antohe, V. A.; Radu, A.; Mátéfi-Tempfli, M.; Attout, A.; Yunus, S.; Bertrand, P.; Dutu, C. A.; Vlad, A.; Melinte, S.; Mátéfi-Tempfli, S. & Piraux, L. (2009). Nanowire-templated microelectrodes for high-sensitivity pH detection. *Appl. Phys. Lett.*, 94, 073118-3.

Ao, X. & He, S. (2005). Negative refraction of left-handed behaviour in porous alumina with infiltrated silver at an optical wavelength. *Appl. Phys. Lett.*, 87, 101112-3.

Avcorp Industries Inc., www.avcorp.com/section.asp?catid=152&subid=165&pageid=185, 2009.

Baibich, M.N.; Broto, J.M.; Fert, A.; Nguyen Van Dau, F.; Petroff, F.; Etienne, P.; Creuzet, G.; Friederich, A. & Chazelas, J. (1988). Giant magnetoresistance of (001)Fe/(001)Cr magnetic superlattices. *Phys. Rev. Lett.*, 61, 2472 - 2474.

Bakonyi, I. & Péter, L. (2009). Electrodeposited multilayer films with giant magnetoresistance (GMR): progress and problems. *Prog. Mater. Sci.*, doi:10.1016/ j.pmatsci.2009.07.001.

Beleggia, M.; Tandon, S.; Zhu, Y. & De Graef, M. J. (2004). On the magnetostatic interactions between nanoparticles of arbitrary shape. J. Magn. Magn. Mater., 278(1-2), 270-284.

Binasch, G.; Grünberg, P.; Saurenbach, F. & Zinn, W. (1989). Enhanced magnetoresistance in layered magnetic structures with antiferromagnetic interlayer exchange. *Phys. Rev.* B, 39, 4828 - 4830.

Campion, A. & Kambhampati, P. (1998). Surface-enhanced Raman scattering. *Chem. Soc. Rev.*, 27, 241 - 250.

Choi, J.; Nielsch, K.; Reiche, M.; Wehrspohn, R.B. & Gösele, U. (2003). Fabrication of monodomain alumina pore arrays with an interpore distance smaller than the lattice constant of the imprint stamp. *J. Vac. Sci. Technol.* B, 21(2), 763-766.

Clime, L.; Ciureanu, P. & Yelon, A. (2006). Magnetostatic interactions in dense nanowire arrays. *J. Magn. Magn. Mater.*, 297, 60-70.

Collins, P. G. & Zettl, A. (1997). Unique characteristics of cold cathode carbon-nanotube-matrix field emitters. *Phys. Rev. B*, 55(15), 9391 – 9399.

Crouse, M. M.; Miller, A. E.; Crouse, D. T. & Ikram, A. A. (2005). Nanoporous alumina template with in situ barrier oxide removal, synthesized from a multilayer thin film precursor. *J. Electrochem. Soc.*, 152(10), D167 – D172.

Das, B. & Singaraju, P. (2005). Novel quantum wire infrared photodetectors. *Infrared Phys. Technol.*, 46, 209 – 218.

Davydov, D. N.; Sattari, P. A.; AlMawlawi, D.; Osika, A.; Haslett, T. L. & Moskovits, M. (1999). Field emitters based on porous aluminium oxide templates. *J. Appl. Phys.*, 86(7), 3983 - 3987.

De Heer, W. A.; Chatelain, A. & Ugarte, D. (1995). A carbon nanotube field-emission electron source. *Science*, 270, 1179 - 1180.

Deng, Z.; Yenilmez, E.; Leu, J.; Hoffman, J. E.; Straver, E. W. J.; Dai, H. & Moler, K. A. (2004). Metal-coated carbon nanotube tips for magnetic force microscopy. *Appl. Phys. Lett.*, 85, 6263 – 6265.

Ebels, U.; Radulescu, A.; Henry, Y.; Piraux, L. & Ounadjela, K. (2000). Spin accumulation and domain wall magnetoresistance in 35 nm Co wires. *Phys. Rev. Lett.*, 84, 983-986.

Evans, P. R.; Yi G. & Schwarzacher W. (2000). Current perpendicular to plane giant magnetoresistance of multilayered nanowires electrodeposited in anodic aluminum oxide membranes. *Appl. Phys. Lett.*, 76, 481 – 483.

Evans, P. R.; Kullock, R.; Hendren, W. R.; Atkionson, R.; Pollard, R. J. & Eng, L. M. (2008). Optical transmission properties and electric field distribution of interacting 2D silver nanorod arrays. *Adv. Func. Mater.*, 18, 1075 – 1079.

Favier, F.; Walter, E.C.; Zach, M. P.; Benter, T. & Penner, R.M. (2001). Hydrogen sensors and switches from electrodeposited palladium mesowire arrays. *Science*, 293, 2227-2231.

Fleischmann, M.; Hendra, P.J. & McQuillan, A.J. (1974). Raman spectra of pyridine adsorbed at a silver electrode. *Chem. Phys. Lett.*, 26(2), 163 – 166.

Fodor, P. S.; Tsoi, G. M. & Wenger, L. E. (2008). Investigation of magnetic interactions in large arrays of magnetic nanowires. *J. Appl. Phys.*, 103, 07B713-3.

Foong, T. R. B.; Sellinger, A. & Xu, X. (2008). Origin of the bottlenecks in preparing anodized aluminium oxide (AAO) templates on ITO glass. *ACS Nano*, 2(11), 2250 – 2256.

Fournier-Bidoz, S.; Kitaev, V.; Routkevich, D.; Manners, I. & Ozin, G.A. (2004). Highly ordered nanosphere imprinted nanochannel alumina (NINA). *Adv. Mater.*, 16, 2193-2196.

Furneaux, R. C.; Rigby, W. R. & Davidson, A. P. (1989). The formation of controlled-porosity membranes from anodically oxidized aluminium. *Nature*, 337, 147 - 149.

Garcia-Martin, J. M.; Thiaville, A.; Miltat, J.; Okuna, T.; Vila, L. & Piraux, L. (2004). Imaging magnetic vortices by magnetic force microscopy: experiments and modelling. *J. Phys. D: Appl. Phys.*, 37, 965 – 972.

García-Vidal, F. J. & Pendry, J. B. (1996). Collective Theory for Surface Enhanced Raman Scattering. *Phys. Rev. Lett.*, 77, 1163 - 1166.

Genov, D.A.; Sarychev, A.K.; Shalaev, M. & Wei, A. (2004). Resonant field enhancements from metal nanoparticle arrays. *Nano Lett.*, 4(1), 153 - 158.

Gingery, D. & Buhlmann, P. (2007). Single-step electrochemical method for producing very sharp Au scanning tunneling microscopy tips. *Rev. Sci. Instrum.*, 78, 113703 - 113.

Gochowski E. (2003). The HDD roadmap. Published online by Hitachi Global Storage Technologies, Inc., http://colossalstorage.net/hdd_technology2003.pdf.

Grimsditch, M.; Jaccard, Y. & Ivan Schuller, K. (1998). Magnetic anisotropies in dot arrays: Shape anisotropy versus coupling. *Phys. Rev.* B, 58, 11539-11543.

Hafner, J. H.; Cheung, C.-L.; Oosterkamp, T. H. & Lieber, C. M. (2001). High-yield assembly of individual single-walled carbon nanotube tips for scanning probe microscopies. *J. Phys. Chem. B*, 105 (4), 743 - 746.

Huang, X.; Tan, L.; Cho, H. & Stadler, B. J. H. (2009). Magnetoresistance and spin transfer torque in electrodeposited Co/Cu multilayered nanowire arrays with small diameters. *J. Appl. Phys.*, 105, 07D128-3.

Hurst, S. J.; Payne, E. K.; Qin, L. & Mirkin, C. A. (2006). Multisegmented one-dimensional nanorods prepared by hard-template synthetic methods. *Angew. Chem. Int. Ed.*, 45, 2672 - 2692.

Jessensky, O.; Müller, F. & Gösele, U. (1998). Self-organized formation of hexagonal pore structures in anodic alumina. *J. Electrochem. Soc.*, 145, 3735 - 3740.

Kabashin, A.V.; Evans, P.; Pastkovsky, S.; Hendren, W.; Wurtz, G. A.; Atkinson, R.; Pollard, R.; Podolskiy, V.A. & Zayats, A. V. (2009). Plasmonic nanorod metamaterials for biosensing. *Nature Mater.*, doi:10.1038/NMAT2546.

Kahl, M. & Voges, E. (2000). Analysis of plasmon resonance and surface-enhanced Raman scattering on periodic silver structures. *Phys. Rev. B*, 61(20), 14078 - 14088.

Kaitsu, I.; Inamura, R.; Toda, J. & Morita, T. (2006). Ultra high density perpendicular magnetic recording technologies. *Fujitsu Sci. Tech. J.*, 42(1), 122 - 130.

Kartopu, G.; Es-Souni, M.; Sapelkin, A.V. & Dunstan, D. (2006). A novel SERS-active substrate system: Template-grown nanodot-film structures. *Phys. stat. sol.* (a), 203(10), R82 - R84.

Kartopu, G.; Yalçin, O.; Es-Souni, M. & Başaran, A. C. (2008a). Magnetization behavior of ordered and high density Co nanowire arrays with varying aspect ratio. *J. Appl. Phys.*, 103, 093915-6.

Kartopu, G.; Es-Souni, M.; Sapelkin, A.V. & Dunstan, D. (2008b). Large-scale, reliable, and robust SERS-active nanowire-substrates prepared using porous alumina templates. *J. Nanosci. Nanotech.*, 8(2), 931 - 935.

Kartopu, G.; Habouti, S. & Es-Souni, M. E. (2008c). Synthesis of palladium nanowire arrays with controlled length and diameter. *Mat. Chem. Phys.*, 107, 226 - 230.

Keller, F.; Hunter, M. S. & Robinson, D. L. (1953). Structural features of oxide coatings on aluminium. *J. Electrochem. Soc.*, 100 (9), 411-419.

Kim, B.; Park, S.; McCarthy, T.J. & Russell, T.P. (2007). Fabrication of ordered anodic aluminum oxide using a solvent-induced array of block-copolymer micelles. *Small*, 3(11), 1869-1872.

Kim, C.; Gu, W.; Briceno, M.; Robertson, I. M.; Choi, H. & Kim, K. (2008). Adv. Mater., Copper nanowires with a five-twinned structure grown by chemical vapor deposition. *Adv. Mater.*, 20(10), 1859 - 1863.

Kim, D.; Bourée, J.-E. & Kim, S. Y. (2009). Calculation of the field enhancement for a nanotube array and its emission properties. *J. Appl. Phys.*, 105, 084315-5.

Kim, K. T.; Sim, S. J. & Cho, S. M. (2006). Hydrogen gas sensor using Pd nanowires electrodeposited into anodized alumina template. *IEEE Sens. J.*, 6(3), 509 - 513.

Kline, T. R.; Tian, M.; Wang, J.; Sen, A.; Chan, M. W. H. & Mallouk, T. E. (2006). Templategrown metal nanowires. *Inorg. Chem.*, 45, 7555 - 7565.

Kolmakov, A.; Zhang, Y.; Cheng, G. & Moskovits, M. (2003). Detection of CO and O_2 using tin oxide nanowire sensors. *Adv. Mater.*, 15(12), 997 - 1000.

Krishnan, R. & Thompson, C.V. (2007). Monodomain high-aspect-ratio 2D and 3D ordered porous alumina structures with independently controlled pore spacing and diameter. *Adv. Mater.*, 19, 988-992.

Kumar, A.; Fähler, S.; Schlörb, H.; Leistner, K. & Schultz L. (2006). Competition between shape anisotropy and magnetoelastic anisotropy in Ni nanowires electrodeposited within alumina templates. *Phys. Rev.* B, 73, 064421-064425.

Lee, J. H.; Wu, J. H.; Liu, H. L.; Cho, J. U.; Cho, M. K.; An, B. H.; Min, J. H.; Noh, S. J. & Kim, Y. K. (2007). Iron-gold barcode nanowires. *Angew. Chem. Int. Ed.*, 46, 3663 - 3667.

Lee, W.; Ji, R.; Gösele, U. & Nielsch, K. (2006). Fast fabrication of long-range ordered porous alumina membranes by hard anodization. *Nature Mater.*, 5, 741 - 747.

Li, A. P.; Müller, F.; Birner, A.; Nielsch, K. & Gösele, U. (1998). Hexagonal pore arrays with a 50–420 nm interpore distance formed by self-organization in anodic alumina. *J. Appl. Phys.*, 84, 6023 - 6026.

Liang, J.; Chik, H. & Xu, J. (2002). Nonlithographic fabrication of lateral superlattices for nanometric electromagnetic-optic applications. *IEEE J. Quantum Electron.*, 8(5), 998 - 1008.

Liu, N.W.; Datta, A; Liu, C.Y. & Wang, Y.L. (2003). High-speed focused-ion-beam patterning for guiding the growth of anodic alumina nanochannel arrays. *Appl. Phys. Lett.*, 82, 1281-1283.

Luo, S. H.; Wan, Q.; Liu, W. L.; Zhang, M.; Di, Z. F.; Wang, S. Y.; Song, Z. T.; Lin, C. L. & Dai, J. Y. (2004). Vacuum electron field emission from SnO_2 nanowhiskers synthesized by thermal evaporation. *Nanotechnology*, 15, 1424 - 1427.

Mallet, J.; Yu-Zhang, K.; Mátéfi-Tempfli, S.; Mátéfi-Tempfli, M. & Piraux, L. (2005). Electrodeposited $L1_0$ Co_xPt_{1-x} nanowires. *J. Phys. D: Appl. Phys*, 38, 909 - 914.

Martin, C. R. (1994). Nanomaterials: A membrane-based synthetic approach. *Science*, 266, 1961 - 1966.

Martin, C. R. & Baker, L. A. (2005). Expanding the molecular electronics toolbox. *Science*, 309, 67 - 68.

Masuda, H. & Fukuda, K. (1995). Ordered metal nanohole arrays made by a two-step replication of honeycomb structures of anodic alumina. *Science*, 268, 1466 - 1468.

Masuda, H.; Yamada, H; Satoh, M.; Asoh, H.; Nakao, M. & Tamamura, T. (1997). Highly ordered nanochannel-array architecture in anodic alumina. *Appl. Phys. Lett.*, 71, 2770-2772.

Mátéfi-Tempfli, S.; Mátéfi-Tempfli, M. & Piraux, L. (2009). Fabrication of nanowires and nanostructures: combining template synthesis with patterning methods. *App. Phys. A*, 96, 603 - 608.

Menon, L. ; Lu, W. T. ; Friedman, A. L. ; Bennett, S. P. ; Heiman, D. & Sridhar, S. (2008). Negative index metamaterials based on metal-dielectric nanocomposites for imaging applications. *J. Appl. Phys.*, 93, 123117-3.

Metzger, R. M.; Konovalov, V. V.; Sun, M.; Xu, T.; Zangari, G.; Xu, B.; Benakli, M. & Doyle, W. D. (2000). Magnetic nanowires in hexagonally ordered pores of alumina. *IEEE Trans. Magn.*, 36(1), 30 - 35.

Mikulskas, I.; Juodkazis, S.; Tomašiūnas, R. & Dumas, J. G. (2001). Aluminum oxide phtonic crystals grown by a new hybrid method. *Adv. Mater.*, 13(20), 1574 - 1577.

Mock, J. J.; Oldenburg, S. J.; Smith, D. R.; Schultz, D. A. & Schultz, S. (2002). Composite plasmon resonant nanowires. *Nano Lett.*, 2(5), 465 - 469.

Murray, B. J.; Walter, E. C. & Penner, R. M. (2004). Amine vapor sensing with silver mesowires. *Nano Lett.*, 4(4), 665 - 670.

Nanosensors, (2008). www.nanosensors.com/Carbon_Nanotube_Probes.pdf

Nicewarner- Peña , S. R.; Freeman, R. G.; Reiss, B. D.; He, L.; Pena, D. J.; Walton, I. D.; Cromer, R.; Keating, C. D.; & Natan, M. J. (2001). Submicrometer metallic barcodes. *Science*, 294, 137 - 141.

Nie, S. & Emory, S.R. (1997). Probing single molecules and single nanoparticles by surfaceenhanced Raman scattering. *Science*, 275, 1102 - 1106.

Nielsch, K.; Müller, F.; Li, A. P. & Gösele, U. (2000). Uniform nickel deposition into ordered alumina pores by pulsed electrodeposition. *Adv. Mater.*, 12(8), 582 - 586.

Nielsch, K.; Choi, J.; Schwirn, K. & Wehrspohn, R. B. & Gösele, U. (2002). Self-ordering regimes of porous alumina: The 10 porosity rule. *Nano Lett.*, 2 (7), 677 - 680.

Nishio, K.; Yanagishita, T.; Hatakeyama, S.; Maegawa, H. & Masuda H. (2008). Fabrication of ideally ordered anodic porous alumina with large area by vacuum deposition of Al onto mold. *J. Vac. Sci. Technol. B*, 26(1), L10-L12.

Ohgai, T.; Hoffer, X.; Gravier, L.; Wegrowe, J.-E. & Ansermet J.-P. (2003). Bridging the gap between template synthesis and microelectronics: spin-valves and multilayers in self organized anodized alumimium nanopores. *Nanotechnology*, 14, 978 - 982.

O'Sullivan, J.P. & Wood, G.C. (1970). The morphology and mechanism of formation of porous anodic films on aluminium. *Proc. Roy. Soc. Lond.*, 200, 5071-5076.

Peng, C.Y.; Liu, C.Y; Liu, N.W.; Wang, H.H.; Datta, A. & Wang, Y.L (2005). Ideally ordered 10 nm channel arrays grown by anodization of focused-ion-beam patterned aluminum. *J. Vac. Sci. Technol. B*, 23(2), 559-562.

Piraux, L.; George, J. M.; Despres, J. F.; Leroy, C.; Ferain, E.; Legras, R.; Ounadjela, K. & Fert, A. (1994). Giant magnetoresistance in magnetic multilayered nanowires. *Appl. Phys. Lett.*, 65(19), 2484 - 2446.

Piraux, L.; Dubois, S.; Fert, A. & Belliard, L. (1998). The temperature dependence of the perpendicular giant magnetoresistance in Co/Cu multilayered nanowires. *Eur. Phys. J. B*, 4(4), 413 - 420.

Pratt, W. P.; Lee, S.-F.; Slaughter, J. M.; Loloee, R.; Schroeder, P. A. & Bass, J. (1991). Perpendicular giant magnetoresistances of Ag/Co multilayers. *Phys. Rev. Lett.*, 66, 3060 - 3063.

Qin, L.; Park, S.; Huang, L. & Mirkin, C. A. (2005). On-wire lithography. *Science*, 309, 113 - 115.

Qin, L.; Zou, S.; Xue, C.; Atkinson, A.; Schatz, G. C. & Mirkin, C. A. (2006). Designing, fabricating, and imaging Raman hot spots. *Proc. Nat. Acad. Sci.*, 103 (36), 13300 - 13303.

Quy, N. V.; Hoa, N. D; Yu, W. J.; Cho, Y. S.; Choi, G. S. & Kim, D. (2006). The use of anodic aluminium oxide templates for triode-type carbon nanotube field emission structures toward mass-production technology. *Nanotechnology*, 17, 2156 - 2160.

Renishaw plc, (2009). AFM-Raman system. http://www.renishaw.com/en/6638.aspx

Richter, H. J. (2007). The transition from longitudinal to perpendicular recording. *J. Phys. D: Appl. Phys.*, 40, R149 - R177.

Rivas, J.; Kazadi Mukenga Bantu, A.; Zaragoza, G.; Blanco, M. C. & López-Quintela, M. A. (2002). Preparation and magnetic behavior of arrays of electrodeposited Co nanowires. *J. Magn. Magn. Mater.*, 249(1-2), 220-227.

Robinson, A. P.; Burnell, G.; Hu, M. & MacManus-Driscoll, J. L. (2007). Controlled, perfect ordering in ultrathin anodic aluminum oxide templates on silicon. *Appl. Phys. Lett.*, 91, 143123-143125.

Sauer, M. G.; Brehm, G.; Schneider, S.; Graener, H.; Seifert, G.; Nielsch, K.; Choi, J.; Göring, P.; Gösele, U.; Miclea, P. & Wehrspohn, R.B. (2005). In situ surface-enhanced Raman spectroscopy of monodisperse silver nanowire arrays. *J. Appl. Phys.*, 97, 024308-6.

Schönenberger, C.; van der Zande, B. M. I.; Fokkink, L. G. J.; Henny, M.; Schmid, C.; Krüger, M.; Bachtold, A.; Huber, R.; Birk, H. & Staufer, U. (1997). Template synthesis of nanowires in porous polycarbonate membranes: Electrochemistry and morphology. *J. Phys. Chem*. B, 101, 5497–5505.

Sommerlatte, J.; Nielsch, K. & Bötter, H. (2007). (In German) Thermoelektrische multitalente. *Physik Journal*, 6(5), 35 - 41.

Strijkers, G. J.; Dalderop, J. H. J.; Broeksteeg, M. A. A.; Swagten, H. J. M. & de Jonge, W. J. M. (1999). Structure and magnetization of arrays of electrodeposited Co wires in anodic alumina. *J. Appl. Phys.*, 86, 5141-5145.

Sun, Z. & Kim, H.K. (2002). Growth of ordered, single-domain, alumina nanopore arrays with holographically patterned aluminium films. *Appl. Phys. Lett.*, 81, 3458-3460.

Sysoev, V. V.; Button, B. K.; Wepsiec, K.; Dmitriev, S. & Kolmakov, A. (2006). Toward the nanoscopic "electronic nose": Hydrogen vs carbon monoxide discrimination with an array of individual metal oxide nano- and mesowire sensors. *Nano Lett.* 6(8), 1584 - 1588.

Tan, L.; McGary, P. & Stadler, B. J. H. (2008). Controlling the angular response of magnetoresistance in Co/Cu multilayered nanowires using Co crystallographic orientation. *J. Appl. Phys.*, 103, 07D503-3.

Tang, X.-T.; Wang, G.-C. & Shima, M. (2007). Layer thickness dependence of CPP giant magnetoresistance in individual CoNi/Cu multilayer nanowires grown by electrodeposition. *Phys. Rev. B*, 75, 134404 - 134413.

The Nobel Foundation, http://nobelprize.org/nobel_prizes/physics/laureates/2007/.

Tresback, J. S.; Vasiliev, A. L. & Padture, N. P. (2005). Engineered metal-oxide-metal heterojunction nanowires. *J. Mater. Res.*, 20(10), 2613 - 2617.

Tresback, J. S. & Padture, N. P. (2008).Low-temperature gas sensing in individual metal–oxide–metal heterojunction nanowires. *J. Mater. Res.*, 23(8), 2047 - 2052.

Valet, T. & Fert, A. (1993). Theory of the perpendicular magnetoresistance in magnetic multilayers. *Phys. Rev.* B, 48, 7099 - 7113.

Vila, L.; Vincent, P.; Pra, L. D.-D.; Pirio, G.; Minoux, E.; Gangloff, L.; Demoustier-Champagne, S.; Sarazin, N.; Ferain, E.; Legras, R.; Piraux, L. & Legagneux, P. (2004). Growth and field-emission properties of vertically aligned cobalt nanowire arrays. *Nano Lett.*, 4(3), 521 - 524.

Wade, T.L. & Wegrowe, J. E. (2005). Template synthesis of nanomaterials. *Eur. Phys. J. Appl. Phys.*, 29(1), 3 - 22.

Wang, H.-H.; Liu, C.-Y.; Wu, S.B.; Liu, N.-W.; Peng, C.-Y.; Chan, T.-H.; Hsu, C.-F.; Wang, J.-K. & Wang, Y.-L. (2006). Highly Raman-enhancing substrates based on silver nanoparticle arrays with tunable sub-10 nm gaps. *Adv. Mater.*, 18, 491 - 495.

Whitney, T. M.; Searson, P. C.; Jiang, J. S. & Chien, C. L. (1993). Fabrication and magnetic properties of arrays of metallic nanowires. *Science*, 261, 1316 - 1319.

Wilson, N. R. & Macpherson, J. V. (2009). Carbon nanotube tips for atomic force microscopy. *Nature Nanotech.*, 4, 483 - 491.

Yao, J.; Liu, Z.; Liu, Y.; Wang, Y.; Sun, C.; Bartal, G.; Stacy, A. M. & Zhang, X. (2008). Optical negative refraction in bulk metamaterials of nanowires. *Science*, 321, 930.

Yenilmez, E.; Wang, Q.; Chen, R. J.; Wang, D. & Dai, H. (2002). Wafer scale production of carbon nanotube scanning probe tips for atomic force microscopy. *Appl. Phys. Lett.*, 80, 2225 - 2227.

Yin, A. J.; Li, J.; Jian, W.; Bennett, A. J. & Xu, J. M. (2001). Fabrication of highly ordered metallic nanowire arrays by electrodeposition. *Appl. Phys. Lett.*, 79, 1039 - 1041.

6

Ferromagnetic Nanowires and Nanotubes

Xiu-Feng Han, Shahzadi Shamaila and Rehana Sharif
Institute of Physics, Chinese Academy of Sciences, Beijing 100190
China

1. Introduction

Nowadays one of the most exciting areas in materials science is the study of nanomaterials due to their potential applications in fields as diverse as optics, electronics, catalysis, magnetism, electrochemistry, information processing and storage, etc. Preparation of inorganic, organic or organic-inorganic hybrid materials in the nanometer scale can be achieved either by physical or chemical methods. In many cases it requires the use of solids presenting voids or cavities in which the material can be synthesized (Ozin, 1992). This field of nanotechnology represents an exciting and rapidly expanding area of research that crosses the borders between physical and engineering sciences. These ideas have driven scientists to develop methods for making nanostructures such as nanowires and nanotubes.
Preparation of magnetic materials in the nanometer scale can be achieved by different methods, such as electrochemistry, nano-print techniques, physical deposition combined with micro-fabrication method etc. These techniques have been developed along with a significantly enhanced fundamental understanding (Cao, 2004; Xia et al., 2003; Huczko, 2000; Burda et al., 2005), though the field is involving rapidly with new synthesis methods and new kinds of nanowires or nanotubes. For the growth of nanowires and nanotubes, evaporation condensation growth has been demonstrated for the synthesis of various oxide nanowires and nanotubes. Similarly, dissolution-condensation method has been used for the synthesis of various metallic nanowires from solutions. Various elementary and compound semiconductor nanowires have been synthesized by vapor-liquid-solid (VLS) growth method (Duan & Lieber, 2000). Substrate ledge or step induced growth of nanowires or nanotubes, has also been under investigation (Zach et al., 2000). Among all these methods, the template-based electrodeposition is a very simple, effective, versatile and a low cost technique for the growth of nanowires and nanotubes of various materials. Particularly, the inexpensive formation of periodically ordered structures (e.g., nanotube and nanowire arrays) with a periodicity lower than 100 nm has triggered extensive activities in research. The present, huge progress in nanotechnology is a direct result of the modern trend towards the miniaturization of devices and the development of specific instrumentation that could visualize the nanoworld and allow surface to be studied at nanoscale resolution (Eftekhari, 2008). Practically all the traditional and modern experimental methods for materials growth are used to grow different nanostructured systems and as well as low dimension devices. The differences among the standard techniques of materials growth and the associated growth mechanisms have given place to two well defined strategies for nanostructures fabrication, i.e., nanophysics and nanochemistry (Ozin, 1992) also identified in the current scientific literature as top-down (Lundstrom, 2003) and bottom-up (Yang, 2003),

respectively. Although, it is impossible to define with absolute clarity the frontiers between nanophysics and nanochemistry to obtain the best performances in nanostructures fabrication but the combination of both strategies has obtained successful results and allows to understand the different properties and factors like higher storage and velocity of information transmission, quantization of the conductance, enhanced mechanical properties, etc. (Alivisatos, 1996; Brus, 1994; Krans et al., 1995), particularly those related with the next generation of nanoelectronic devices (Dobrzynski, 2004). The strong reduction of the dimensions and precise control of the surface geometry of nanostructured materials has resulted in the occurrence of novel and unique magnetic and magnetization properties.

Ferromagnetic nanowires and nanotubes exhibit unique and tunable magnetic and magnetization properties due to their inherent shape anisotropy. Current interest in research on ferromagnetic nanowires and nanotubes is stimulated by their applications in different fields such as magnetism, optics, electronics (Li et al., 1999), spintronics, electrochemistry, magnetic catalyzer or absorbent, magneto- or bio-sensors, micro-electromechanical systems (MEMS), future ultra-high-density magnetic recording media (Sun et al., 2000) and biotechnology (Escrig et al., 2007). An ideal ultrahigh density recording medium would have a nanostructure with magnetically isolated small grains. The ultrahigh density magnetic recording with 1 bit down to nanosize is touching the superparamagnetic limit. To overcome this limitation, the possible method is either to increase the effective anisotropy of material or to increase the thickness from nanodots to nanocylinders. Since longitudinal recording may have difficulty achieving acceptable thermal stability from 40 to 100 Gbit/in^2, perpendicular recording media is now being seriously considered for storage at 100-1000 Gbit/in^2. To sustain even 100 Gbit/in^2, either the recording media must possess an average grain and magnetic cluster domain size near 10 nm, possess high coercivity of 5-10 kOe to resist bit demagnetization, and simultaneously allow only 10% signal amplitude loss in 10 years. The small diameter, single domain nanowires of Ni, Co fabricated into the pores of porous anodic alumina (Thurn-Albrecht et. al., 2000; Nielsch et. al., 2001) has been found to be suitable for the above purpose. The high aspect ratio of the nanowires results in enhanced coercivity and suppresses the onset of the 'superparamagnetic limit', which is considered to be very important for preventing the loss of magnetically recorded information among the nanowires. Suitable separation among the nanowires is maintained to avoid the interwire interaction and magnetic dipolar coupling.

This chapter gives a review about ferromagnetic nanocylinders (nanowires and nanotubes) presented by other researchers during the previous 10 years including our most recent results (Han et. al., 2009; Shamaila et. al., 2009; Sharif et. al., 2008). Anodized aluminum oxide (AAO) and track etched polycarbonate (PC) membranes have been used widely to prepare ferromagnetic elemental and alloy nanocylinders while electrochemical depositions have been presented as major template synthetic strategies. This chapter addresses to, (i) various electrodeposited ferromagnetic elemental Fe, Co, Ni, and alloy NiFe, CoFe, CoPt, CoFeB and CoCrPt nanowire and nanotube arrays, in both AAO and PC membranes with different diameters and lengths, (ii) the investigations of these nanocylinders as function of geometrical parameters, (iii) a systematic discussion of the relationships among their structure, the comparison of structural, magnetic and magnetization reversal properties of ferromagnetic nanowires with that of corresponding nanotubes. The results show that the electrodeposition technique allows to systematically varying the length and ratio of internal to external diameter (thickness) of the nanowires and nanotubes. The magnetization switching of ferromagnetic cylinder is influenced by the ratio of internal to external radii (thickness) of the ferromagnetic solid cylinder due to their geometry dependent magnetic properties.

2. Materials and experimental techniques

2.1 Membrane as template

Porous membranes are generally employed in filtration technologies for the separation of different species (polymers, colloids, molecules, salts, etc.), depending on their pore radii that may vary from µm to nm. Although they exhibit in general heterogeneous porosity, a few of them can be prepared with well-defined shape pores of a narrow distribution of diameters such as nanochannel array on glass (Tonucci et al., 1992), radiation track-etched mica (Possin, 1970), mesoporous materials (Wu & Bein, 1994), porous silicon by electrochemical etching of silicon wafer (Fan et al., 1999), zeolites (Enzel et al., 1992) and carbon nanotubes (Guerret et al., 1994; Ajayan et al., 1995). Bio-templates are also explored for the growth of nanowires (Knez et al., 2003) and nanotubes (Gasparac et al., 2004). Among these it is worth mentioning track-etch membranes and anodized Al_2O_3 membranes. PC membranes are made by bombarding a nonporous polycarbonate sheet, typically 6 to 20 µm in thickness, with nuclear fission fragments to create damage tracks, and then chemically etching these tracks into pores (Fleisher et al., 1975). In these radiation track etched membranes, pores have a uniform size as small as 10 nm, but they are randomly distributed. Pore densities can be as high as 10^9 pores/cm^2. Polycarbonate (PC) track-etch membranes (commercially available from Nucleopore, Poretics, Millipore) show cylindrical pores, mainly perpendicular to the membrane sheet although they are tilted up to 34°. Track-etch mica membranes present higher chemical and thermal stability with diamond-like cross-section pores. The commonly used alumina membranes with uniform and parallel pores are made by anodic oxidation of aluminum sheet in solutions of sulfuric, oxalic, or phosphoric acids (Furneaux et al., 1989; Despic & Parkhuitik 1989). The pores can be arranged in a regular hexagonal array and densities as high as 10^{11} pores/cm^2 can be achieved (AlMawiawi et al., 1991). Pore size ranging from 5 nm to 100 µm can be made (AlMawiawi et al., 1991; Foss et al., 1992). Anodic Al_2O_3 membranes are prepared by electrochemical oxidation of Al producing pores of asymmetric structure. Whatman (Anodisc) and Merck (Anotec) commercially sell anodic Al_2O_3 membranes but they are restricted to a very limited range of pore diameters. Consequently, many researchers prepare their own templates. Recently, the preparation of polycrystalline and monocrystalline pore arrays with large interpore distance in anodic Al_2O_3 has been reported (Li et al., 2000).

2.2 Anodization for AAO

Anodization of aluminium in acidic solutions leads to a nanoporous alumina membrane (Jessensky et al., 1998; Sullivan & Wood et al., 1970; Li et al., 2008; Singaraju et al., 2006, Du et al., 1999). Different acids like sulphuric acid, oxalic acid and phosphoric acids can be used for anodiziation. Practically, sulphuric acid is used to prepare the AAO templates of small pore diameter ranging from 3 nm to about 50 nm, phosphoric acid is used to fabricate templates with large pore diameter (≥ 60 nm). Oxalic acid is used to make AAO templates with medium pore diameter. Anodic oxide templates has been fabricated as follows: The high purity (99.999%) Aluminium (Al) foil was ultrasonically degreased in trichloroethylene for 5 min., and etched in 1.0 M NaOH for 3 min. at room temperature (RT). It was then electropolished in a mixed solution of $HClO_4 : CH_3CH_2OH = 1 : 4$ (by volume) for 3 min. with a constant potential of about 12 volts (V). To obtain highly ordered pores, a two-step anodization was employed. In the first anodization step the Al foil was anodized at 0 °C and 40 V dc in 0.3 M oxalic acid for about 12 h to form textures on Al surface. The formed aluminum oxide layer was then removed by immersing anodized Al into a mixed solution of 0.4 M chromic acid and 0.6 M phosphoric acid solution at 60 ° C. Subsequently, the samples were reanodized for different periods of time under the same anodization conditions as in the first step. These self assembled anodic aluminium oxide (AAO) templates were used to fabricate different types of nanowires and nanotubes by electrochemical deposition method (Shamaila et al., 2008a; Shamaila et al., 2008b; Shamaila et al., 2009b; Sharif et al. 2007).

2.3 Electrochemical deposition

Most of the studies reported in literature are based on two types of membranes: polymer ion track membranes and anodic alumina. Both present a number of advantages which makes them suitable for the fabrication of high aspect ratio nanostructures, namely nanowires and nanotubes. The method of filling of the pores, thus of fabricating the nanostructures are various but the most employed one is electrochemical deposition (Toimil-Molares et al., 2004; Sima et al., 2004) also known as electrodeposition. Electrodeposition was employed for the preparation of metallic and semiconductor nanowires (Enculescu, 2006). Electrochemical deposition was used about 40 years ago in filling pores in mica with metals by Possin (Possin, 1970; Possin, 1971). Electroless deposition was also used to nanoporous membranes allowing the preparation of hollow structures in contrast with electrochemical deposition which leads in most of the cases to rod-like deposits (Bercu et al., 2004). The template approach represents an interesting path towards preparation of nano objects with controlled morphological properties mainly due to the fact that by appropriate choosing of host templates the shape and dimension of the prepared structures are precisely determined (Martin et al., 1994; Fert & Piraux 1999; Enculescu et al., 2003). The template materials have some certain requirements to get the desired pore or channel size, morphology, size distribution and density of pores, like, the compatibility of template materials with the processing conditions, the internal pore walls should be wet by depositing materials or solution and the deposition should start from the bottom or one end of the template channels and proceed from one side to another.

2.4 Fabrication of nanowires and nanotubes

Ferromagnetic nanowires and nanotubes have been fabricated in three kind of templates, self assembled anodic aluminum oxide (AAO) (home made), commercially available AAO membranes and polycarbonate (PC) membranes. For nanotubes, the AAO templates were preannealed in air at 100 °C in order to remove moisture from the AAO templates. A conductive layer of different thickness is sputtered on one side of AAO and PC template to serve as working electrode for nanowires and nanotubes. For nanotubes, the layer was so thin that this just covered the pore walls of the templates, leaving the orifices open. Electrodeposition was performed in a three-electrode cell under constant voltage at room temperature, where the sputtered conducting layer served as the working electrode, saturated calomel electrode (SCE) reference and a graphite pole was used as counter electrode. The electrolytes for electrodeposition consisted of their respective salts in deionized water as mentioned in the references (Shamaila et al., 2008a; Shamaila et al., 2008b; Shamaila et al., 2009a; Shamaila et al., 2009b; Sharif et al. 2006; Sharif et al. 2007; Sharif et al. 2008; Sharif et al. 2008, Liu et al. 2009). The different lengths of nanowires and nanotubes were obtained by adjusting the time of electrodeposition. The different wall thicknesses for nanotubes were obtained by adjusting the thickness of working electrode layer. Co/Cu multilayer nanowires and nanotubes have also been fabricated in homemade AAO templates by electrodeposition method.

2.5 Characterization

The morphology and size of nanowires and nanotubes was characterized by scanning electron microscopy (SEM). The composition was analyzed by induced coupling plasma atomic emission spectrometer (ICP) combined with chemical analysis. The structural analysis of ferromagnetic nanowires and nanotubes is done by transmission electron microscopy (TEM) and X-Ray diffraction (XRD) spectroscopy. Magnetic properties of the samples were tested by a vibrating sample magnetometer (VSM) and superconducting quantum interference device (SQUID).

3. Morphology of templates and ferromagnetic nanowires and nanotubes

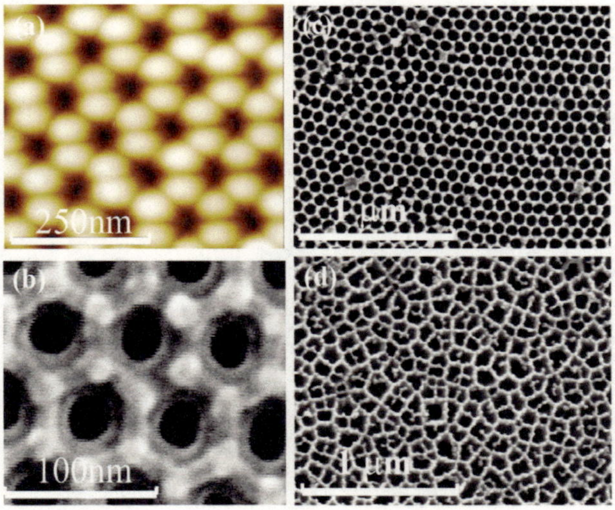

Fig. 1. (a) AFM top view (0.5x0.5 μm^2); (b-d) SEM image of (b) homemade AAO templates (anodized in 0.3M oxalic acid solution) with diameter, d ~ 40 nm, (c) homemade (anodized in 5%H_3PO_4 solution) and (d) commercial AAO templates with diameter, d ~ 200 nm.

Figure 1 shows (a) AFM top view (1x1 μm^2) (b) SEM view of homemade anodic alumina (AAO) template prepared by two-step anodization in 0.3 M oxalic acid, with diameter equal to 40 nm; (c) homemade AAO template anodized in 5% phosphoric acid and (d) Commercial AAO templates, with diameter equal to 200 nm. The fabricated AAO t emplates contain self-assembled uniform pore arrays with quasi-hexagonal ordering. The average centre-to-centre spacing (D_i) and pore diameter (d) depend on anodization conditions and the electrolyte used for anodization. Generally, pore diameter of AAO film is increased if the anodized voltage is large, and length (L) of the pores is increased with time for anodization.

Figure 1a-d shows that our home made AAO templates are highly uniform and ordered as compare to the commercially available anodic alumina membranes as shown in Figure 1d.

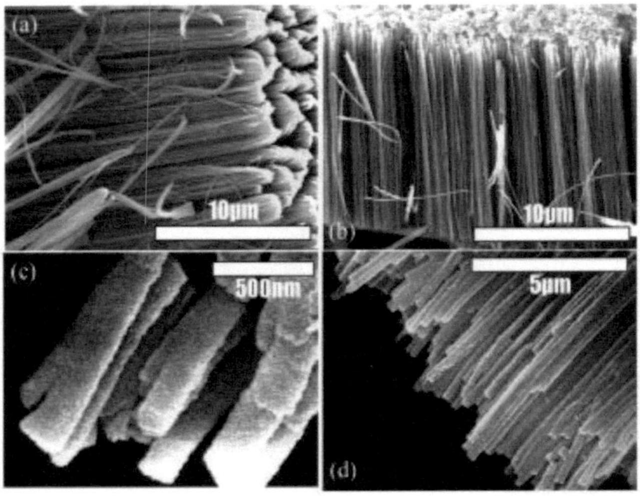

Fig. 2. SEM images of isolated (a) Co, (b) Ni, (c) CoPt, and (d) CoCrPt, Nanowires separated from AAO template by dissolving the alumina layer in NaOH aqueous solution.

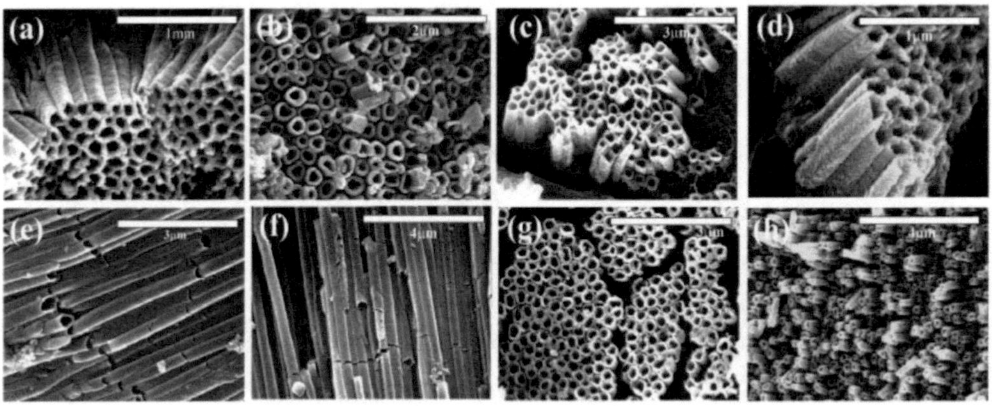

Fig. 3. SEM images of isolated (a) Co, (b) NiFe, (c) Ni, (d) CoFe, (e) CoPt, (f) Fe, (g) CoFeB, and (h) CoCrPt nanotubes, separated from AAO template by dissolving the alumina layer in NaOH aqueous solution.

By using two-step anodization in different electrolytes, highly ordered and uniform AAO templates can be fabricated with a wide range of diameter and length. These self-assembled AAO templates can be used to fabricate Co, Ni, CoPt, CoFe, CoFeB etc. nanowire arrays by electrochemical deposition method (Shamaila et al., 2008a; Shamaila et al., 2008b; Shamaila et al., 2009b; Sharif et al. 2007). Figure 2a-d shows the SEM images of isolated (a) Co, (b) Ni, (c) CoPt, and (d) CoCrPt nanowires separated from AAO template by dissolving the alumina layer in NaOH aqueous solution. The average diameter *d* of Co and Ni nanowires shown in Figure 2 is ~ 40 nm and that of CoPt and CoCrPt NWs is ~ 200 nm. These SEM images reveal that nanowires of several materials with different diameters can be fabricated by using the templates of different diameters.

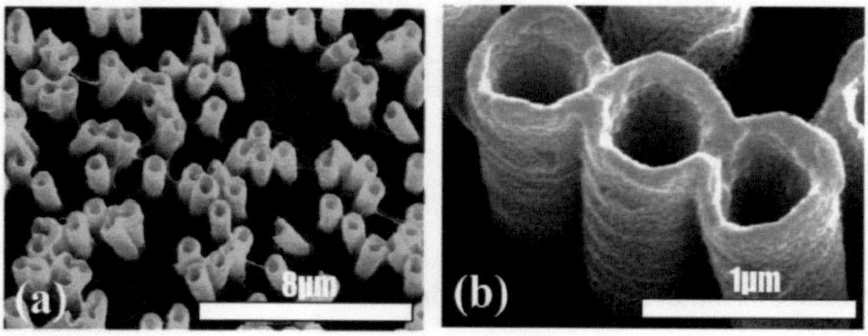

Fig. 4. SEM images of isolated (a) Fe, (b) Co Nanotubes separated from PC template by dissolving the PC template in chloroform.

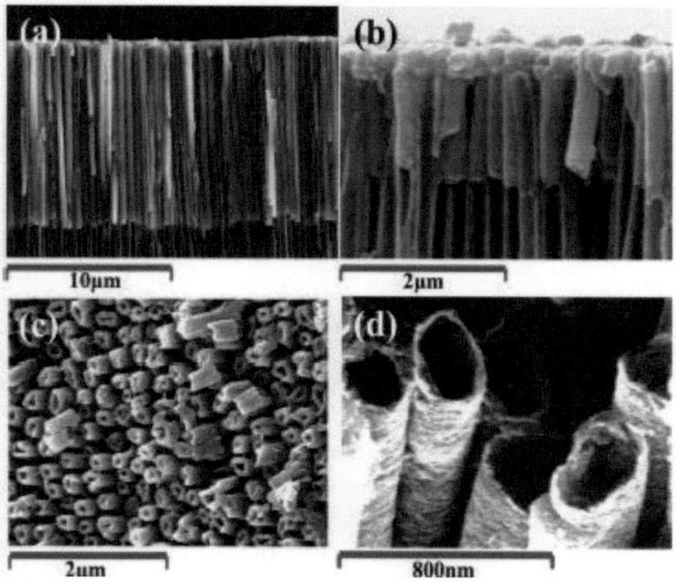

Fig. 5. (a,b) SEM images of nanowires in AAO template with different Lengths (L) (a) L ~12 µm, (b) L ~1 µm. (c,d) SEM images of nanotubes with different wall thickness t_w for nanotubes (c) t_w ~ 60 nm, and (d) t_w ~ 20 nm.

Figure 3a-h shows the SEM images of nanotubes with different materials, (a) Co (b) NiFe, (c) Ni, (d) CoFe, (e) CoPt, (f) Fe, (g) CoFeB, and (h) CoCrPt nanotubes. These nanotubes were fabricated in AAO templates and for SEM images these were separated by dissolving the alumina layer in NaOH aqueous solution. The average outer diameter (*d*) of these nanotubes is ~ 200 nm. Figure 4a and b shows the SEM images of (a) Fe and (b) Co nanotubes fabricated in polycarbonate (PC) membrane as template with average outer diameter (*d*) of ~ 400 nm. SEM images of nanotubes reveal that nanotubes of several materials can be fabricated by using different templates of different diameters.

Figure 5a-d shows the SEM images of nanocylinders with (a and b) different lengths (*L*) for nanowires (a) L~12 µm, (b) L~ 1 µm and (c and d) different wall thickness t_W for nanotubes (c) t_W ~ 60 nm, and (d) t_W ~ 20 nm. Since the nanocylinders are characterized geometrically by their length (*L*), external and internal diameter (*d* & *a*) respectively, and wall thickness (t_W), where t_W =0 gives nanowire and t_W > 0 gives nanotube geometry. This t_W makes the nanotubes distinct from that of nanowires and strongly affects the magnetization reversal mechanism and thereby, the overall magnetic behaviour (Escrig et al., 2008). These SEM images reveal that nanowires with wide range of diameter and lengths, and nanotubes with different external and internal diameter, length and wall thickness can be fabricated by low cost electrodeposition method.

4. Structural characterizations of ferromagnetic nanowires and nanotubes

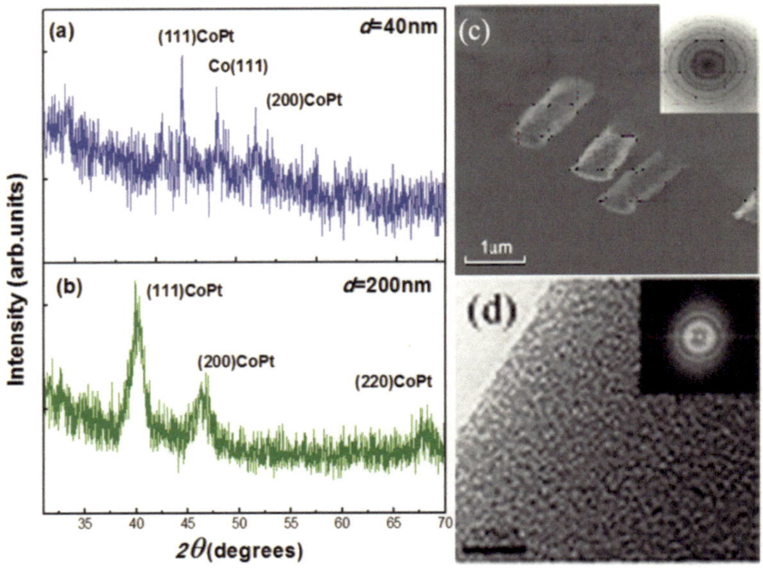

Fig. 6. (a,b) XRD pattern of aligned CoPt nanowires in the AAO templates of diameter (a) 40 nm and (b) 200 nm. (c) TEM image of Co nanotubes, inset is the diffraction pattern of the nanotubes showing its fcc crystalline structure, (d) HRTEM images of the CoFeB nanowire arrays, the inset is SAED image.

The structure analysis of ferromagnetic nanowires and nanotubes is done by HRTEM and selected area electron diffraction (SAED) patterns after releasing the nanowires and nanotubes from the templates (Shamaila et al., 2009b; Sharif et al., 2006; Sharif et al., 2007; Sharif et al., 2006). The XRD is also performed for the structure analysis of nanocylinders (Shamaila et al., 2008a; Shamaila et al., 2008b; Shamaila et al., 2009b; Sharif et al., 2006; Sharif et al., 2008). Figure 6a and b shows the XRD pattern of aligned CoPt nanowires in the AAO templates of diameter (a) 40 nm and (b) 200 nm. In the as-synthesized conditions, the XRD patterns show that the samples of CoPt nanowires comprise of fcc phase with the identifiable diffraction peaks, namely (111) and (200) of which the (111) reflection is the most intense. For 200 nm CoPt nanowires (220) peak has also been detected, together with the diffraction peaks (111) and (200). These XRD results suggest that the CoPt nanowires are of fcc polycrystalline structure with randomly oriented grains. Figure 6c is the TEM image of Co nanotubes. The inset shows diffraction pattern of the nanotubes showing its fcc crystalline structure. Figure 6d exhibits the HRTEM images of the CoFeB nanowire arrays, the inset is SAED image which show amorphous phase for CoFeB nanowire arrays. Structural analysis explores that nanowires and nanotubes of crystalline and amorphous structure for different kinds of nanocylinders. Structural analysis understanding can be used to fabricate crystalline or amorphous nanowires and nanotubes according to the required applications.

5. Magnetic and magnetization properties of nanowires and nanotubes

No.	Composition	$H_{c//}$ (Oe)	$H_{c\perp}$ (Oe)	$SQ_{//}$	$SQ_{\perp}$	Ref
1	Fe	89	125	0.02	0.05	In this work
2	Co	863	300	0.57	0.11	Shamaila et al.2009b
3	Ni	230	105	0.22	0.06	Han et al.2003
4	$Ni_{86}Fe_{14}$	769	313	0.65	0.12	Hao et al.2001
5	$Co_{94}Fe_6$	66	111	0.01	0.03	Sharif et al.2008b
6	$Co_{90}Pt_{10}$	105	80	0.02	0.04	In this work
7	$Co_{91}Fe_7B_3$	208	265	0.08	0.07	Sharif et al.2008b
8	$Co_{75}Cr_{13}Pt_{12}$	121	233	0.09	0.32	Shamaila et al.2009a

Table 1. Ferromagnetic nanowires in AAO template (L > 10 μm)

No.	Composition	$H_{c//}$ (Oe)	$H_{c\perp}$ (Oe)	$SQ_{//}$	$SQ_{\perp}$	Ref
1	Fe	364	163	0.09	0.35	In this work
2	Co	115	75	0.03	0.05	In this work
3	Ni	111	102	0.16	0.24	In this work
4	$Ni_{93}Fe_7$	57	63	0.02	0.25	In this work
5	$Co_{94}Fe_6$	107	192			In this work
6	$Co_{90}Pt_{10}$	125	177	0.06	0.09	In this work
7	$Co_{80}Fe_{17}B_3$	353	108	0.04	0.4	In this work
8	$Co_{75}Cr_{13}Pt_{12}$	251	135	0.03	0.13	Shamaila et al.2009a

Table 2. Ferromagnetic nanotubes in AAO template (L > 10 μm)

Table I shows the magnetic properties of metal nanowires like Fe, Co, and Ni and alloy NiFe, CoFe, CoPt, CoFeB, and CoCrPt nanowire arrays fabricated by electrodeposition in AAO template. Table II shows the magnetic properties of metal nanotubes like Fe, Co, and Ni and alloy NiFe, CoFe, CoPt, CoFeB, and CoCrPt nanotube arrays fabricated by electrodeposition in AAO template (Shamaila et al., 2008a; Shamaila et al., 2008b; Shamaila et al., 2009a; Shamaila et al., 2009b; Sharif et al., 2006; Sharif et al., 2007; Sharif et al., 2008a; Sharif et al., 2008b). Table III shows the magnetic properties of metal nanotubes like Fe, Co, and Ni nanotube arrays fabricated by electrodeposition in polycarbonate (PC) template. The magnetic properties of pure and alloy nanowire and nanotube arrays in AAO and PC templates can be compared from these tables. Because of the intrinsic properties difference between different metals and their alloys, the demagnetization field ($2\pi M_s$) of individual cylinders (for infinite cylinder) and dipolar interaction among the nanocylinders will be different. Therefore their properties are different from each other.

No.	Composition	$H_{c//}$ (Oe)	$H_{c\perp}$ (Oe)	$SQ_{//}$	$SQ_{\perp}$	Ref
1	Fe	145	149	0.05	0.22	Sharif et al.2008a
2	Co	158	197	0.11	0.37	Sharif et al.2008a
3	Ni	80	127	0.06	0.31	Sharif et al.2008a

Table 3. Ferromagnetic nanotubes in Polycarbonate (PC) template (L ~ 6 μm)

5.1 Easy axis of ferromagnetic nanowires

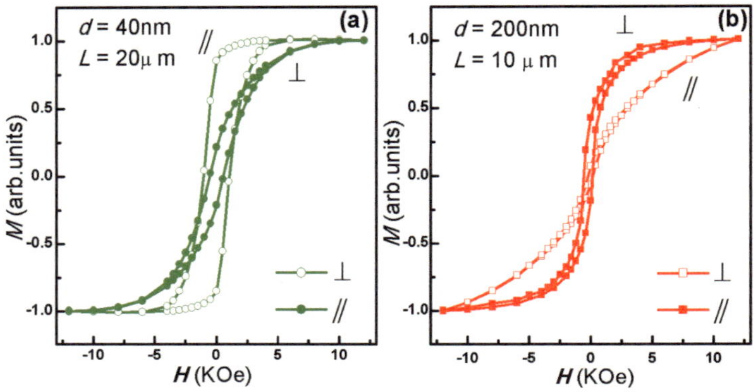

Fig. 7. M-H curves for CoPt nanowire arrays of diameter (a) 40 nm and (b) 200 nm shows the crossover of easy axis as a function of diameter.

Shape and geometry dependent magnetization behaviour of the CoPt nanowires and nanotubes is explained here. Diameter and length have strong effects on the magnetic properties of nanocylinder. The typical room temperature magnetic hysteresis (M-H) curves for 40 and 200 nm CoPt nanowires, with the external field applied parallel and perpendicular to the nanowire axis, are shown in Figure 7a and b. The difference between the perpendicular and parallel M-H curves defines the uniaxial anisotropy for CoPt nanowire arrays. For 40 nm CoPt nanowires (Figure 7a) the remanent squareness (SQ) in the parallel (//) geometry is larger than that of the perpendicular (⊥) geometry attributing to the parallel alignment of the magnetic easy axis along the wire axis (Mallet et al., 2004). For 200 nm CoPt nanowires as shown in Figure 7b, SQ in the perpendicular geometry is larger than that of the parallel geometry. Therefore the easy axis of magnetic anisotropy favours to be aligned perpendicular to the wire axis. This variation in the alignment of easy axis for the two diameters shows a crossover of easy axis of magnetization as a function of diameter. The values of magnetic parameters H_c, SQ and alignment of easy axis for different diameters and lengths are presented in Table IV.

Comparison of the parameters, given in Table IV shows the effect of the diameter and length on the magnetic properties of CoPt nanowire and nanotube arrays. The alignment of easy axis can also be specified by the sign of difference of saturation fields (ΔH_s) here $\Delta H_s = H_{s//} - H_{s\perp}$ (Ciureanu et al., 2005). Here $H_{s//}$ is the saturation field when magnetic field is applied parallel to the nanocylinder axis and $H_{s\perp}$ is the saturation field when the magnetic field is applied perpendicular to the nanocylinder axis. Negative sign of ΔH_s indicates the easy axis along the nanocylinder axis and positive sign shows the orientation of easy axis perpendicular to the nanocylinder axis.

No	Diameter (nm)	Length (μm)	$H_{c//}$ (Oe)	$H_{c\perp}$ (Oe)	$SQ_{//}$	$SQ_{\perp}$	ΔH_s	Easy axis	Ref.
1	40	2	2033	575	0.93	0.15	-2164.45	//	Shamaila et al.2009a
2	40	20	1051	555	0.84	0.21	-3375.59	//	Shamaila et al.2009a
3	200	0.8	652	303	0.38	0.14	-717.45	//	Shamaila et al.2009a
4	200	10	189	141	0.07	0.18	4018.07	⊥	Shamaila et al.2009a

Table 4. Effect of Length on the easy axis of ferromagnetic $Co_{90}Pt_{10}$ nanowires in AAO template

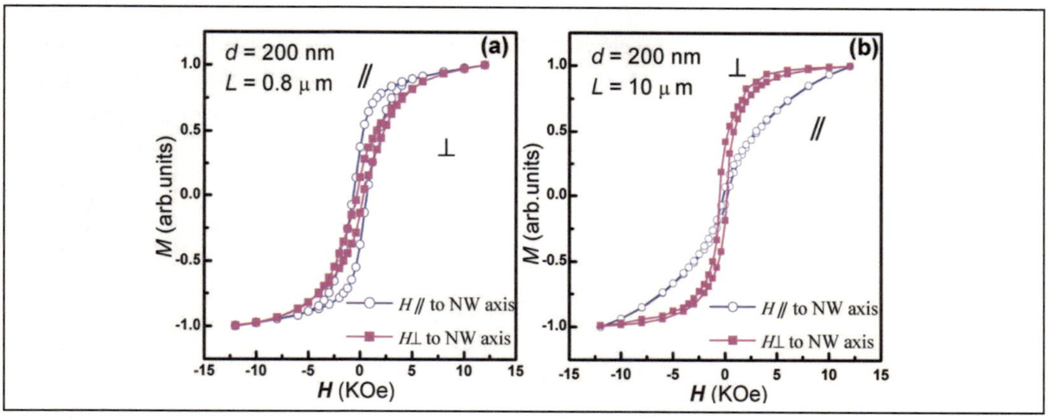

Fig. 8. M-H curves for CoPt nanowire arrays with length (a) 0.8 μm and (b) 10 μm shows the cross over of easy axis as a function of length.

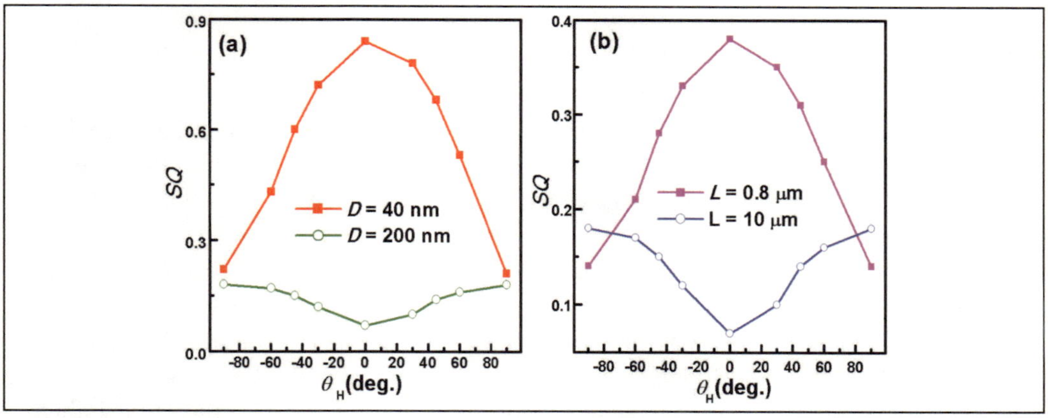

Fig. 9. Angular dependence of remanent squareness (SQ(θ)) (a) with different diameters and (b) with different lengths, of CoPt nanowires where θ is the angle between the field direction and the wire axis.

Figure 8a-b show the effect of length on the magnetic properties of CoPt nanowire arrays for diameter $d \sim 200$ nm. Figure 8a-b show that parallel magnetization is favoured for very short nanowires whereas; perpendicular magnetization is favoured for long nanowires. Thus, in CoPt nanowires, it is observed that when the wire length is larger than a critical value, the parallel wire axis crosses over from easy to hard as also was observed by others for Co nanowire arrays (Rivas et al., 2002). This crossover is clear from the shape of M-H curves as well as from the sign of Delta H_s (Table 4) of these samples. This effect can be used to turn the parallel wire axis from easy to hard, by modifying the length. Additional evidence for the crossover of easy axis is provided by the angular dependence of the SQ of CoPt nanowire arrays measured as a function of wire diameter and length (Figure 9). SQ(θ) curves show bell-shaped or otherwise bell-shaped behaviour corresponding to the easy axis of their magnetization. Bell-shaped curves for sample with d = 40 nm and otherwise bell-shaped curve for sample with d = 200 nm confirms the crossover of easy axis from parallel to perpendicular as a function of diameter (Figure 9a). Furthermore, bell-shaped behaviour of sample with d = 200 nm, and length = 0.8 µm with the easy axis parallel to the wire axis and otherwise bell-shaped curve for sample with length = 10 µm with easy axis perpendicular to the wire axis shows the crossover of easy axis as a function of length. Figure 9a and b also show that values for SQ for samples with d = 40 nm are larger than sample of 200 nm.

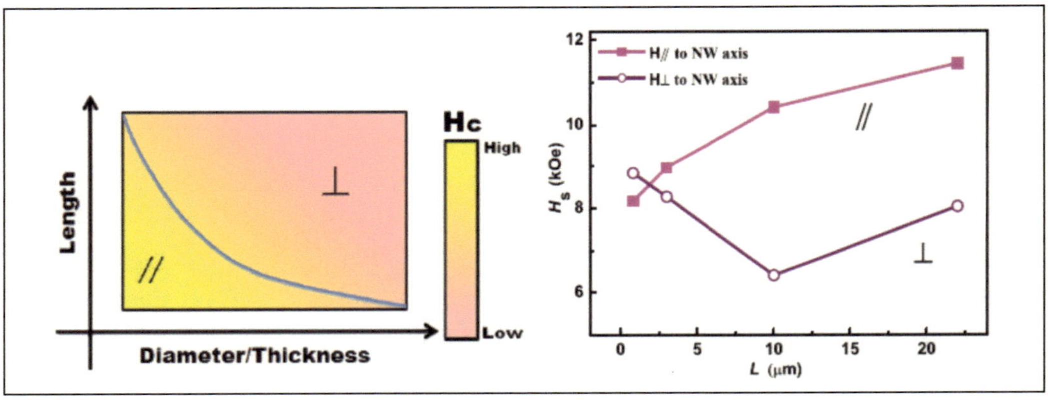

Fig. 10. (a) Variations in the values of $H_{c//}$ and $H_{c\perp}$. (b) Saturation field Vs length of CoPt nanocylinders when H is applied parallel and perpendicular to the axis.

Figure 10a represents the trend of variation of $H_{c//}$ and $H_{c\perp}$ as a function of diameter, length and tube wall thickness. Figure 10b shows the variation of H_s with the length of nanocylinders in parallel and perpendicular geometry. For very small length of nanocylinders, $H_{s\perp}$ is larger than $H_{s//}$ resulting in -ve value of ΔH_s. Whereas, for length larger than a critical length, $H_{s//}$ is larger than $H_{s\perp}$ giving the +ve value of ΔH_s. The phenomenon of critical length has been explained later in the discussion. The overall anisotropic field (H_k) for nanocylinders is mainly determined by following three contributions: (1) the shape anisotropy field ($2\pi M_s$) which will induce a magnetic easy axis parallel to the nanocylinder axis; (2) magnetostatic dipole interaction field among the cylinders which will induce a magnetic easy axis perpendicular to the nanocylinder axis; (3) the magnetocrystalline anisotropy field (H_a). The total effective anisotropic field is given as follows (Han et al., 2003):

$$H_k = 2\pi M_s - 6.3\pi M_s r^2 L / D^3 + H_a \quad (1)$$

Where M_s is the saturation magnetization, r is the radius of the wire, L is the length and D is the interwire distance. The second term in Eq.1 is the total dipole field acting on one wire due to all other wires. Eq.1 predicts that as the wire length increases, H_k linearly decreases to zero when

$$L = L_c = 2D^3 / 6.3r^2 \quad (2)$$

while neglecting the contribution of H_a. For $L > L_c$, H_k is negative pointing out that there is a crossover of easy axis for magnetization from parallel to perpendicular to the axis. For CoPt nanowire arrays with d = 200 nm, r = 100 nm and D = 250 nm; $L_c \sim 1$ μm is calculated. The shape anisotropy field is weak in this case therefore when $L > L_c$, H_k is negative and crossover of easy axis of magnetization from parallel to perpendicular to nanowire axis is observed.
The orientation of easy axis of magnetization can be determined by the total energy $\sum E$ in the parallel and perpendicular geometries where $\sum E$ can be obtained by taking into account the magnetostatic interaction energy E_{mi}, the demagnetization energy E_{de}, the magnetocrystalline anisotropy E_k, where $E_{mi/\!/} > E_{mi\perp}$, $E_{de/\!/} < E_{de\perp}$, $E_{k/\!/} < E_{k\perp}$ (Gao et al., 2006). For large diameters with small interwire distances, the dipolar interaction is increased, $\sum E_{/\!/} \sim \sum E_{\perp}$ and the easy axis favors to be aligned perpendicular to the wire axis (Rivas, 2002). For small diameters, $\sum E_{/\!/} < \sum E_{\perp}$ and the easy axis favors to be aligned along the cylinder axis. It should also be pointed that other reasons cannot be excluded such as the orientation of crystallographic axes and grain size of the nanocylinders. A wide range of values of H_c (from $H_{c/\!/}$ = 2032 Oe for d = 40 nm to $H_{c/\!/}$ = 100 Oe for d = 200 nm) and SQ (from $SQ_{/\!/}$ = 0.93 for d = 40 nm to $SQ_{/\!/}$ = 0.07 for d = 200 nm) is observed depending upon different factors involved in the magnetic properties of the CoPt nanocylinders. The results discussed here reveal that ferromagnetic nanocylinders with easy axis of magnetization either parallel or perpendicular to the nanocylinder axis having the desired values of H_c and SQ for perpendicular recording media can be obtained by modifying the diameter, length or geometry of the nanocylinders.

5.2 Magnetization reversal mechanism

The two most common magnetization reversal modes can be modeled by coherent rotation or curling. Generally for magnetic nanowires the magnetization reversal mechanism depends upon the diameter of the nanowires. For a specific material, the critical diameter for the transition from coherent rotation to non-coherent rotation is given by

$$d_c = 2.08 (A^{1/2} / M_s) \quad (3)$$

where A is the exchange stiffness and M_s is the saturation magnetization (Zeng et al., 2002).

5.3 Coherent rotation model

In presence of coherent rotation mode in one isolated nanocylinder, the micro-spin configuration in the system should be uniformly magnetized with the external applied field as shown in Figure 11a. If the nanocylinders are of soft magnetic material, the crystal anisotropy can be neglected safely. Thus, only the shape anisotropy is contributed to the reversal mechanism. The magnetization can be easily modeled by $m_z = \cos(\theta)$. Hence the total energy is given by

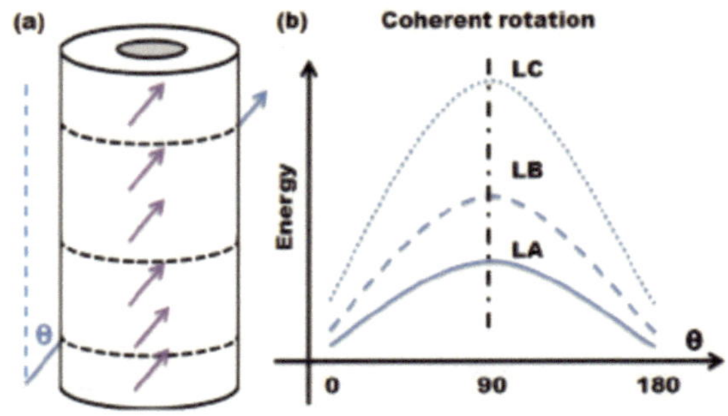

Fig. 11. (a) Coherent rotation mode. The z-direction is parallel along the nanocylinder. θ is the angle between the parallel axis and external field. (b) The relative energy model of nanocylinders in different lengths, where $L_C > L_B > L_A$. The maximum of energy is at 90°.

$$E(\theta) = \pi\mu_0 M_s^2 H R^2 (1-\beta^2)[\sin^2\theta + (3\cos^2\theta - 1)N_z] / 4 \tag{4}$$

Where

$$N_z = \frac{2/H}{1-\beta^2} \int \frac{1-e^{-qH}}{q^2} \left[J_1(qR) - \beta J_1(\beta qR)\right]^2 dq \tag{5}$$

The relative energy differences of the different lengths can be calculated by the above equations. As shown in the Figure 12b, the shape anisotropy is increasing with the length of isolated nanocylinder. However, the experimental results demonstrate that the easy axis of nanocylinder will be changed by the length from parallel along the z-direction to the x-y panel (perpendicular). The coherent mode is not suitable for realistic experiment, which suggests that more complex rotation mechanism should be included. Furthermore the d_c of nanocylinders, for the occurrence of coherent rotation, is in a very small range due to the distribution of energy along the diameter, length and tube wall thickness.

5.4 Non-coherent rotation model

The rigorous micromagnetic simulations have been performed for single, isolated nanowires and elongated particles (Seberino & Bertram, 1999; Ferre et al., 1997; Hinzke & Nowak, 2000). In these cases, the magnetostatic interaction of neighbouring wires was generally omitted from the simulations. Instead of attempting to mimic an infinitely extended array of wires using more or less plausible simplifications, a magnetostatically coupled ensemble consisting of a comparatively small number of wires is modeled without making simplified assumptions concerning the magnetic structure or the dipolar fields (Hertel, 2001). Furthermore, the more serious micromagnetic simulations show that the switching in nanowires will be under more realistic mode of nuclear deformation in array properties, by considering a real array of nanowires. Generally the noncoherent magnetization reversal is concluded as to be curling switching mechanism in ferromagnetic nanocylinders.

5.5 Curling rotation model

For nanowires with diameter larger than the critical diameter, the magnetization reversal process can be described by the curling mode, and H_c decreases with increasing diameter of the nanowires. For example the critical diameter for CoPt nanowires was calculated as d_c=14 nm (Mallet et al., 2004). The equation describing the dependence of H_c on the diameter of nanowires in the curling mode has been given as follows (Mallet et al., 2004),

$$H_c = \frac{2\pi kA}{M_s}\frac{1}{r^2} + \frac{2K_u}{M_s} \quad (6)$$

where r is the radius of the nanowire and k is a constant related to the shape of the material (1.08 for an infinite cylinder.) K_u is the uniaxial anisotropy constant. To certain extent, the relation of H_c and d_c can also be generalized for nanotubes under curling mode.

5.6 Angular dependent rotation mode

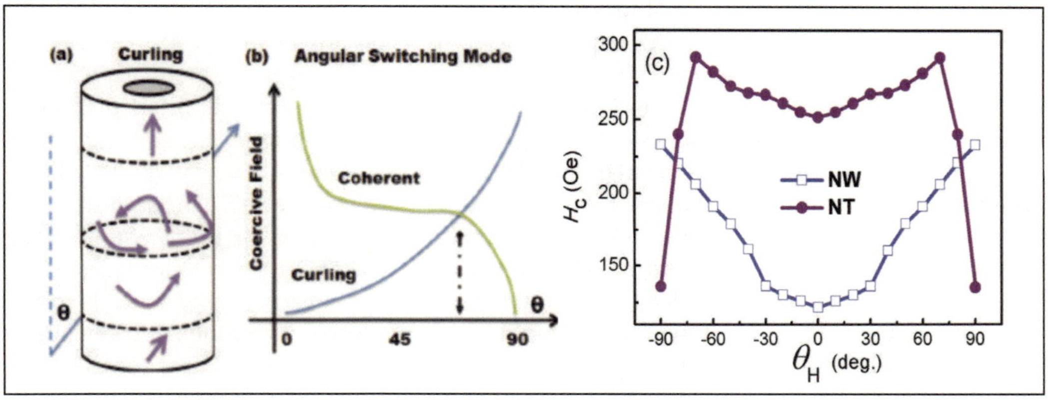

Fig. 12. (a) Angular dependence of curling rotation mode in presence of exchange, magnetostatic field and crystal anisotropy for nanotube. (b) The relative coercive field for angular rotation under curling (dark) and coherent (light) mode. (c) Angular dependence of coercivity [$H_c(\theta)$] of CoCrPt nanowires and nanotubes where θ is the angle between the field direction and the nanocylinder axis.

Although the magnetic behaviour of nanowires has been intensely investigated, tubes have received less attention, in spite of the additional degree of freedom they present; not only the length L and radius r can be varied, but also the thickness of the wall, t_w. Changes in thickness are expected to strongly affect the mechanism of magnetization reversal, and thereby, the overall magnetic behaviour (Escrig et al., 2007, Sharif et. al., 2008a, Sui et al., 2004). In fact, experimental evidence speaks in favour of coherent rotation (Wernsdorfer et al., 1997) and curling (Wirth et al., 1999) in nano-scale particles with relatively small aspect ratios, but neither the observed coercivities nor activation volumes support delocalized reversal for elongated nanowires (Skomski et al., 2000).

Angular dependence of coercive field in presence of exchange, magnetostatic field and crystal anisotropy has been given in Figure 12a & b. Figure 12c shows angular dependence of H_c at room temperature for CoCrPt nanowires and nanotubes. Two possible reversal modes, coherent and curling, give different angular dependence of coercivity. Since d ~ 200 nm of our samples is much greater than critical diameter (d_c) (Sun et al., 2005) for coherent rotation, therefore, reversal mechanism is expected to occur through curling rotation. The curling mode of reversal mechanism for nanocylinders predicts that

$$H_c = \frac{a(1+a)}{\sqrt{a^2+(1+2a)\cos^2\theta}} H_k \tag{7}$$

where a =-1.08 $(d_c / d)^2$ (Han et al., 2003). Eq.7 describes that H_c increases as angle (θ) increases, whereas coherent rotation mode predicts that H_c decreases as angle increases (Han et al., 2003, Sun et al., 2005). For CoCrPt nanowires, H_c increases with increasing angle θ from 0° to ± 90° representing curling mode of reversal mechanism. H_c of nanotubes initially increases with increasing angle up to θ = ± 70°, in good agreement with the curling model; however, above this critical angle H_c decreases abruptly expressing an M-type variation. M-type curve for nanotubes (Figure 12c) reveal that at large angles, coherent rotation is dominant, while curling happens only for small angles ($\theta \leq 70°$). The distinct geometry of nanotubes presents two dynamic configurations of magnetic moments with the applied field. When the field angle is small the magnetic moments will align preferably parallel to the tube axis and reversal will take place by curling rotation. At large field angles the moments will align perpendicular to tube axis and coherent reversal mode will be observed. Different alignment of moments and surface effects due to t_w in nanotubes are causes of transition from curling to coherent reversal mechanism for higher angles. The transition angle depends on d and t_w of nanotubes as previously proved theoretically for other nanotubes (Escrig et al., 2008). Our experimental data is in accordance with the trend given in Figure 12a & b. However, the angular dependence of various ferromagnetic nanocylinders to understand the insights of magnetization switching requires some further explanations. Further understanding describes that the curling rotation mechanism can be adopted more generally for nanowires only whereas for nanotubes curling mechanism is influenced by some other phenomenon which are explained in detail discussion given below.

5.7 Non-coherent model for nanotubes

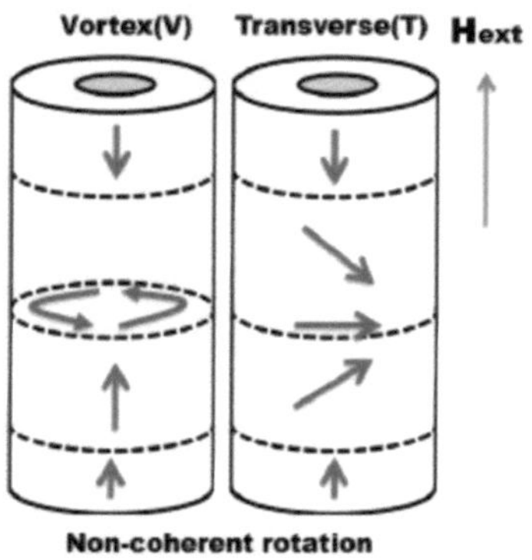

Fig. 13. (a) Simple non-coherent vortex rotation model. (b) Simple transverse rotation model. The z-direction is parallel along the tube structure. The x-direction is perpendicular to the z-direction in paper panel. Then, the y-direction is the outer one towards the paper panel.

The underline physics of the non-coherent rotation is in the dynamic field-driven domain wall motion in the isolated nanocylinder. Considering the simplest cases of parallel switching in the isolated nanotubes, two switching mechanisms can be idealized for the propagation of domain wall in ferromagnetic nanotubes, one is vortex domain wall and another is transverse domain wall motion, as shown in Figure 13a and b and in reference (Landeros et al., 2007). Starting from the equations presented by Landeros for the vortex mode *V*, the magnetization is a little complex spin closure structure along the tube wall. For the transverse mode *T*, the magnetization of domain wall is the several coherent rotations in y-z panel. In both cases, the switching along z-direction is the domain motion at one end of the tube and propagates toward the other, as illustrated in Figure 13. Using the equations presented by Landeros (Landeros et al., 2007), the zero-field energy barrier as well as the width of the domain for each reversal model as a function of the tube wall thickness can be calculated. For the vortex rotation model the micro-spin configuration could be modeled by,

$$m(z) = \begin{cases} \hat{z}, & 0 \le z \le z_w - w/2 \\ m_\phi(z)\hat{\phi} + m_z(z)\hat{z}, & z_w - w/2 < z < z_w + w/2 \\ -\hat{z}, & z_w + w/2 \le z \le L. \end{cases} \quad (8)$$

Where $m_z(z)=\cos\Theta(z)$, $\Theta(z) = \Pi((z-z_w)/w+1/2)$ and $m_x = -\, m_\phi(z)\sin\phi$, $m_y = --m_\phi(z)\cos\phi$. The transverse domain is only different at the description of magnetization within the region of transverse domain wall. Which can be described as $m(z)=m_x(z)\hat{x}+m_x(z)\hat{z}, m_z(z)=\cos\Theta(z)$. The z_w is the position at the centre of domain wall. Then, by considering the crystal anisotropy, shape anisotropy and exchange interactions, the energy barrier can be calculated by integral of the spacial magnetization structures. The general tendency of the calculation results have been reported (Landeros et al., 2007).

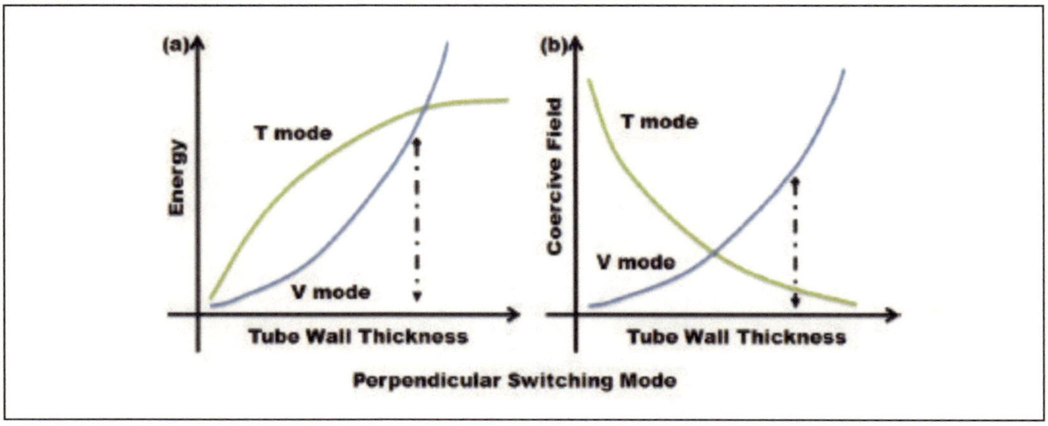

Fig. 14. (a) z-direction switching energy barrier versus tube wall thickness. (b) Coercivity field of z-direction switching versus different tube thickness.

Considering the results in Figure 14a, the relative energy differences of vortex and transverse mode suggested a transition point. The transition is taken place after vortex overcomes the transverse energy versus tube thickness or nanowire diameter. The experimental results are not far below the transition point, as the arrow indicates in the

Figure 14a, while, the Figure 14 implicates an increase of coercivity under vortex mode. But the cross over in Figure 14b does not mean the transition from vortex switching to transverse. In our experimental results presented in the tables and above discussion, we usually observe vortex-like curling mode.

In perfect ellipsoids of revolution subject to a field parallel to the long axis, magnetization reversal starts by coherent rotation or curling, although there remains a remote possibility of a buckling mode (Aharoni, 1996). The transition between the two modes depends on the radius of the ellipsoid similar to that of d_c in case of nanocylinders. Although the above explanations can generally model our experimental results, the more rigorous switching modes in case of presenting nanocylinder array properties is a little different in dynamic mechanism as will be discussed below.

5.8 Magnetostatic interactions of ferromagnetic nanocylinder array

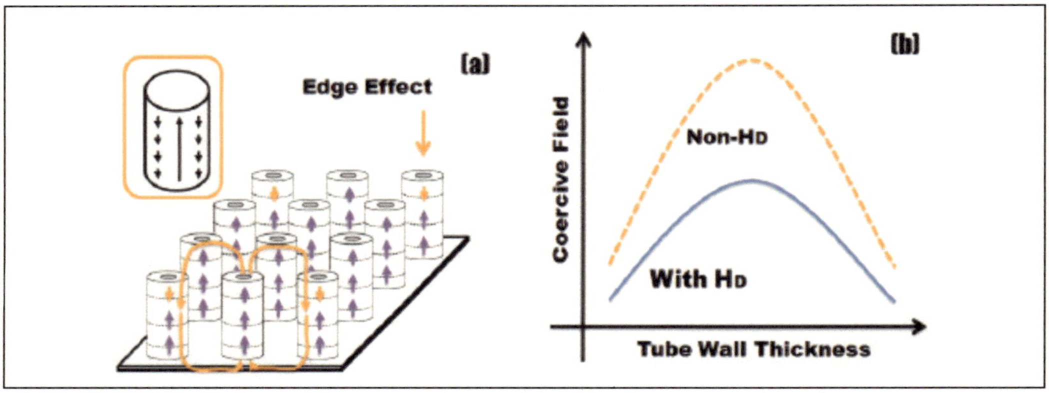

Fig. 15. (a) The model of nanocylinder array properties. The inset in figure a, is the possible demagnetized nanowire micro-spin configuration. (b) The model of diamagnetic field influenced in array of nanotube. The dash line is without magnetic interactions. The solid line is taken with interactions into concern.

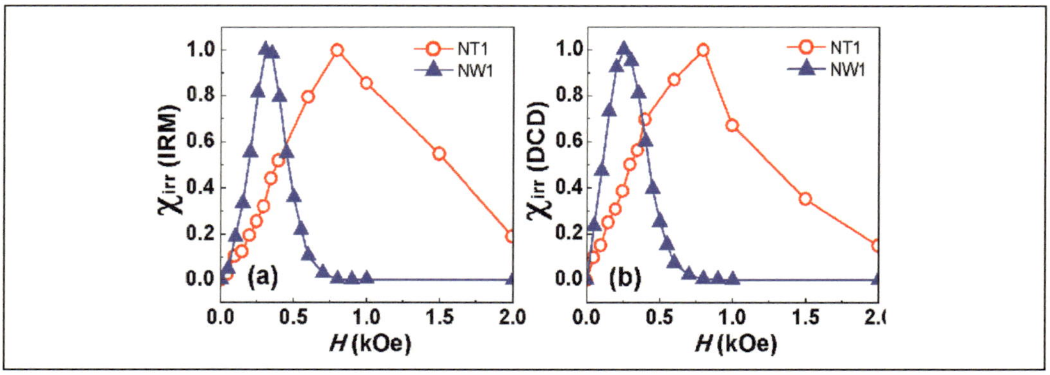

Fig. 16. Normalized $\chi_{irr} = d(M/M_s)/dH$ obtained from the derivative of (a) IRM curves (b) DCD curves, for CoCrPt NWs and NTs.

Taking the array interactions into concern, one can easily imagine that the interactions starting from one nanotube is considered as the demagnetization field influenced on the others as shown in the Figure 15a. In principle, the array interaction will make the switching coercivity different in the centre of the array from that at the corner. In the above model and discussion, only the self-demagnetization field limited in the nanocylinder is taking into concern. The direct influence in experiment is indicated by the changing of easy axis versus length. Thus, the coercivity (H_c) of a nanocylinder among the array can be separated as intrinsic, H_{int}, and stray field correction, H_{stray}. As suggested by Escrig (Escrig et al., 2007a; Escrig et al., 2007b; Escrig et al., 2008), the tendency of parallel switching coercivity field can reach its maximum due to the competition of vortex and transverse mode (shown in the Figure 15b). The stray field is proved to cause a largely reduction of the coercivity in nanocylinder array.

The comparison of experimental results for (χ_{irr}) curves of nanocylinders, which are derivatives of remanence (DCD and IRM) curves with respect to field, is given in the Figure 16a and b. The coercive point on these curves, i.e. remanent coercivity (H_r) at which remanence falls to zero is more appropriate for characterization of media rather than magnetization coercivity. The remanence curves are contributed by the moments which are unable to overcome the E_B for reversal mechanism and magnetization component recorded in such case is due only to irreversible changes (Uren et. al., 1988). A very thin tube behaves as a rolled-up thin film, in which the magnetic moments always tend to remain within the plane of the array. Hence the interaction among nanotubes could be larger than that of nanowires. As indicated by our experiment, Figure 16a and b, the field in χ_{irr} (DCD) or χ_{irr} (IRM) curve for nanotubes are almost three times larger than that of nanowires and the curve is wider for NTs. Furthermore, when tubes of large wall thicknesses approach the case of wires, the surface effects are less crucial, but interactions among diametrically opposed regions become more important.

In another point of view, we ascribe such difference between calculations and experimental results to the interaction of each tube with the stray fields produced by the array—an effective anti-ferromagnetic coupling between neighbouring tubes, which reduces the coercive field as previously demonstrated in the case of nanowires (Hertel et. al., 2001; Escrig et. al. 2008). In these interacting systems, at finite temperatures, the process of magnetization reversal can be viewed as the overcoming of a single energy barrier (E_B). In an array with all the nanotubes initially magnetized in the same direction, the magneto-static interaction between neighbouring tubes favours the magnetization reversal of some of them. A reversing field aligned opposite to the magnetization direction lowers the energy barrier, thereby increasing the probability of switching. Thermal fluctuations can allow the magnetization of a sample to surmount the E_B and switch from one stable direction to the other. A reversing field aligned in opposite direction from magnetization direction acts to lower the E_B, thereby increasing the probability of switching. The dependence of applied field on E_B is often described by the expression

$$E_B = U(1-H/H_{c0})^m \tag{9}$$

where U is E_B at zero applied field, H is applied field, and H_{c0} is the field needed to overcome barrier at zero temperature (Sun et. al., 2005; Zeng et. al., 2005). Due to many uncertain parameters, the energy barrier in nanotubes and nanowires can be modelled more realistically only by micromagnetic simulations (Zeng et. al., 2005), especially for the array properties.

6. Concluding remarks

6.1 Magnetic recording media

By touching the limitation of super paramagnetism in 1 Tbit / in^2 of the magnetic recording media, the dilemma between the grain size of the recording media and the coercivity is the main concern in this field. The media fabricated by the nano-technique, such as Electron Beam Lithography (EBL), Focus ion Beam System (FIB) and Nano-imprint, can meet the need of controllable grain size at very high cost. However, even nano-fabrication is still not stable for the grain size under and below 100 nm. While the electro-deposition offers low cost and is an easy way to fabricate nanowires and nanotubes below 100 nm. It is one of few methods that can overcome the geometrical restrictions of inserting metals into very deep nanometric recesses, making it the best method for nanowire and nanotube fabrication. By controlling the length of nanotubes and wall thickness, the vortex domain wall can be formed in the nanotube arrays, which can largely reduce the interaction between array elements. And compared to the nanowire, the array interaction can be reduced to 1/3 according to the experimental results in this work. Since the magnetic properties can be changed by geometry to further improve the recording density, the fabrication of nanocylinders is comparable with nano-rings (Figure 17) which are fabricated by nano-fabrication methods. The further dedicated adjustments of the deposition time could make the chemical deposition goes down to 40 nm or even smaller.

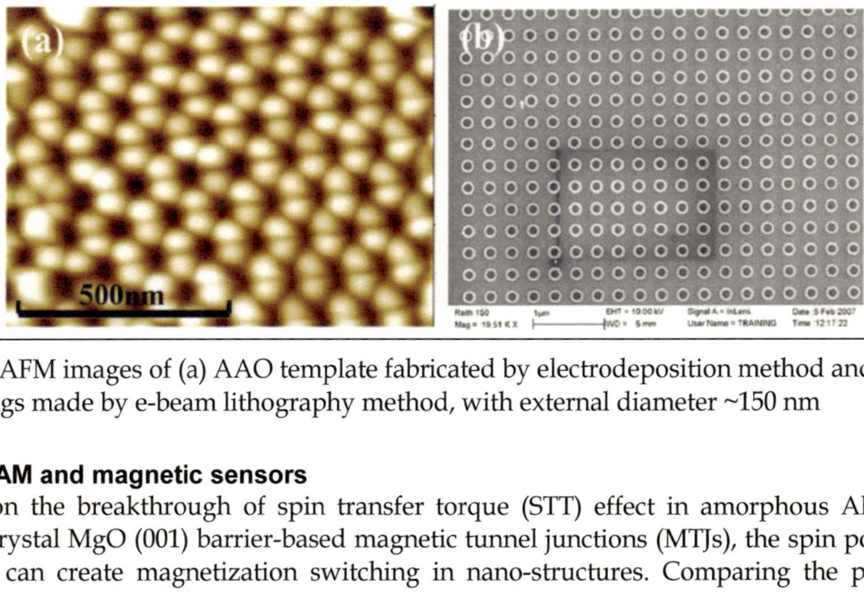

Fig. 17. AFM images of (a) AAO template fabricated by electrodeposition method and (b) nanorings made by e-beam lithography method, with external diameter ~150 nm

6.2 MRAM and magnetic sensors

Based on the breakthrough of spin transfer torque (STT) effect in amorphous Al-O and single crystal MgO (001) barrier-based magnetic tunnel junctions (MTJs), the spin polarized current can create magnetization switching in nano-structures. Comparing the previous switching properties driven by magnetic field, the information process can be much easier. To achieve such nano-phenomenon in nanocylinders is a challenging task; the preliminary results have been shown in the Figure 18. Figure 18a is Magneto-resistance (MR) curve of [Co(10nm)/Cu(15nm)]-100 multilayer nanowire array deposited in AAO template with diameter of ~ 300 nm and length ~ 60 μm at room temperature. MR ratio is obtained about 0.6%. MR-H curve of [Co(15nm)/Cu(15nm)]-240- multilayer nanowire arrays with diameter ~60 nm, length ~ 8 μm and MR ratio ~9.5% is shown in Figure 18b. However, since the effective length of multilayers was only 2.5 μm and it has Cu electrode on both sides, MR can get a considerable value ~ 12 %, if the resistance of Cu electrode can be cut off. TEM was used to characterize the morphology of multilayer nanowires, as shown in Figure 18c and d. TEM images show the clear interface between Co and Cu. They offer attractive potential to serve diverse applications, in particular, for high-density magnetic recording devices, magnetic random access memory and magnetic field sensors. However, average physical properties can only be obtained in nanowire arrays. For physical interest and technology applications, the best method is to control single nanowire and nanotube and investigate their interesting properties, like magnetic-electro transport properties and fabrication of ultra-small magnetic sensor. Different kinds of methods have been used to make electrodes for a single nanowire. Figure 18e and f show one of the methods adopted in our group to handle a single nanowire by Focus ion beam (FIB). First, large electrodes were patterned by UV-lithography, then the solution containing separated nanowires was dropped onto the substrate and then electrodes were made by using FIB. The process of nanowire transfer of a single nanowire has been shown in SEM images, Figure 18e, nanowire was picked up and Figure 18f, was transferred to the destination by micro-tip of FIB. They offer attractive potential to serve diverse applications, in particular, for high density magnetic recording devices, magnetic random access memory and magnetic field sensors.

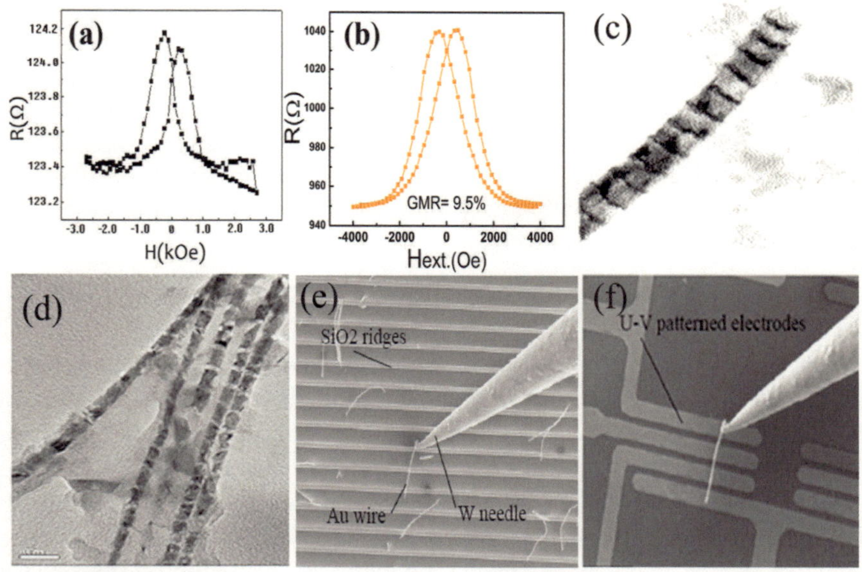

Fig. 18. (a,b) MR curves and (c,d) TEM image of of (a,c) [Co(15nm)/Cu(15nm)]-240 (b,d) [Co(10nm)/Cu(15nm)]-100 multilayer nanowires. (e,f) SEM image of nanowire transfer process using micro-tip of Focus Ion Beam (FIB).

6.3 Biomagnetics

A growing area of application for nanotubes is particularly promising in the use of nanoengineered magnetic particles to selectively manipulate and probe biological systems. Since this field of biomagnetics is growing rapidly, and there is already a broad range of applications including cell separation (Moore et al., 1998; Escrig et al., 2007), drug-delivery systems (Martin & Kohli, 2003), biosensing (Baselt et al., 1998), studies of cellular function

(Alenghat et al., 2000) as well as a variety of potential medical and therapeutic uses (Alenghat et al., 2000). Martin and co-workers have done many works on the biomedical delivery (Hillebrenneret al., 2006), such as drug-delivery systems (Martin & Kohli, 2003), molecular separation (Lee et al., 2002), single-DNA sensing (Fan et al., 2005).

The literature has been filled with nanomaterials based solutions to biomedical, bioengineering, and pharmaceutical problems. Nanowire sensors with single molecule selectivity have been engineered to detect specific chemicals, proteins, and complementary DNA, a critical function for genomics research (Ball, 2005). Researchers at Harvard University recently used iron oxide containing magnetic nanoparticles to control calcium intake in live cells, demonstrating for the first time a physical rather than chemical means of controlling cellular function (Mannix et al., 2008). Scaffolds made of biodegradable nanowires have been shown to repair brain damage and repair vision in animals, or coax neurons into forming engineered patterns (Bullis, 2006; Llinas et al., 2005; Sharma et al., 2007). Neural tissue has been coerced to live happily on computer chips and respond to electrical inputs. The pharmaceutical industry has already released new drugs based on nanotechnologies for slow release and local treatment (Foranet al., 2005; Lieberet al., 2004; Patolsky et al., 2006; Patolsky et al., 2007).

6.4 Other applications

The growth of nanowires and nanotubes will be precipitated by a burst of science and engineering break through occurring most rapidly in the last few years. Nanobatteries made from paper or nanowires are promising power necessary for nanomachines derived from solar energy, blood flow, waste vibrations, and even urine (Wang et al., 2007; Lee, 2005). A mat of nanowires with the touch and feel of paper can be an important new tool in the cleanup of oil and other organic pollutants (Yuan, 2008). Furthermore, the nanocylinders can be used as a kind of micro-wave absorber.

7. Summary

Recent trends demonstrate that interest in nanomaterials and their unique behaviour is increasing rapidly even in the presence of some doubts to the commercial applicability of nanotechnologies. Various new, exciting and potential applications require that researchers should gain better understanding and control of the nanostructure tools to implement them in applications. This article provides an overview on a variety of methods that have been developed for generating the nanostructures. It has been focused to understand the fundamentals of ferromagnetic nanowires and nanotubes, their synthesis and properties so that these could be realistically applied to new magnetic recording and electronic applications. Two systems were examined in detail through fabrication and experiment, according to the theoretical calculations, and investigations into published literature. Synthesis and properties measurements demonstrated the routes to create high quality materials as well as provided test systems for understanding basic growth mechanisms.

Ferromagnetic nanowires and nanotubes were synthesized through electrodeposition method which is an efficient method to synthesize high quality, and uniformly distributed nanocylinders in nanoporous templates. One of the main advantages of this method is that by controlling the deposition parameters, it is possible to get the desired length, diameter, tubewall thickness, as well as the morphology and size distribution. The density of the nanocylinder can be designed by modifying the template pores according to the desired application.

This article also gives a review on the structural and magnetic properties of ferromagnetic nanowire and nanotube arrays including elemental Fe, Co, Ni and their alloys such as NiFe, CoFe, CoPt, CoFeB, and CoCrPt using AAO and PC template. Geometry dependent magnetic anisotropy for nanotubes and nanowires has been discussed in details in experiment and theoretically, especially for angular dependent coercivity (H_c) and remanence squareness (SQ). With the variety of length, wall thickness and diameter, it shows different magnetic switching modes. In our nanowire and nanotube arrays, we can easily tune the direction of the easy axis from parallel to perpendicular or from perpendicular to parallel direction. It is a good way for application in perpendicular recording media in the near future and the magnetic states can also be controlled by changing length, wall thickness and diameter. Desired geometry of nanocylinders with a wide range of diameter, thickness and length can be adopted for potential applications such as quite short nanowires (nanoparticles) and nanotubes (nanorings) for patterned recording media (Albrecht et al., 2002), perpendicular STT-MRAM and nanoring MRAM (Wen et al., 2007).

8. References

Aharoni, A. (1966). *Phys. Status Solidi*, Vol. 16, p. 1

Ajayan, P.M.; Stephan, O.; Redlich, Ph. & Colliex, C. (1995), *Nature*, Vol. 375, p. 564

Albrecht, M.; Moser, A.; Rettner, C.T.; Anders, S.; Thomson, T. & Terris, B.D. (2002). *Appl. Phys. Lett.*, Vol. 80, p. 3409

Alenghat, F.J.; Fabry, B.; Tsai, K.Y.; Goldmann, W.H. & Ingber, D.E. (2000). *Biochem. Biophys. Res. Commun.*, Vol. 277, p. 93

Alivisatos, A. P. (1996). *Science*, Vol. 271, p. 933

AlMawiawi, D.; Coombs, N. & Moskovits, M. (1991). *J. Appl. Phys.*, Vol. 70, p. 4421

Ball, P. (2005). Nanowire sensors pass drugs test. *Nature Nanozone News*

Baselt, D.R.; Lee, G.L.; Natesan, M.; Metzger, S.W.; Sheehan, P.E. & Colton, R.J. (1998). *Biosens. Bioelectron.*, Vol. 13, p. 731

Bercu, B.; Enculescu, I. & Spohr, R. (2004). *Nuclear Instruments and Methods in Physics, B.*, Vol. 225, p. 497

Brus, L. (1994). *J. Phys. Chem.*, Vol. 98, p. 3575

Bullis, K. (2006). *Technology Rev.*, Vol. 109, p. 16

Burda, C.; Chen, X.; Narayanan, R. & El-Sayed, M. A. (2005). *Chem. Rev.*, Vol. 105, p. 1025

Cao, G. Z. (2004); *Nanostructures and nanomaterials: Synthesis, properties and applications*, Imperial College Press, London

Ciureanu, M.; Beron, F.; Clime, L.; Ciureanu, P.; Yelon, A.; Ovari, T.A.; Cochrane, R.W.; Normandin, F. & Veres, T. (2005). *Electrochim. Acta*, Vol. 50, p. 4487

Despic, A.; & Parkhuitik, V.P. (1989). *Modern Aspects of Electro- chemistry, Plenum*, New York Vol. 20

Dobrzynski, L. (2004). *Phys. Rev. B.*, Vol. 70, p. 193307

Du, Y.; Cai, W.L.; Mo, C.M. & Chen, J. (1999). *Appl. Phys. Lett.*, Vol. 72, p.2951

Duan, X. & Lieber, C. M. (2000). *Adv. Mater.*, Vol. 12, p. 298

Eftekhari, A. (2008). *Nanostructured Materials in Electrochemistry*, Wiley-VCH,

Enculescu, I.; & Nanomater. J. (2006). *Biostructures*, Vol. 1, p. 15

Enculescu, I.; Siwy, Z.; Dobrev, D.; Trautmann, C.; Molares, T.M.E.; Neumann, R.; Hjort, K.; Westerberg, L. & Spohr, R. (2003). *Appl. Phys. A.*, Vol. 77, p. 751

Enzel, P.; Zoller, J. J. & Bein T. (1992). *J. Chem. Soc. Chem. Commun.*, Vol. 633

Escrig, J.; Landeros, P.; Altbir, D. & Vogel, E. E. (2007). *J. Magn. Magn. Mater.*, Vol. 310, p. 2448

Escrig, J.; Landeros, P.; Altbir, D.; Vogel, E.E. & P. Vargas, (2007). *J. Magn. Magn. Mater.*, Vol. 308, p. 233

Escrig, J.; Lavin, R.; Palma, J. L.; Denardin, J. C.; Altbir, D.; Cortes, A. & Gomez, H. (2008). *Nanotechnology*, Vol. 19, p. 075713

Fan, R.; Karnik, R.; Yue, M.; Li, D.; Majumdar, A. & Yang, P. (2005). *Nano Lett.*, Vol. 5, p. 1633

Fan, S.; Chapline, M.G.; Franklin, N.R.; Tombler, T.W.; Cassell, A.M. & Dai, H. (1999). *Science*, Vol. 283, p. 512

Ferre, R.; Ounadjela, K.; George, J.; Piraux, L. & Dubois, S. (1997). *Phys. Rev. B*, Vol. 56, p. 14066

Fert, A. & Piraux, L. (1999). *J. Magn. Magn. Mater.*, Vol. 200, p. 338

Fleisher, R.L.; Price, P.B. & Walker, R.M. (1975). *Nuclear Tracks in Solids: Principles and Applications*, University of California Press, Berkeley

Foran, J.R.H.; Steinman, S.; Barash, I.; Chambers, H.G. & Lieber, R.L. (2005). *Dev. Med. Child Neur.*, Vol. 47, p. 713

Foss, C.A.; Tierney, M.J. & Martin, C.R. (1992). *J. Phys. Chem.*, Vol. 96, p.9001

Furneaux, R.C.; Rigby, W.R. & Davidson, A.P. (1989). *Nature*, Vol. 337, p. 147

Gao, T.R.; Yin, L.F.; Tian, C.S.; Lu, M.; Sang, H.; Zhou, S.M. (2006). *J. Magn. Magn. Mater.* Vol. 300, p. 471

Gasparac, R.; Kohli, P.; Trofin, M.L. & Martin, C.R. (2004). *Nano Lett.*, Vol. 4, p. 513

Guerret-Piecourt, C.; Bouar, Y.L.; Loiseau, A. & Pascard, H. (1994). *Nature*, Vol. 372, p. 761

Han, G.C.; Zong, B.Y.; Luo, P. & Wu, Y.H. (2003). *J. Appl. Phys.*, Vol. 93, p. 9202

Han, X.F.; Shamaila, S.; Sharif, R.; Chen, J.Y.; Liu, H.R. & Liu, D.P. (2009). Structural and Magnetic Properties of Various Ferromagnetic Nanotubes, *Adv. Mater.* Vol. 21, p. 1-6.

Hao, Z.; Shaoguang, Y.; Gang, N.; Liang, Y.D. & Wei, D.Y. (2001). *J. Magn. Magn. Mater.* Vol. 234, p. 454

Hertel, R. (2001). *J. Appl. Phys.*, Vol. 90, p. 5752

Hillebrenner, H.; Buyukserin, F.; Stewar J.D. & Martin, C.R. (2006). *Nanomedicine*, Vol. 1, p. 39

Hinzke, D. & Nowak, U. (2000). *J. Magn. Magn. Mater.*, Vol. 221, p. 365

Hitachi, http://www.hitachigst.com

Huczko, A. (2000). *Appl. Phys. A*, Vol. 70, p. 365

Jessensky, O.; Müller F. & Gösele, U. (1998). *Appl. Phys. Lett.*, Vol. 72, p. 1173

Knez, M.; Bittner, A.M.; Boes, F.; Wege, C.; Jeske, H. & Kern K. (2003). *Nano Lett.*, Vol. 3, p. 1079

Krans, J.M.; van Rutenbeek, J.M.; Fisun, V.V.; Yanson, I.K. & Jongh, L.J. (1995). *Nature*, Vol. 375, p. 767

Landeros, P.; Allende, S.; Escrig, J.; Salcedo, E.; Altbir, D. & Vogel, E.E. (2007). *Appl. Phys. Lett.*, Vol. 90, p. 102501

Lee, K. B. (2005). *J. Micromech. and Microengine*, Vol. 15, p. S210

Lee, S.B.; Mitchell, D.T.; Trofin, L.; Nevanen, T.K.; Söderlund, H. & Martin, C.R., (2002). *Science*, Vol. 296, p. 2198

Li, A. P.; Müller F. & Gösele, U. (2000). *Electrochem. Solid-State Lett.*, Vol. 3, p. 131

Li, D.D.; Jiang, C.H.; Ren, X.; Long, M. & Jiang, J.H. (2008). *Mat. Lett.*, Vol. 62, p. 3228

Li, J.; Papadopoulos, C. & Xu, J. M. (1999). *Appl. Phys. Lett.* Vol. 75, p. 367

Lieber, R.L.; Steinman, S.; Barash, I.A. & Chambers, H. (2004). *Muscle and Nerve.*, Vol. 29, p. 615

Liu, H.R.; Shamaila, S.; Chen, J.Y.; Sharif, R.; Lu, Q. F; & Han, X.F. (2009). Magnetization Reversal Mechanism for CoFeB Ferromagnetic Nanotube Arrays. *Chin. Phys. Lett.*, Vol. 26, No. 7, pp. 077503.

Llinas, R.R.; Walton, K.D.; Nakao, M.; Hunter, I. & Anquetil, P.A. (2005). *J. Nanoparticle Research*, Vol. 7, p. 111 Vol.

Lundstrom, M. (2003). *Science*, Vol. 299, p. 210

Mallet, J.; Yu-Zhang, K.; Chien, C.L.; Eagleton, T.S. & Searson, P.C. (2004). *Appl. Phys. Lett.* Vol. 84, p. 3900

Mannix, R.J.; Kumar, S.; Cassiola, F.; Montoya-Zavala, M.; Feinstein, E.; Prentiss, M. & Ingber, D.E. (2008). *Nature Nanotech.*, Vol. 3, p. 36

Martin, C. R. & Kohli, P. (2003). *Nat. Rev. Drug Discovery.*, Vol. 2, p. 29

Martin, C.R.; (1994). *Science*, Vol. 266, p. 1961

Moore, L.R.; Zborowski, M.; Sun, L. & Chalmers, J.J. (1998). *J. Biochem. Biophys. Methods.*, Vol. 37, p. 11

Nielsch, K.; Wehrspohn, R.B.; Fischer, S.F.; Kronmiller, H.; Kirsehner J. & Gosele, U. (2001) *Mater. Res. Soc. Symp. Proc.*, Vol. 9, p. 636

Ozin, G.A. (1992). *Adv. Mater.* 4, 612

Patolsky, F.; Timko, B.P.; Yu, G.H.; Fang, Y.; Greytak, A.B.; Zheng, G.F. & Lieber, C.M. (2006). *Science*, Vol. 313, p. 1100

Patolsky, F.; Timko, B.P.; Zheng, G.F. & Lieber, C.M. (2007). *MRS Bulletin*, Vol. 32, p. 142

Possin, G.E. (1970). *Rev. Sci. Instrum.*, Vol. 41, p. 772

Possin, G.E. (1971). *Physica*, Vol. 55, p. 339

Rivas, J.; Bantu, A.K.M.; Zaragoza, G.; Blanco, M.C.; Lo'pez-Quintela M.A. (2002). *J. Magn. Magn. Mater.*, Vol. 249, p. 220

Seberino, C. & Bertram, H. (1999). *J. Appl. Phys.*, Vol. 85, p. 5543

Shamaila, S.; Liu, D.P.; Sharif, R.; Chen, J.Y.; Liu, H.R. & Han, X.F. (2009a). Electrochemical fabrication and magnetization properties of CoCrPt nanowires and nanotubes *Appl. Phys. Lett.* Vol. 94, pp. 203101_1-3

Shamaila, S.; Sharif, R.; Chen, J.Y.; Liu, H.R. & Han, X.F. (2009b). Magnetic Field Annealing Dependent Magnetic properties of Co1-xPtx Nanowire Arrays. *J. Magn. Magn. Mater.* Vol. 321, pp. 3984-3989, ISSN: 0304-8853

Shamaila, S.; Sharif, R.; Riaz, S.; Khaleeq-ur-Rahman, M. & Han, X.F. (2008a). Fabrication and magnetic characterization of Co_xPt_{1-x} nanowire Arrays. *Appl. Phys. A.*, Vol. 92, pp. 687-691, ISSN: 0947-8396 (Print) 1432-0630 (Online)

Shamaila, S.; Sharif, R.; Riaz, S.; Ma, M.; Khaleeq-ur-Rahman, M. & Han, X. F. (2008b). Magnetic and magnetization properties of electrodeposited fcc CoPt nanowire arrays. *J. Magn. Magn. Mater.*, Vol. 320, pp. 1803-1809, ISSN: 0304-8853

Sharif, R.; Shamaila, S.; Ma, M.; Yao, L.D.; Yu, R.C.; Han, X.F. & Khaleeq-ur-Rahman, M. (2008a). Magnetic switching of ferromagnetic nanotubes. *Appl. Phys. Lett.,* Vol. 92, pp. 032505_1-3.

Sharif, R.; Shamaila, S.; Ma, M.; Yao, L.D.; Yu, R.C.; Han, X.F.; Wang, Y. & Khaleeq-ur-Rahman M. (2008b). Magnetic and microstructural characterizations of CoFe and CoFeB nanowires. *J. Magn. Magn. Mater.* 320, 1512, ISSN: 0304-8853

Sharif, R.; Zhang, X.Q.; Shahzadi, S.; Jiang, L.X.; Han, X.F. & Kim, Y.K. (2006). Effect of magnetic field annealing upon Co-rich nanowires. *IEEE Trans. Magn.*, Vol. 42, pp. 2778-2780

Sharif, R.; Zhang, X.Q.; Shamaila, S.; Riaz, S.; Jiang, L.X. & Han X.F. (2007), Magnetic and magnetization properties of CoFeB nanowires. *J. Magn. Magn. Mater.* Vol. 310, pp.e830-e832, ISSN: 0304-8853

Sharma, H.S.; Ali, S.F.; Dong, W.; Tian, Z.R.; Patnaik, R.; Patnaik, S.; Sharma, A.; Boman, A.; Lek, P.; Seifert, E. & Lundstedt, T. (2007). *8th Int. Neuroprotection Soc. Meet.* Vol. 1122, p. 197

Sima, M.; Enculescu, I.; Trautmann, C. & Neumann, R. (2004) *Adv. Mater.*, Vol. 6, p. 121

Sima, M.; Enculescu, I.; Visan, T.; Spohr, R.; Trautmann, C. (2004). *Molecular Crystals and Liquid Crystals,* Vol. 418, No. 21, p. 749

Singaraju, P.; Venkat, R.; Kanakala, R. & Das, B. (2006). *Eur. Phys. J. Appl. Phys.*, Vol. 35, p. 107

Skomski, R.; Zeng, H.; Zheng, M. &Sellmyer, D.J. (2000). *Phys. Rev. B,* Vol. 62, p. 3900

Sui, Y.C.; Skomski, R.; Sorge, K.D. & Sellmyer, D.J. (2004). *Appl. Phys. Lett.*, Vol. 84, p. 1525

Sullivan, J.P.O'. & Wood, G.C. (1970). *Proc. R. Soc, London,* Vol. 317, p. 511

Sun, L.; Hao, Y.; Chien, C.L. & Searson, P.C. (2005). *IBM J. Res. and Dev.*, Vol. 49, p. 79

Sun, S.; Murray, C. B.; Weller, D.; Folks, L. & Moser A. (2000), *Science,* Vol. 287, p. 1989

Thurn-Albrecht, T. et. al., (2000). *Science,* Vol. 290, p. 2126

Toimil-Molares, M.E.; Chtanko, N.; Cornelius, T.W.; Dobrev, D.; Enculescu, I.; Blick, R.H. & Neumann, R. (2004). *Nanotechnology,* Vol. 15, p. S 201

Tonucci, R.J.; Justus, B.L.; Campillo, A.J. & Ford, C.E. (1992). *Science,* Vol. 258, p. 783

Uren, S.; Walker, M.; O'Grady, K. & Chantrell, R. W. (1988). IEEE Trans. Magn. Vol. 24, p. 1808.

Wang, X.D.; Song, J.H.; Liu, J. & Wang, Z.L. (2007). *Science,* Vol. 316, p. 102

Wen, Z.C.; Wei, H.X. & Han, X.F. (2007). Patterned nanoring magnetic tunnel junctions. *Appl. Phys. Lett.*, Vol. 91, pp. 122511_1-3

Wernsdorfer, W.; Orozco, E.B.; Hasselbach, K.; Benoit, A.; Bar-bara,B.; Demoncy, N.; Loiseau, A.; Pascard, H. & Mailly, D. (1997). *Phys. Rev. Lett.*, Vol. 78, p. 1791

Wirth, S.; Molnar, S.V.; Field, M. & Awschalom, D.D. (1999). *J. Appl. Phys.,* Vol. 85, p. 5249

Wu, C.G. & Bein, T. (1994), *Science,* Vol. 264, p. 1757

Xia, Y. N.; Yang, P.; Sun, Y.; Wu, Y.; Mayers, B. & Gates, B. (2003). *Adv. Mater.,* Vol. 15, p. 353

Yang, P. (2003). *Nature,* Vol. 425, p. 243

Yuan, J. (2008). *Nature Nanotechnology,* Vol. 3, p. 332

Zach, M. P.; Ng, K.H. & Penner, R.M. (2000). *Science,* Vol. 290, p. 2120

Zeng, H.; Skomski, R.; Menon, L.; Liu, Y.; Bandyopadhyay, S. & Sellmyer, D.J. (2002). *Phys. Rev. B,* Vol. 65, p. 134426

7

Extracting Individual Properties from Global Behaviour: First-order Reversal Curve Method Applied to Magnetic Nanowire Arrays

Fanny Béron[1], Louis-Philippe Carignan[2], David Ménard[2] and Arthur Yelon[2]
[1]Universidade Estadual de Campinas (UNICAMP),
[2]École Polytechnique de Montréal
[1]Brazil
[2]Canada

1. Introduction

The behaviour of a group cannot be simply reduced to the sum of the behaviour of individuals (Le Bon, 1895). Examples abound in nature: school of fish or swarms of insect act like a single individual, but with emergent properties, i.e. properties which none of the individuals in the group possess, but which appear when these individuals are grouped into a system. A key element characterising a crowd is that the individuals should know that they are part of an ensemble (Le Bon, 1895). In other words, there must be interactions between them, on a short or long range. Otherwise they only constitute a system of several isolated individuals without emergent properties. Whatever the system, it is usually much easier to access its global behaviour, compared to the behaviour of individuals and their method of communication. However, the understanding of the global behaviour, in order to be able to control it subsequently, necessarily requires the knowledge of individual behaviour and of their interactions.

According to these criteria, the magnetic behaviour of structures composed of magnetic entities embedded in a non-magnetic matrix must be considered as a crowd: entities "talking" to each other via exchange or dipolar interactions depending upon the distance that separates them. Therefore, the behaviour of the global (or collective) structure may differ from that of the local (or individual) entities. This difference will be accentuated in the presence of non-uniformity (geometric, structural, etc.) of the entities. To access the magnetostatic properties of a system (coercivity, remanence, interactions, etc.), one usually uses a magnetometer to measure the major hysteresis curve. It is possible to use the same experimental setup to obtain the local magnetostatic properties of the system, by measuring multiple minor hysteresis curves, called first-order reversal curves (FORC) (Mayergoyz, 1985).

This technique can be particularly efficient and powerful in the case of highly interacting systems, like ferromagnetic nanowire arrays (Fig. 1). The large dipolar interaction field between the wires, associated with small interwire distance, strongly affects the overall array behaviour. Array properties (high anisotropy, resonance frequency in the gigahertz, etc.) make them promising candidates for high frequency devices (Saib et al., 2005), high-density magnetic memories (Almawlawi et al., 1991; Ross, 2001) and sensors (Lindeberg &

Hjort, 2003). Moreover, with a multilayer structure (i.e. alternation of magnetic and non-magnetic nanodiscs), controlling the ratio between the magnetic and non-magnetic nanodiscs provides easy control of the effective anisotropy of the array (Tang et al., 2007; Carignan et al., 2007). This possibility makes the multilayer nanowire array particularly interesting for high frequency devices (Ye et al., 2007), especially circulators (Saib et al., 2001), although the initial goal of using multilayer nanowires was to obtain giant magnetoresistance devices (Blondel et al., 1994, Piraux et al., 1994; Piraux et al., 2007).

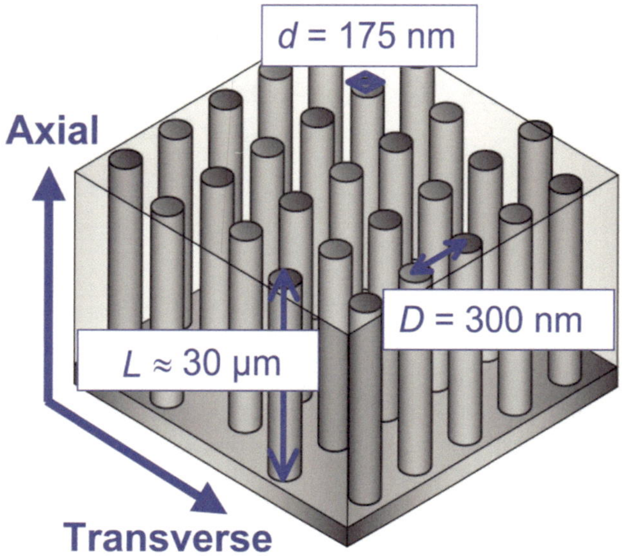

Fig. 1. Schematic of a ferromagnetic nanowire array. The geometric dimensions (diameter *d*, interwire distance *D* and length *L*) are indicated, as are the two principal directions (axial and transverse) for the application of an external magnetic field.

Despite the advantages of the FORC method for nanowire array characterisation, few studies exist on the subject (Spinu et al., 2004; Béron et al., 2006; Lavín et al., 2008; Peixoto & Cornejo, 2008; Béron et al., 2008b). This is mainly due to the lack of interpretative models suitable for highly interacting systems. This confines the analysis of the FORC result, which graphically represents the magnetisation reversal, to qualitative conclusions. This problem motivated the elaboration of the physical analysis model (Béron et al., 2008a), which is based on physically meaningful hypotheses. It is therefore possible to extract the intensity and spatial distribution of the interaction field at saturation, the coercivity distribution of individual nanowires (Béron et al., 2008b) and a quantitative evaluation of the overall magnetic anisotropy (Béron et al., 2007). In addition, insights about the magnetisation reversal mechanism can also be obtained. Here, we demonstrate the utility of FORC characterisation of nanowire arrays by showing how to extract these local properties from a global measurement. The versatility of the FORC method is demonstrated by presenting experimental examples of nanowire arrays with different saturation magnetisations (Ni = 490 emu/cm^3, $Co_{94}Fe_5B_1$ = 1500 emu/cm^3), structures (uniform or multilayer) and applied field directions (parallel or perpendicular to the nanowire axis).

For systems such as arrays of nanoparticles with high dispersion of such properties as particle size, spacing and geometry, it may be more difficult to analyse the effects of these parameters, and to develop the kind of physical models described here. For systems such as sheets of transformer steel, it may even be difficult to identify physically meaningful objects whose properties and interactions may be modelled. However, we believe that the FORC method is sufficiently powerful, and the physical modelling approach presented here is sufficiently promising, that effort in this direction is likely to be rewarded.

2. First-order reversal curve (FORC) method

Application of the FORC method requires that the system exhibits hysteretic behaviour. A phenomenon is called hysteretic when, for a given value of input, there exist several possible output values. The principal consequence is the division into two paths between the saturation points, depending on the direction of variation of the input. The set of the two branches is called hysteresis cycle or major hysteresis curve, while the term hysteresis area refers to the interval between the saturation points, i.e. where the two path are not superposed (Fig. 2). The remanence represents the state for a null input, while the coercivity is the input value needed for the system to give a null output. Finally, the slope of the hysteresis curve is called susceptibility (χ). The easy axis of a system refers to the easiest direction to magnetise, for which the energy is a minimum in the absence of an input. This implies the presence of some anisotropy in the system (shape, crystalline, surface, etc.)
From a mathematical point of view, a hysteretic phenomenon can be described as a transducer for which the input-output relationship has multiple branches, a branch transition occurring for each applied extreme input value. A hysteresis mathematical model thus needs to keep in memory the previous extreme applied input values (Mayergoyz, 1985).

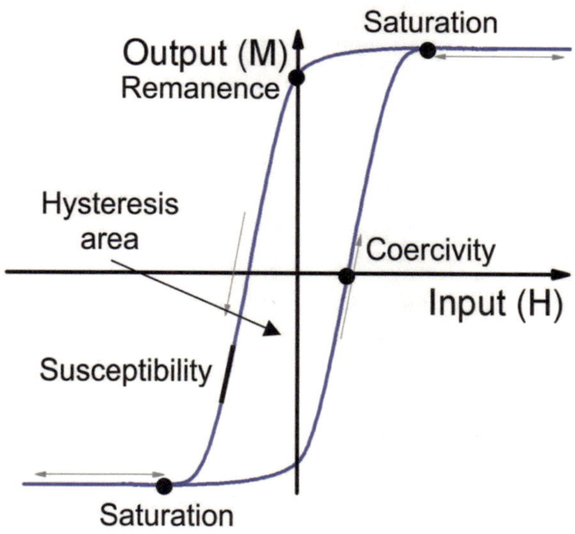

Fig. 2. Characteristics of a hysteresis curve.

The nature of the physical phenomena exhibiting a hysteresis is exceedingly diverse: magnetic, mechanical, optical, etc. However, because of its mathematical nature, the following model and the conclusions derived from it remain valid regardless of the physical phenomenon. One just needs to adjust the input and output variables with the correct parameters, for example, the applied magnetic field and the magnetisation in case of magnetic materials, or the applied tension and the deformation generated for a plastic material.

2.1 Measurement and calculations

The FORC method was based, at its origin, on the classical Preisach model (Preisach, 1938), a mathematical model where hysteresis can be modelled as a set of elementary processes, or operators, called hysterons and characterized by two parameters, H_c and H_u (Fig. 3). It is important to note that these hysterons do not necessarily have a physical significance, and so, will subsequently be called "mathematical hysterons" (Mayergoyz, 1985). Indeed, when applying the FORC method to a physical system, one does not need to use the classical Preisach model to interpret the result. Several different approaches are possible, either based on physical hypothesis (ex. physical analysis model, Béron et al., 2008a) or on mathematical distributions (ex. moving Preisach model, Della Torre, 1966).

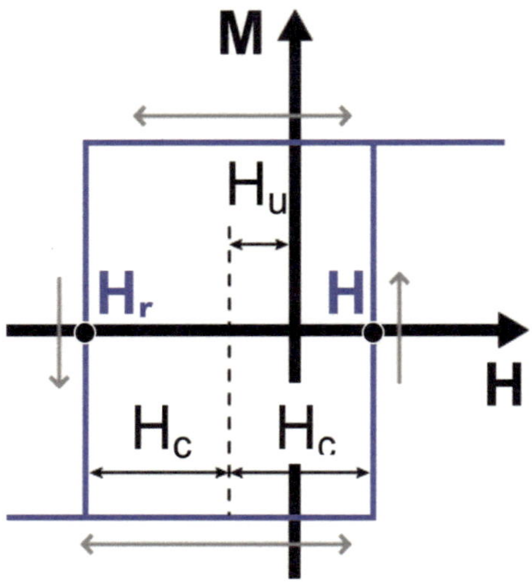

Fig. 3. Mathematical hysteron from the classical Preisach model. The magnetisation switches down, abruptly, at $H_r = -(H_c + H_u)$, and switches back up for a certain $H = (H_c - H_u)$ value. It is completely described by the H_c and H_u parameters, which respectively represent the coercivity and the bias field of the mathematical hysteron.

The goal of any FORC measurements is to retrieve the H_c and H_u parameters of each mathematical hysteron of the system. To achieve this, the system is first positively saturated, in order to put all the magnetisation of the hysterons in the "up" position. The input variable (in our case, the external applied field H) is then lowered until a point called reversal field (H_r), which switches "down" the magnetisation of some hysterons, depending upon their H_c and H_u parameters. Then, the field H is increased again and the magnetisation M is measured. The difference of magnetisation between the applied field and the reversal field is directly proportional to the amount of hysterons that switched back "up". This kind of minor hysteresis curve is called first-order reversal curve. The information about all hysterons in the system, the FORC distribution ρ_{FORC}, can be obtained by generalising the process, i.e. applying a second-order mixed derivative of M on a set of FORCs beginning at different H_r (Fig. 4) (Mayergoyz, 1985):

$$\rho_{FORC}(H,H_r) = -\frac{1}{2}\frac{\partial^2 M(H,H_r)}{\partial H \partial H_r} \quad (H > H_r) \tag{1}$$

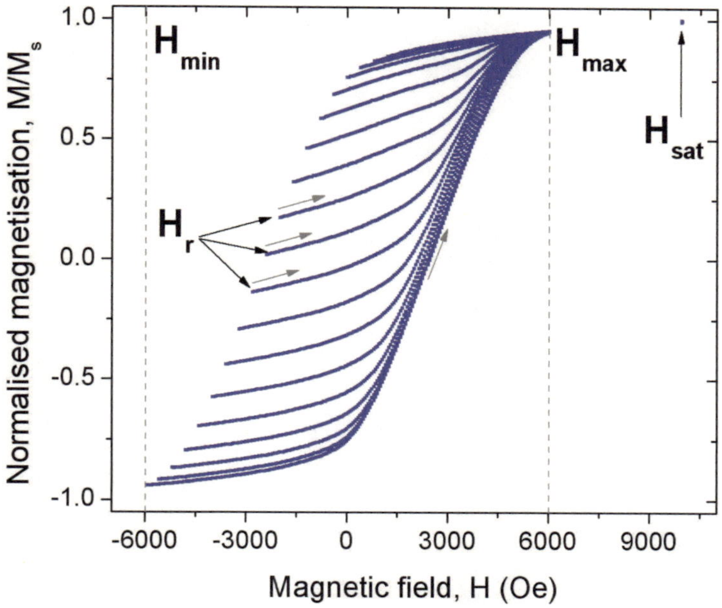

Fig. 4. Set of experimental first-order reversal curves. For practical reasons, only one curve out of four is shown. (CoFe nanowire array, d = 15 nm, D = 55 nm, L = 1.6 μm, axial direction, ΔH = 50 Oe, ΔH_r = 100 Oe)

In order to obtain the complete representation of the magnetisation reversal, the reversal field values cover the hysteretic area of the major hysteresis curve, from H_{min} to H_{max}. To accelerate the acquisition time, no measurement is taken between H_{max} and the saturation field. Also, experimentally, one should introduce a pause in the measurement at the saturation and reversal fields, long enough in order to avoid any magnetic viscosity effect. For metallic nanowires, 5 to 10 seconds proved to be sufficient (Béron, 2008). The choice of the experimental field steps ΔH (between consecutive data on the FORC) and ΔH_r (between different FORC) is the result of a trade-off between the FORC result precision and the measuring time. In order to optimise the experiment, these field steps can be varied according to the precision needed in different regions, but the $\Delta H_r/\Delta H$ ratio should be kept between 1 (high hysteresis susceptibility) and 2.5 (low hysteresis susceptibility) (Béron et al., 2006).

Before calculating equation (1), the FORCs are extrapolated in a way which minimises the discontinuity at $H = H_r$, created by the lack of data taken in the $H < H_r$ region (Béron et al., 2007). The FORC distribution then gives only and completely the information related to the irreversible processes (mathematical hysterons with $H_c \neq 0$). The characterisation of the reversible processes can be done through the calculation of a reversibility indicator η (Fig. 5) (Béron et al., 2007):

$$\eta(H = H_r) = \frac{\chi_{FORC}(H = H_r)}{\chi_{Hyst}(H = H_r)} \tag{2}$$

A value of η = 0 corresponds to a completely irreversible process and of η = 1 to a fully reversible process.

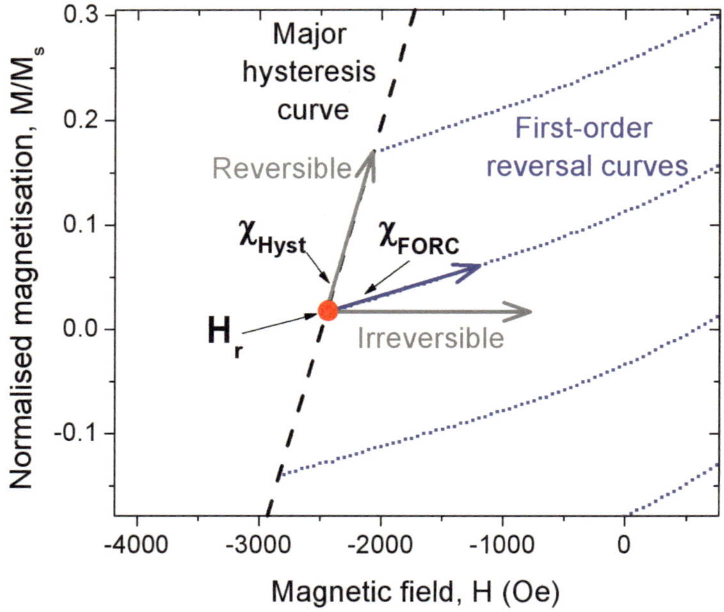

Fig. 5. Calculation of the reversibility indicator η at a given reversal field H_r. (Zoom of Fig. 4)

2.2 Analysis of results

Information about both reversible and irreversible processes can be combined in a graphical representation called "FORC result" (Fig. 6) (Béron, 2008). It consists of a contour plot of the FORC distribution ρ_{FORC}, with a scale going from blue (minimum ρ_{FORC}) to red (maximum ρ_{FORC}), which directly indicates the statistical distribution of the mathematical hysterons. In order to read the H_c and H_u mathematical hysteron parameters directly from the graph, it is convenient to execute a change of coordinates, to define a coercive field axis (H_c) and an interaction field axis (H_u) (see Fig. 3):

$$H_c = \frac{H - H_r}{2} \qquad H_u = -\frac{H + H_r}{2} \tag{3}$$

In the classical Preisach model approach, the H_c and H_u cross-sections of the FORC distribution directly yield the coercivity and interaction field distributions. Even if this remains valid only for systems that respect the classical Preisach model requirements (which is not the case for most physical systems), the coordinate change remains appropriate; the FORC distribution generally spreads along the H_c and H_u axes. The reversibility indicator η is added to the FORC result as a greyscale strip, ranging from white (fully reversible behaviour) to black (fully irreversible behaviour). Two quantitative parameters can be defined on the FORC result: the H_c position of the FORC distribution maximum is called FORC coercivity (H_c^{FORC}), while its half-width along the H_u axis is called global interaction field (ΔH_u) (Fig. 6) (Béron et al., 2008b). They will later be used for the FORC analysis along with the physical analysis model.

The FORC result analysis is generally the most difficult step in the use of the FORC method. The three major problems are 1) the FORC distribution represents the statistical distribution

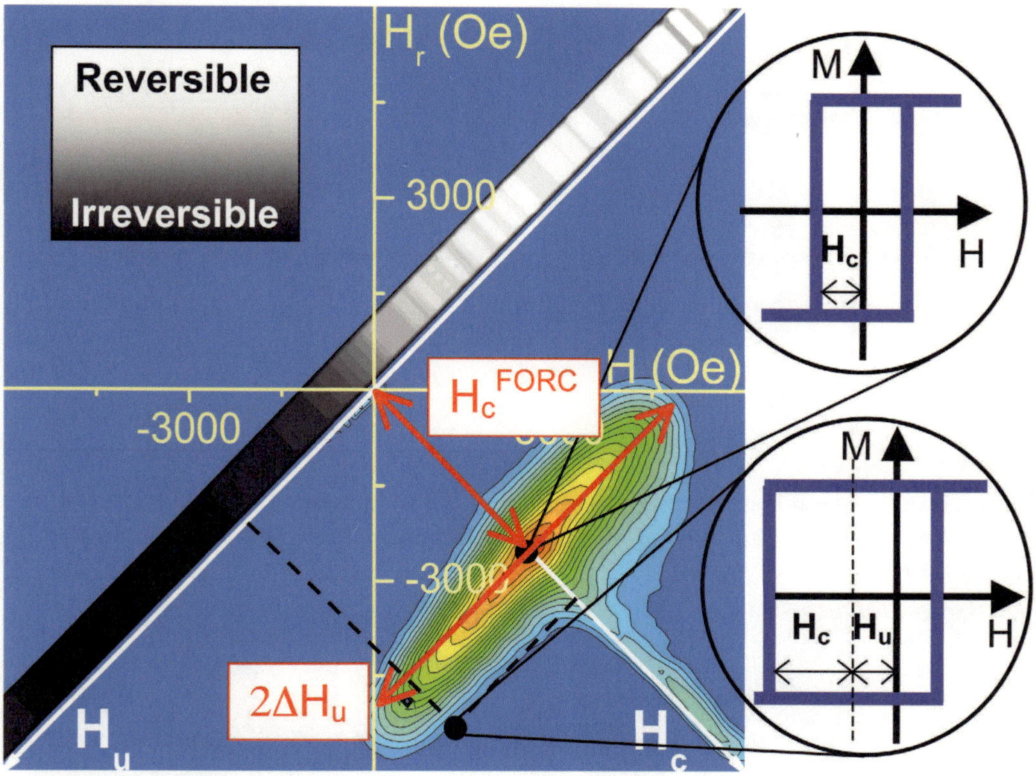

Fig. 6. FORC result (calculated from the set of FORCs shown in figure 4). Two mathematical hysterons are represented, illustrating the use of the H_c and H_u axes to read the coercivity and bias field of the FORC distribution. The parameters H_c^{FORC} and ΔH_u are also indicated.

of entities that do not necessarily have a physical meaning (the mathematical hysterons), 2) the FORC distribution can be deformed by the presence of state-dependent interaction field during the reversal, such as a mean interaction field, and 3) different physical systems can exhibit identical FORC results. Several approaches to FORC interpretation, designed to overcome these problems, exist. One is called the physical analysis model (Béron et al., 2008a), which allows us to establish some quantitative relationships between the characteristics of a physical system and of the FORC result. It is achieved by using physically meaningful hysterons, representing the supposed magnetic behaviour of the system entities, coupled with an interaction term, to simulate the global behaviour of the system. For example, the behaviour of each nanowire can theoretically be represented by the coherent rotation model (Stoner & Wohlfarth, 1948). The nanowire is approximated as an infinite cylinder where all the spins remain parallel to each other during the magnetisation reversal. This defines two different hysterons, depending upon the applied field direction, that can be used in the physical analysis model (Fig. 7).

The FORC behaviour strongly depends upon the type of hysterons. For example, for easy axis hysterons, a coercivity distribution will spread the FORC distribution along the H_c axis (Fig. 8a). With an interaction field (and no coercivity distribution), the result depends upon its direction relative to the saturation magnetisation. An antiparallel interaction field (opposite to the magnetisation) will elongate the FORC distribution along the H_u axis (Fig. 8b), while a

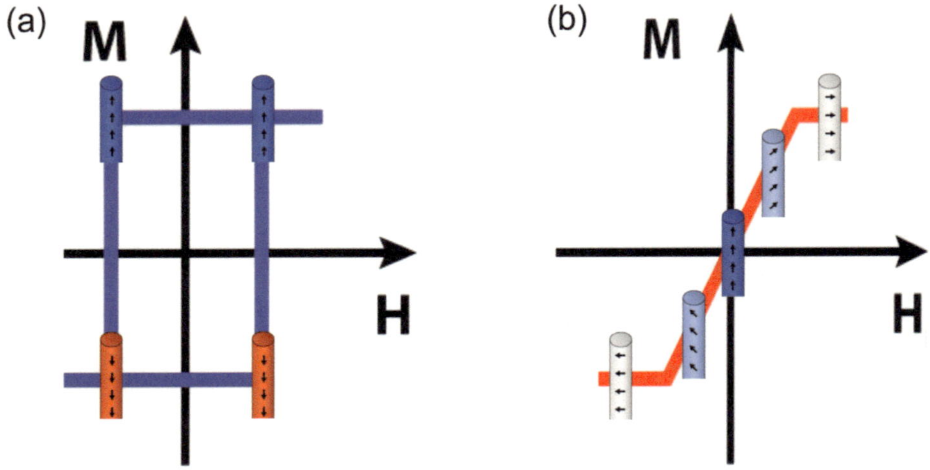

Fig. 7. Hysterons based on the theoretical behaviour of an individual nanowire, depending upon the applied field direction. (a) Axial applied field: easy axis hysteron (b) Transverse applied field: hard axis hysteron

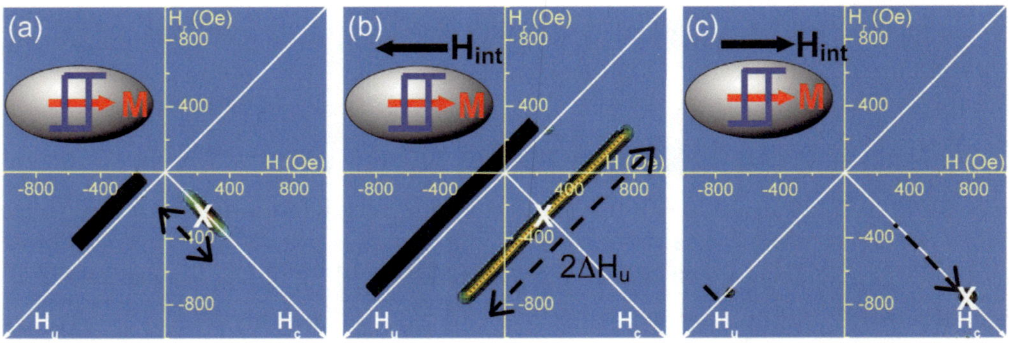

Fig. 8. Simulated FORC results with easy axis hysterons. The H_c^{FORC} parameter is indicated as a white X. (a) Normal coercivity distribution, no interaction field (b) Antiparallel mean interaction field, no coercivity distribution (c) Parallel mean interaction field, no coercivity distribution

parallel interaction field will have as consequence to translate it toward higher values on the H_c axis (Fig. 8c) (Béron et al., 2008a). These simulations showed that the shape of the FORC distribution, if clear and intuitive for a simple system (only one characteristic, i.e. fully reversible or irreversible, coercivity distribution or mean interaction field), quickly becomes more complex and far from the direct interpretation based on the mathematical hysterons.

From a quantitative point of view, the average individual coercivity of easy hysterons can be extracted from the FORC result by taking the H_c coordinate at $H_u = 0$ of the maximum of the FORC distribution, i.e. the H_c^{FORC} parameter, at $H_u = 0$. This remains valid for systems with coercivity distribution (no interaction field, Fig. 8a) or with antiparallel interation field (no coercivity distribution, Fig. 8b). In the limit of total absence of both coercivity and interaction field distributions, the half-width of the FORC distribution elongation (ΔH_u) is equal to the absolute value of the coefficient parameter k ($k < 0$) of a mean interaction field, which yields the value of interaction field at saturation ($M = M_s$) (Fig. 8b):

$$H_{\text{int}} = k\frac{M}{M_s} \tag{4}$$

Finally, in the presence of a mean parallel interaction field ($k > 0$) and no coercivity distribution, the H_c^{FORC} value is the sum of the individual coercivity and the coefficient k. (Fig. 8c) (Béron et al., 2008a).

3. Fabrication of nanowire arrays

Nanowire arrays may be fabricated by electrodeposition of metallic ions into the cylindrical pores of a dielectric template (Masuda & Fukuda, 1995). This deposition technique is economic, fast, versatile and does not require high vacuum facilities. A three electrodes configuration is typically used, the sample being used as the working electrode while the counter electrode is in platinum (Pt) (Fig. 9). The applied potential is given with respect to that of the reference electrode, usually a saturated calomel, because of the stability of its potential.

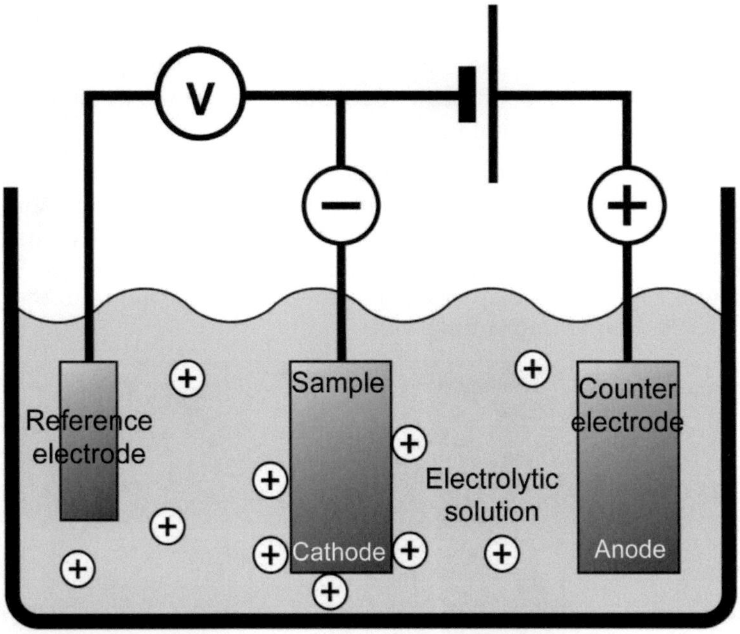

Fig. 9. Schematic of an electrolytic cell used in the electrodeposition of the ferromagnetic nanowires.

The nanoporous dielectric template is typically made of anodised aluminium (alumina) (O'Sullivan & Wood, 1970), ion bombarded polycarbonate (Ferain & Legras, 2003) or diblock copolymer (Thurn-Albrecht et al., 2000). It is important that the pores be of constant diameter, parallel to each other, and perpendicular to the template surface. The most widely used template is currently alumina, mainly due to the ease of tailoring the template geometry (pore diameter, length and interpore distance) in laboratory, to the hexagonal arrangement of the pores, and to the rigidity of the template. They are fabricated in a two-step anodisation process (Fig. 10) (Masuda & Fukuda, 1995; Pirota et al., 2004; Zhao et al.,

2007). A cleaned aluminium substrate is first slowly anodised in an acidic bath under constant voltage. Since the pore order is getting more regular during this process, the initial layer of alumina (aluminium oxide) is chemically etched, before a second anodisation is performed, under the same conditions as the first. Finally, a chemical etch can enlarge the pores, if needed.

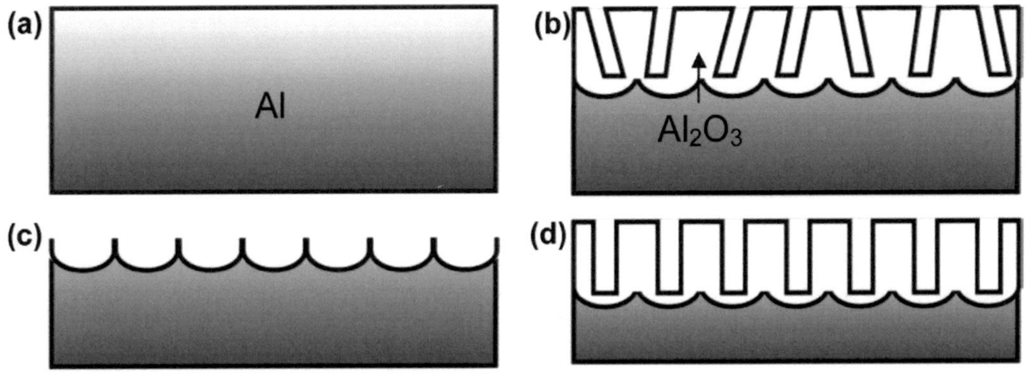

Fig. 10. Schematic of the two-step anodisation process for fabricating nanoporous anodised aluminium templates. (a) Electropolished Al substrate (b) First anodisation (b) Etching of the alumina layer (d) Second anodisation

The samples discussed here were all fabricated in commercially available alumina templates (thickness = 60 μm, pore diameter = 175 nm, interpore distance = 300 nm, pore density = 10^9 pores/cm^2, figure 11a). However, due to the filtration purpose of these templates, the pores split into multiple smaller pores in the first 500 to 700 nm (Fig. 11b). This region was therefore removed by mechanical polishing before magnetic characterisation, in order to make the magnetic behaviour more uniform.

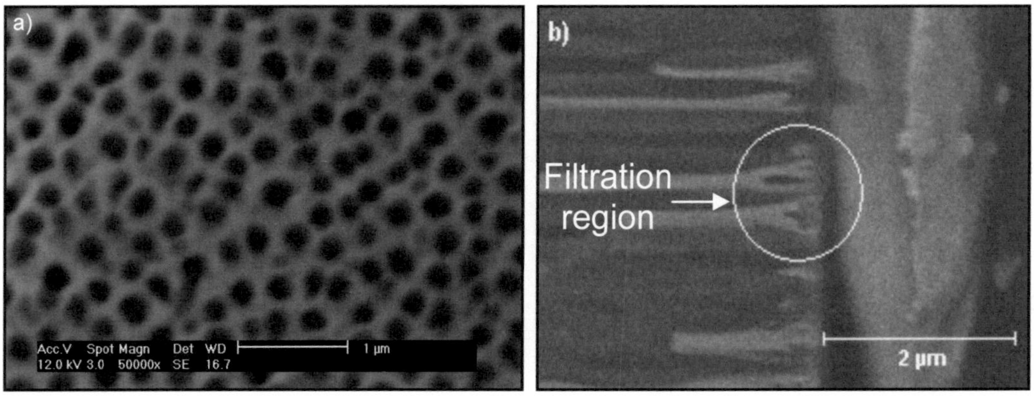

Fig. 11. Commercial alumina template (Anodisc™ 0.02 μm) (a) Top view (b) Cross-sectional view of the pore bottom, showing the filtration region, removed before magnetic characterisation.

The pulse-current electrodeposition technique was used for the fabrication of the uniform nanowires. This technique consists of sending short current pulses (8 ms), during which the deposition occurs, followed by a pause (152 ms) (Ciureanu et al., 2005) (Fig. 12a). It avoids the formation of composition gradients along the nanowires (Nielsh et al., 2000), prevents formation of hydrogen in the pores and therefore allows uniform filling of the pores. Neither composition (Ni and CoFeB) exhibited large magnetocrystalline anisotropy, since the CoFeB was amorphous and the magnetocrystalline anisotropy of Ni is low. Multilayer nanowires can also be fabricated in an electrolytic bath containing all the elements to be deposited (Blondel et al., 1994; Piraux et al., 1994). In this case, the potential is alternated between two values, where the deposition of each material is favoured for one value of potential (Alper et al., 1993) (Fig. 12b). The non-magnetic metal is kept dilute in the solution to avoid excessive inclusion in the magnetic layer. Ni/Cu nanowires were fabricated by this method (-1 V for Ni, -0.56 V for Cu), varying the Cu/Ni ratio from 0.3 to 1.17 (Carignan et al., 2007).

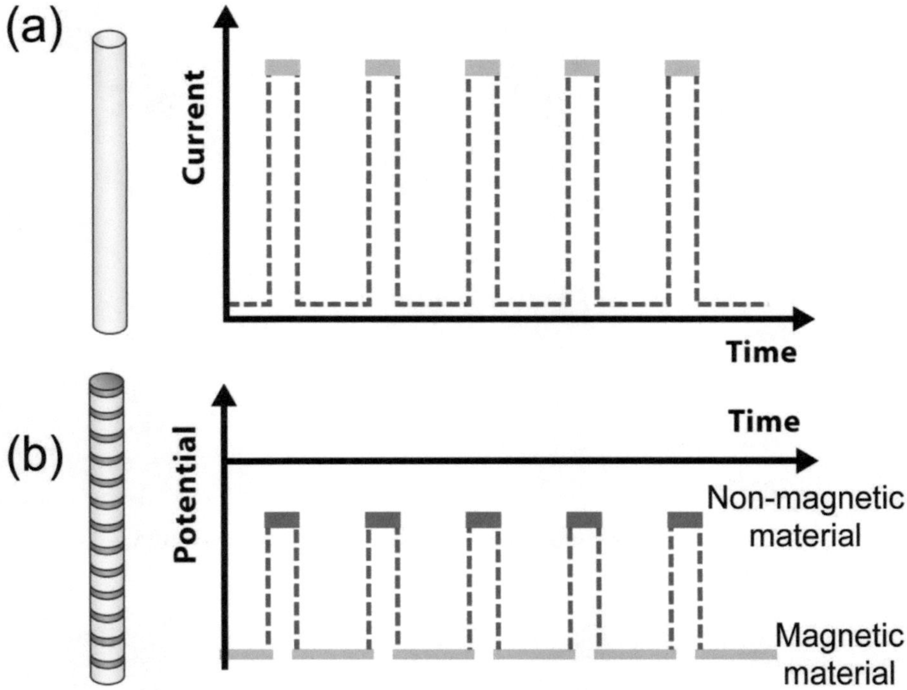

Fig. 12. Programmed electrical parameters for the electrodeposition process. (a) Pulse-current electrodeposition (b) Alternating potential. The grey lines represent the deposition of the different materials.

4. Experimental FORC results

The experimental FORC results for four different nanowire arrays, two uniform (Ni and CoFeB) and two multilayer, are presented in Figs 13-16 as typical examples. Despite the fact that the nanowire arrays have the same geometry, one may easily see that they exhibit different magnetic behaviour. They were therefore chosen in order to exhibit the versatility of the FORC method. They are all characterised by the presence of both reversible and irreversible processes, which complicates their quantitative analysis.

In an axial applied field, the FORC distribution of elongated cylinders always exhibits the same kind of shape, i.e. narrow in the H_c direction and highly elongated along the H_u axis. According to the coherent rotation model, each nanowire could theoretically be represented by an easy axis hysteron, due to the large shape anisotropy. Therefore, the interpretation of the FORC distribution shape, according to the coherent rotation model, would lead to the conclusion that all the nanowires have almost the same coercivity, with the geometric non-uniformity leading to a narrow coercivity distribution. They are subject to a high antiparallel interaction field, created by the dipolar interactions, as seen from the elongated distribution along the H_u axis. However, it will be seen in section 5 that the coherent rotation model fails to account for the low coercivity of the arrays and for several features of the FORC results. The FORC distribution shape obtained under a transverse applied field varies significantly, suggesting radically different magnetic behaviour depending upon the nanowire array properties.

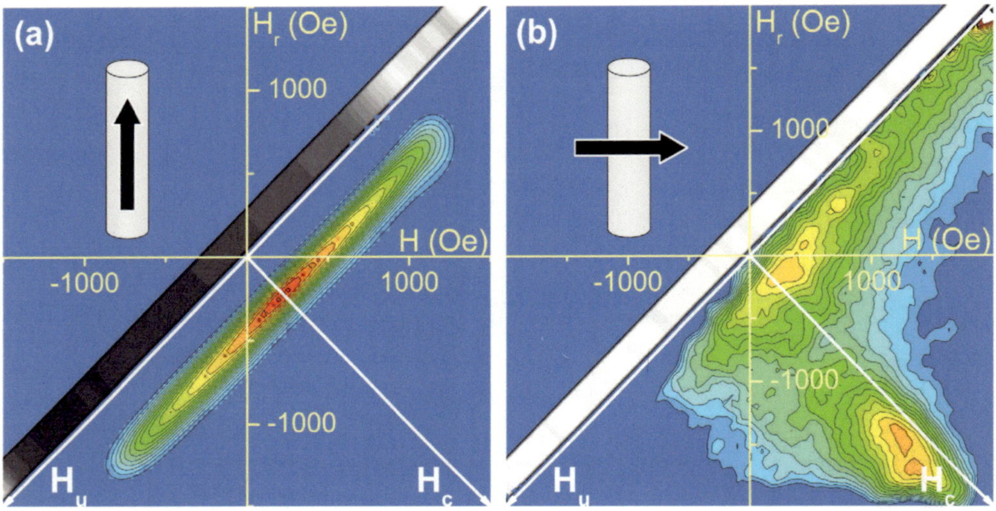

Fig. 13. FORC results for a Ni nanowire array (L = 19 μm, axial easy axis). Applied field (a) axial and (b) transverse.

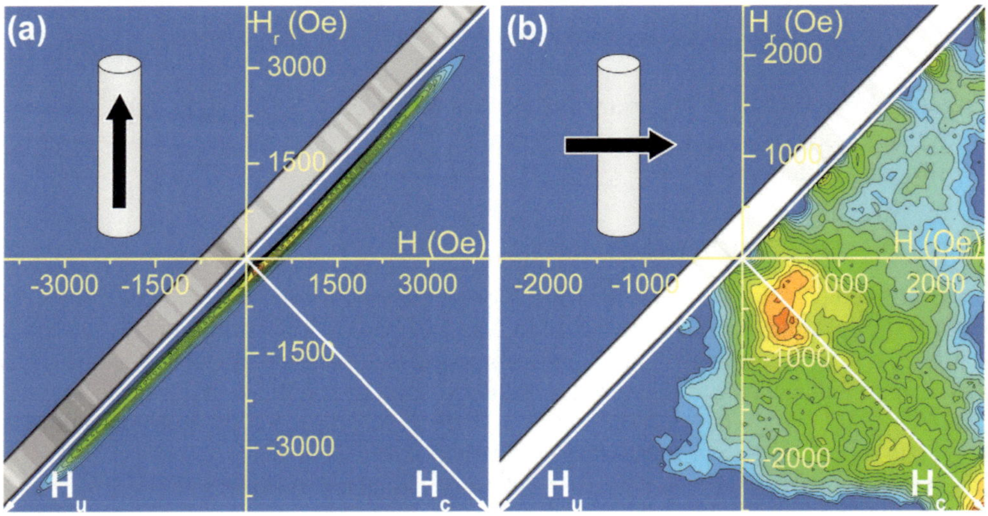

Fig. 14. FORC results for a CoFeB nanowire array (L = 25 μm, axial easy axis). Applied field (a) axial and (b) transverse.

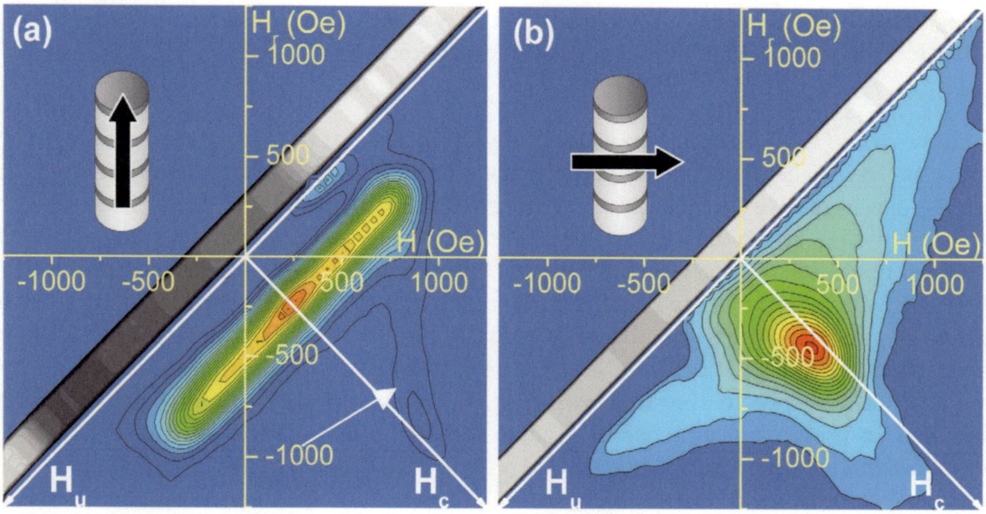

Fig. 15. FORC results for a Ni/Cu nanowire array (L = 32.5 μm, Ni nanodisc thickness = 50 nm, Cu nanodisc thickness = 15 nm, Cu/Ni thickness ratio = 0.3, axial easy axis). Applied field (a) axial and (b) transverse.

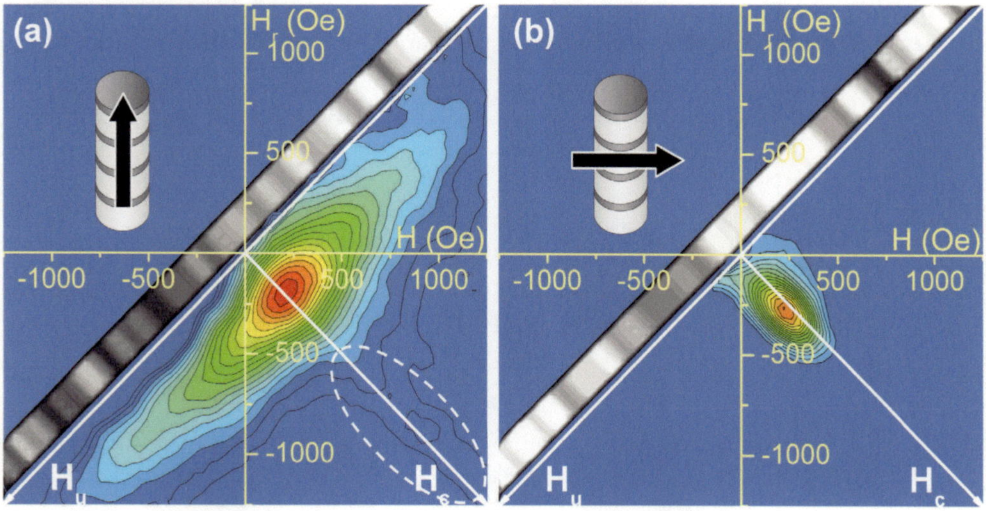

Fig. 16. FORC results for a Ni/Cu nanowire array (L = 3.25 μm, Ni nanodisc thickness = 30 nm, Cu nanodisc thickness = 35 nm, Cu/Ni thickness ratio = 1.17, transverse easy axis). Applied field (a) axial and (b) transverse.

5. Extracting individual nanowire properties

5.1 Individual nanowire coercivity/interaction field

In strongly interacting systems, like ferromagnetic nanowire arrays, the extraction of the interaction field and of individual coercivities have to be done together. This is because the interaction field modifies the FORC distribution, which can then no longer be analysed using the classical Preisach model, in which the coercive and interaction fields are independent. In the specific case of nanowire arrays, both FORC results obtained with axial and with transverse applied fields should be considered simultaneously.

Axial applied field

When the axial FORC distribution is qualitatively equivalent to that presented in Fig. 8b, as is the case for all the experimental distributions presented above (Figs 13-16), it suggests a narrow coercivity distribution and a net antiparallel interaction field. This allows us to use H_c^{FORC} as a first approximation to the average coercivity of the individual nanowires. However, two cases may occur for which this approximation is no longer adequate: 1) if the coercivity distribution is large enough to disturb the FORC distribution, inducing a second branch (Fig. 15a, shown by the arrow) or enlarging it along the H_c axis (Fig. 16a); 2) if both the reversibility indicator and the antiparallel interaction field are large (Fig. 14a); their simultaneous presence in a system will shift the entire FORC distribution towards higher coercivity, in proportion to these two values (Béron, 2008). In the later case, the H_c^{FORC} approximation should be corrected before being used as the average individual nanowire/nanodisc coercivity.

When H_c^{FORC} satisfies the conditions for it to represent the individual wire coercivity, the net value of the axial interaction field at saturation is then well described by ΔH_u, defined as the half-width at half-height of the uniform part of the ΔH_u cross-section (Fig. 17). Good agreement (Béron et al., 2008b) was found between the experimental ΔH_u values of uniform nanowires (Figs 13-14a) and the predicted values of interaction field at saturation from a micromagnetic model (Clime et al., 2006) and from an effective field model (Carignan et al., 2007).

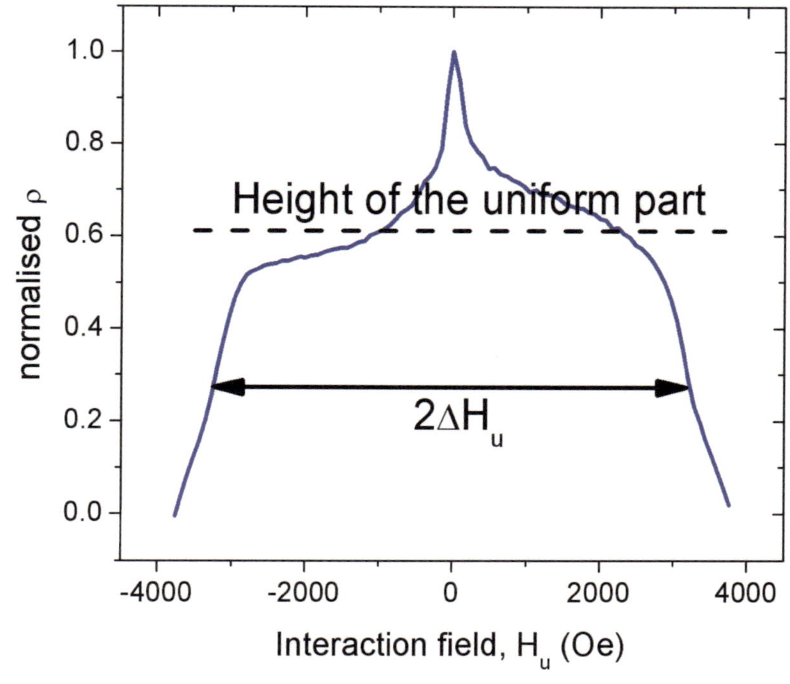

Fig. 17. Axial FORC distribution cross-section along the H_u axis and its respective ΔH_u value. (CoFeB nanowire array, see Fig. 14a)

If we assume a uniform, antiparallel interaction field, described by Eq. (4), we expect a FORC distribution that is straight and parallel to the H_u axis and with a flat H_u cross-section (Fig. 18a). Spatial non-uniformity in the interaction field induces some deviations in this distribution. Three kinds of deviation can be easily identified.

First, a spatial distribution of interaction field caused by the array border can induce a central peak in the H_u cross-section (Fig. 18b). Most of the nanowires are submitted to a dipolar field that can be treated as a mean interaction field. However, the array is not infinite. The nanowires located at the array borders, where the lack of neighbours induces a smaller interaction field, can complete their reversal when this field is still relatively weak. This results in a peak at the centre of the cross-sections along H_u (present for all the experimental results). This hypothesis was confirmed by noting that the amplitude of the central peak is significantly reduced when very large samples are measured. Therefore, characterisation should always be carried out using the largest area possible, in order to avoid this effect.

Second, if the antiparallel interaction field decrease allows the preponderance of a parallel interaction field (for example, the dipolar field between nanodiscs within a multilayer nanowire), then, in addition to the central peak, another distribution can appear, reflecting this parallel interaction field effect (ex.: distribution of higher coercivity, circled on figure 16a).

Finally, for arrays of radius significantly larger than the nanowire length and for which the condition $L/2D >> 1$ is satisfied, then the interaction field decreases near the nanowire extremities (Carignan et al., 2007). Therefore, depending upon the applied field direction (positive or negative), the magnetisation reversal is favoured either by a maximal or minimal value of interaction field, inducing a non-linearity in the FORC distribution (Fig. 18c). This non-linearity is experimentally more visible in the case of CoFeB nanowires (Fig. 14a) than for Ni (Fig. 13a), because higher saturation magnetisation leads to larger interaction fields.

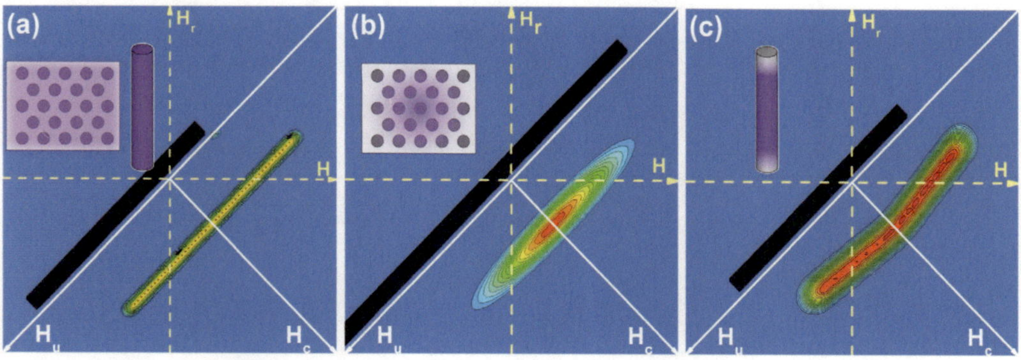

Fig. 18. Simulated FORC results for several configurations of spatial distribution of interaction field. (a) Uniform mean interaction field (b) Distribution in the array (c) Distribution along the nanowires

Transverse applied field

With a transverse applied field on an infinite array, the interaction field at saturation is theoretically parallel to M, and equal to the half of the axial interaction field (Carignan et al., 2007). This value has been successfully used to fit the transverse major hysteresis curves of uniform nanowires (Béron, 2008). For FORC distributions that are narrow on the H_c axis, as in figures 15b and 16b, the transverse individual coercivity is indirectly accessible through use of the analysis illustrated in Fig. 8c. Subtracting the half value of the axial ΔH_u from the transverse H_c^{FORC} value allows us to retrieve the coercivity of individual nanodiscs in multilayer nanowires (Béron et al., 2008c).

5.2 Effective nanowire array anisotropy

In cases where the sum of the minimum values of the reversibility indicator η (Béron et al., 2007, see fig. 19a) in axial and transverse applied field is around 1, an indication of the

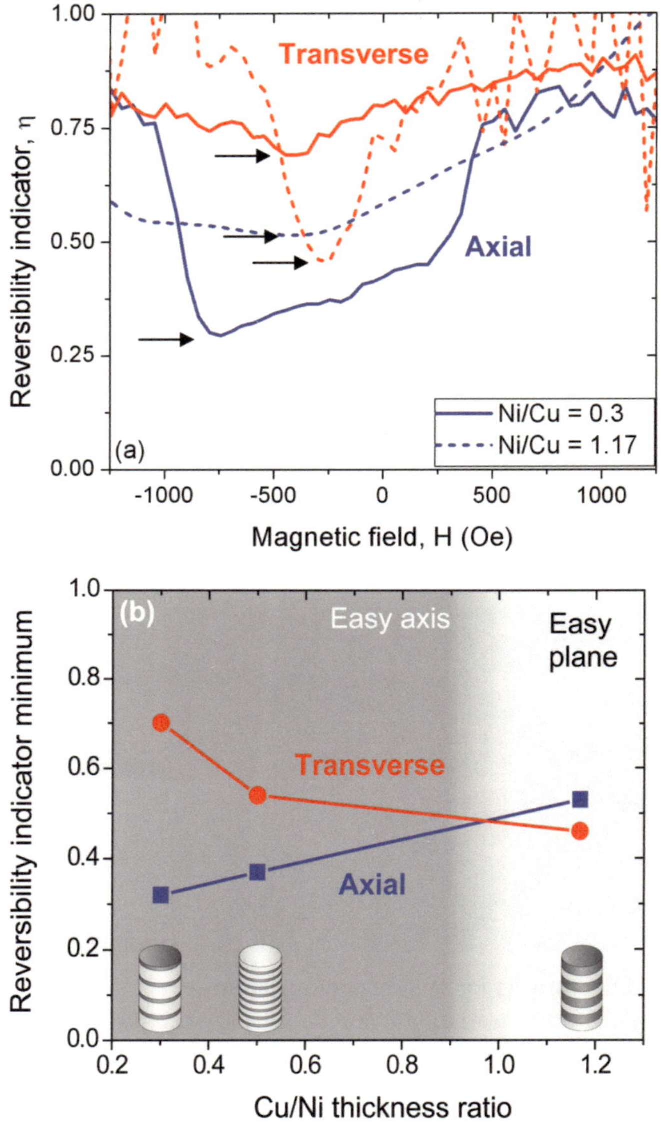

Fig. 19. (a) Reversibility indicator as a function of the reversal field for multilayer nanowires. The position of the minimum is indicated. (b) Reversibility indicator minimum as a function of the Cu/Ni thickness ratio. The cross-over indicates the change of easy axis direction.

overall anisotropy of the nanowire array can be obtained by the FORC method. Otherwise, other processes contribute to the reversibility indicator and therefore the minimum value of η cannot be used to adequately characterise the effective nanowire array anisotropy. When the sum is around 1, the reversibility being lower along an easy direction, the lowest η minimum value indicates an easy direction. A cross-over between the η minimum value curves, occurring when the array is magnetically isotropic, implies a change in the easy axis direction (Fig. 19b) (Béron et al., 2008c). This result for multilayer nanowires is in agreement with those found from the major hysteresis curves and ferromagnetic resonance measurements, as well as demagnetisation factor modelling (Carignan et al., 2007).

5.3 Magnetisation reversal process

Quantitative knowledge of the interaction field, average individual coercivity and overall array anisotropy, along with the FORC distribution shape, can help us determining the magnetisation reversal process. However, one has to remember that the FORC method does not directly give the type of magnetisation reversal occurring in the sample and so has to be used as a tool which suggests a probable mechanism, since systems reversing by different means can exhibit the same FORC result.

Axial applied field

For uniform nanowire arrays and axial applied field, coherent rotation reversal of individual nanowires leads to purely irreversible behaviour, with a unique value of H_c^{FORC} equal to the shape anisotropy field (as in Fig 8b). Another possible reversal mechanism is by nucleation-propagation of a domain wall, which has been observed for the reversal of individual nanowires (Wernsdorfer et al., 1996; Hertel, 2001) (Fig. 20a). Then, the FORC result can exhibit some reversibility, a coercivity lower than the shape anisotropy field and a second distribution of coercivity near saturation, associated with annihilation of the domain walls (Béron, 2008) (circled on Fig. 20b). The FORC results presented in Figs 13a and 14a exhibit these three characteristics and, therefore, nucleation-propagation magnetisation reversal mechanism is more likely in these cases than coherent rotation.

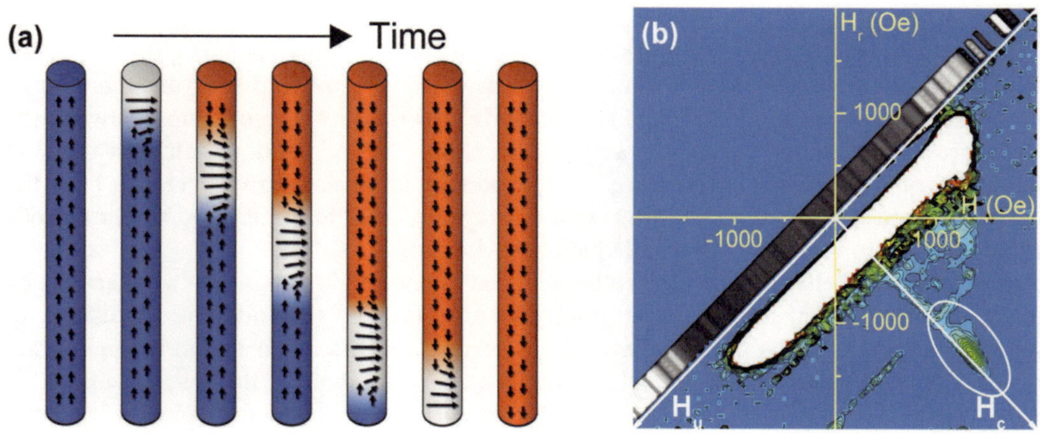

Fig. 20. (a) Schematic of the nucleation-propagation of a domain wall. (b) Ni axial FORC result, where the scale has been adjusted in order to show the second distribution of lower intensity (circled), generated by nucleation-propagation reversal.

Transverse applied field

For an applied field transverse to the nanowire axis, the coherent rotation model predicts fully reversible behaviour for a uniform individual nanowire, being along a hard direction, and no other FORC distribution. The existence of irreversible processes in Figs 13b and 14b, even in low proportion (1 - 4%), indicates that at least one other mechanism is present. This process could be the irreversible reversal of the domain walls created between regions decoupled enough to allow the magnetisation to rotate coherently freely toward one or the other nanowire extremity (Fig. 21) (Henry et al., 2002; Hertel, 2001). Other phenomena, such as closure domains at nanowire extremities, could also contribute to the non-null FORC distribution.

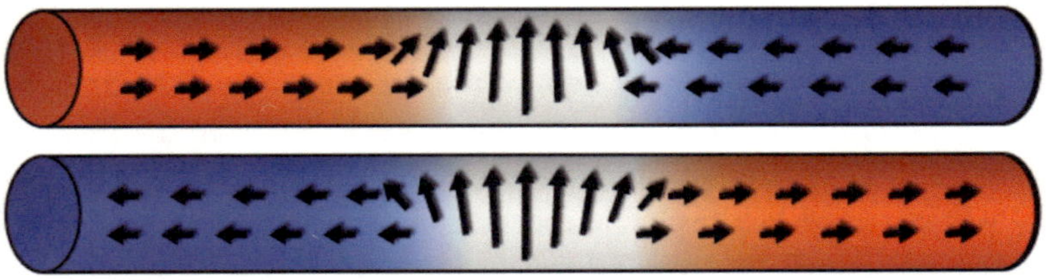

Fig. 21. Schematic of 180° domain walls created during transverse magnetisation reversal. The irreversible reversal could account for the irreversible FORC distribution in Figs. 13b and 14b.

Other types of FORC results are possible and have to be analysed individually, based on the available information concerning the transverse magnetic behaviour of the array. For example, for multilayer nanowires, the FORC result presented in Fig. 15b was attributed to a mixture of coherent and incoherent rotation, the favourable conditions for one or the other changing during the reversal, while Fig. 16b was interpreted as an abrupt, irreversible reversal, caused by field applied along an easy direction (Béron et al., 2008c).

6. Conclusion

The unique structure of ferromagnetic nanowire arrays gives them interesting properties for various applications, ranging from high-density memory to high frequency devices. They may be fabricated by electrodeposition, a process which is fast, cheap and versatile. From a magnetic point of view, their behaviour is mainly governed by a competition between the shape anisotropy (an individual property of each nanowire) and the dipolar interaction field (an effect of the network). Therefore, it is important to be able to characterise both the individual and global properties of nanowire arrays, in order to understand their magnetic behaviour and to implement them adequately in devices.

This experimental characterisation can be achieved in one single step on the whole array, by using the very promising first-order reversal curve (FORC) method. The acquisition of several minor hysteresis curves allows us to record and discriminate the magnetisation reversal of each nanowire, ultimately yielding a representation of the reversal of all the nanowires. From this representation (the FORC result) and the help of the physical analysis model (based on predicted behaviour), one can extract valuable information, usually difficult to obtain experimentally. The half-width of the FORC distribution elongation (ΔH_u) along the interaction field axis (H_u) gives the value of the interaction field at saturation, while a spatial distribution of this field modifies its shape. The spatial non-homogeneity can be in the array (border effect), which induces a central peak in the FORC distribution, or along the nanowires (extremity effect), which breaks the FORC distribution linearity along the H_u axis. An approximation of the average individual coercivity of the nanowires can be obtained by taking the position of the maximum of the FORC distribution on the coercivity axis (H_c) (H_c^{FORC}). This approximation remains valid if the coercivity distribution is narrow, and both the reversibility indicator and the antiparallel interaction field are not too large. Also, in the presence of a parallel interaction field, its saturation value has to be subtracted from H_c^{FORC} in order to reflect the individual coercivity. Finally, a quantitative evaluation of the effective anisotropy of the array is accessible through the minimum value of the reversibility indicator function of the reversal field, if the sum from axial and transverse applied fields is close to 1.

The versatility of the FORC method has been demonstrated here by discussing experimental results for nanowire arrays with different compositions (Ni and CoFeB), and thus different saturation magnetisation, and different structures (uniform and multilayer), all under both axial and transverse external applied fields. However, the application range of the FORC method is not confined to the two principal directions (axial and transverse) of the array. The physical analysis model can also be employed for the study of the magnetic behaviour as a function of the applied field angle (Béron et al., 2009a). The tridimensional and highly reversible magnetisation reversal process in this case led to the development of a novel variant of the FORC method, vector FORC, where both magnetisation components, parallel and perpendicular to the applied field direction, are used. This allows us to adequately follow and characterise reversible magnetisation rotation in three dimensions (Béron et al., 2009b).

7. References

Almawlawi D.; Coombs N. & Moskovits M. (1991). Magnetic properties of Fe deposited into anodic aluminum oxide pores as a function of particle size. *Journal of Applied Physics*, Vol. 70, No. 8, October 1991, 4421-4425, 0021-8979

Alper, M.; Attenborough, K.; Hart, R.; Lane, S.J.; Lashmore, D.S.; Younes, C. & Schwarzacher, W. (1993). Giant magnetoresistance in electrodeposited superlattices. *Applied Physics Letters*, Vol. 63, No. 15, October 1993, 2144-2146, 0003-6951

Béron, F.; Clime, L.; Ciureanu, M.; Ménard, D.; Cochrane, R.W. & Yelon A. (2006). First-order reversal curves diagrams of ferromagnetic soft nanowire arrays. *IEEE Transactions on Magnetics*, Vol. 42, No. 10, October 2006, 3060-3062, 0018-9464

Béron, F.; Clime, L.; Ciureanu, M.; Ménard, D.; Cochrane, R.W. & Yelon A. (2007). Reversible and quasireversible information in first-order reversal curve diagrams. *Journal of Applied Physics*, Vol. 101, No. 9, May 2007, 09J107, 0021-8979

a - Béron, F.; Ménard, D. & Yelon, A. (2008). First-order reversal curve diagrams of magnetic entities with mean interaction field: A physical analysis perspective. *Journal of Applied Physics*, Vol. 103, No. 7, April 2008, 07D908, 0021-8979

b - Béron, F.; Clime, L.; Ciureanu, M.; Ménard, D.; Cochrane, R.W. & Yelon A. (2008). Magnetostatic interactions and coercivities of ferromagnetic soft nanowires in uniform length arrays. *Journal of Nanoscience and Nanotechnology*, Vol. 8, No. 6, June 2008, 2944-2954, 1533-4880 c - Béron, F.; Carignan, L.-P.; Ménard, D. & Yelon, A. (2008). Magnetic Behavior of Ni/Cu Multilayer Nanowire Arrays Studied by First-Order Reversal Curve Diagrams. *IEEE Transactions on Magnetics*, Vol. 44, No. 11, November 2008, 2745-2748, 0018-9464

Béron, F. (2008). *Propriétés magnétostatiques de réseaux de nanofils via les courbes de renversement du premier ordre*, Ph. D. thesis, Université de Montréal, Montréal, Canada

a - Béron, F.; Carignan, L.-P.; Pirota, K.R.; Knobel, M.; Ménard, D. & Yelon, A. (2009). Magnetization reversal of ferromagnetic nanowire arrays studied by angular FORC measurements, 2009 IEEE International Magnetics Conference (INTERMAG), Sacramento, USA, May 2009

b - Béron, F.; Pirota, K.R.; Knobel, M.; Ménard, D. & Yelon, A. (2009). Vector first-order reversal curve (FORC) results, International Conference on Magnetism 2009 (ICM), Karlushe, Germany, July 2009

Blondel, A.; Meier, J.P.; Doudin, B. & Ansermet, J.-P. (1994). Giant Magnetoresistance of Nanowires of Multilayers, *Applied Physics Letters*, Vol. 65, No. 23, December 1994, 3019-3021, 0003-6951

Carignan, L.-P.; Lacroix, C.; Ouimet, A.; Ciureanu, M.; Yelon, A. & Ménard, D. (2007) Magnetic anisotropy in arrays of Ni, CoFeB, and Ni/Cu nanowires. *Journal of Applied Physics*, Vol. 102, No. 2, July 2007, 023905, 0021-8979

Ciureanu, M.; Béron, F.; Clime, L.; Ciureanu, P.; Yelon, A.; Ovari, T.A.; Cochrane, R.W.; Normandin, F. & Veres, T. (2005). Magnetic properties of electrodeposited CoFeB thin films and nanowire arrays, *Electrochimica Acta*, Vol. 50, No. 2, August 2005, 4487-4497, 0013-4686

Clime, L.; Béron, F.; Ciureanu, P.; Ciureanu, M.; Cochrane, R.W. & Yelon A. (2006). Characterization of individual ferromagnetic nanowires by in-plane magnetic measurements of arrays. *Journal of Magnetism and Magnetic Materials*, Vol. 299, No. 2, April 2006, 487-491, 0304-8853

Della Torre, E. (1966). Effect of Interaction on the Magnetization of Single Domain Particles, *IEEE Transactions on Audio and Electroacoustics*, Vol. AU-14, No. 2, June 1966, 86-92, 0018-9278

Ferain, E. & Legras, R. (2003). Track-etch templates designed for micro- and nanofabrication. *Nuclear Instruments & Methods in Physics Research*, Vol. 208,August 2003, 115-122, 0168-583X

Henry, Y.; Iovan, A.; George, J.-M. & Piraux, L. (2002). Statistical analysis of the magnetization processes in arrays of electrodeposited ferromagnetic nanowires, *Physical Review B*, Vol. 66, No. 18, November 2002, 184430, 1098-0121

Hertel, R. (2001). Micromagnetic simulations of magnetostatically coupled nickel nanowires, *Journal of Applied Physics*, Vol. 90, No. 11, December 2001, 5752-5758, 0021-8979

Hertel, R. & Kirschner, J. (2004). Magnetization reversal dynamics in nickel nanowires, *Physica B*, Vol. 343, No. 1-4, January 2004, 206-210, 0921-4526

Lavín, R.; Denardin, J.C.; Escrig, J.; Altbir, D.; Cortés, A. & Gómez, H. (2008). Magnetic characterization of nanowire arrays using first order reversal curves, *IEEE Transactions on Magnetics*, Vol. 44, No. 11, November 2008, 2808-2811, 0018-9464

Le Bon, G. (1895). *Psychologie des foules*, Les Presses universitaires de France, 978-2130542971 Paris, France

Lindeberg, M. & Hjort, K. (2003). Interconnected nanowire clusters in polyimide for flexible circuits and magnetic sensing applications, *Sensors and Actuators A*, Vol. 105, No. 2, July 2003, 150-161, 0924-4247

Masuda, H. & Fukuda, K. (1995). Ordered metal nanohole arrays made by a 2-step replication of honeycomb structures of anodic alumina, *Science*, Vol. 268, No. 5216, June 1995, 1466-1468, 0036-8075

Mayergoyz, I.D. (1985). Hysteresis models from the mathematical and control-theory points of view, *Journal of Applied Physics*, Vol. 57, No. 8, August 1985, 3803-3805, 0021-8979

Nielsh, K.; Müller, F.; Li, A-P. & Gösele, U. (2000). Uniform nickel deposition into ordered alumina pores by pulsed electrodeposition, *Advanced Materials*, Vol. 12, No. 8, April 2000, 582-586, 0935-9648

O'Sullivan, J.P. & Wood, G.C. (1970). Morphology and mechanism of formation of porous anodic films on aluminum, Proceedings of the Royal Society of London, serie A, Vol. 317, No. 1731, July 1970, 511-543, 1471-2954

Peixoto, T.R.F. & Cornejo, D.R. (2008). Characterizing magnetic interactions in Ni nanowires by FORC analysis, *Journal of Magnetism and Magnetic Materials*, Vol. 320, No. 14, July 2008, E279-E282, 0304-8853

Piraux, L.; Renard, K.; Guillemet, R.; Mátéfi-Tempfli, S.; Mátéfi-Tempfli, M.; Antohe, V.A.; Fusil, S.; Bouzehouane, K. & Cros, V. (2007). Template-grown NiFe/Cu/NiFe nanowires for spin transfer devices, *Nano Letters*, Vol. 7, Septembre 2007, 2563-2567, 1530-6984

Piraux, L.; George, J.M.; Despres, J.F.; Leroy, C.; Ferain, E.; Legras, R.; Ounadjela, K. & Fert, A. (1994). Giant magnetoresistance in magnetic multilayered nanowires, *Applied Physics Letters*, Vol. 65, No. 19, November 1994, 2484-2486, 0003-6951

Pirota, K.R.; Navas, D.; Hernandez-Velez, M.; Nielsch, K.& Vazquez, M. (2004). Novel magnetic materials prepared by electrodeposition techniques: arrays of nanowires and multi-layered microwires, *Journal of Alloys and Compounds*, Vol. 369, No. 1-2, April 2004, 18-26,

Preisach, F. (1938). Über die magnetische Nachwirkung. *Zeitschrift für Physik*, Vol. 94, 1938, 277-302, 0722-3277. A presentation of the classical Preisach model is accessible in section 1.1 of Mayergoyz, I. (2003). *Mathematical models of hysteresis and their applications*, Elsevier, 978-0124808737, New York, USA

Ross C.A. (2001). Patterned magnetic recording media, *Annual Review of Materials Research*, Vol. 31, 2001, 203-235, 1531-7331

Saib, A.; Darques, M.L.; Piraux, L.; Vanhoenacker-Janvier, D. & Huynen, I. (2005). An Unbiased Integrated Microstrip Circulator Based on Magnetic Nanowired Substrate, *IEEE Transactions on Microwave Theory and Techniques*, Vol. 53, No. 6, June 2005, 2043-2049, 0018-9480

Saib, A.; Vanhoenacker-Janvier, D.; Raskin, J.-P.; Crahay, A. & Huynen, I. (2001). Microwave tunable filters and nonreciprocal devices using magnetic nanowires, Proceedings of the 2001 1st IEEE Conference on Nanotechnology (IEEE-NANO 2001), pp. 260-265, 0-7803-7215-8, Maui, USA, October 2001, IEEE, Piscataway, USA

Spinu, L.; Stancu, A.; Radu, C.; Li, F. & Wiley, J.B. (2004). Method for magnetic characterization of nanowire structures, *IEEE Transactions on Magnetics*, Vol. 40, No. 4, July 2004, 2116-2118, 0018-9464

Stoner, E.C. & Wohlfarth, E.P. (1948). A mechanism of magnetic hysteresis in heterogeneous alloys, *Philosophical Transactions of the Royal Society of London*, Vol. A240, 1948, 599-642, 0080-4649. Reprinted in *IEEE Transactions on Magnetics*, Vol. 27, No. 4, July 1991, 3475- 3518, 0018-9464

Tang, X.T.; Wang, G.C. & Shima, M. (2007). Magnetic layer thickness dependence of magnetization reversal in electrodeposited CoNi/Cu multilayer nanowires, *Journal of Magnetism and Magnetic Materials*, Vol. 309, No. 2, February 2007, 188-196, 0304-8853

Thurn-Albrecht, T.; Schotter, J.; Kästle, C.A.; Emley, N.; Shibauchi, T.; Krusin-Elbaum, L.; Guarini, K.; Black, C.T; Tuominen, M.T. & Russell, T.P. (2000). Ultrahigh-density nanowire arrays grown in self-assembled diblock copolymer templates, *Science*, Vol. 290, No. 5499, December 2000, 2126-2129, 0036-8075

Wernsdorfer, W.; Doudin, B.; Mailly, D.; Hasselbach, K.; Benoit, A.; Meier, J.; Ansermet, J.-P. & Barbara, B. (1996). Nucleation of magnetization reversal in individual nanosized nickel wires, *Physical Review Letters*, Vol. 77, No. 9, August 1996, 1873-1876, 0031-9007

Ye, B.; Li, F.; Cimpoesu, D.; Wiley, J.B.; Jung, J.-S.; Stancu, A. & Spinu, L. (2007). Passive high-frequency devices based on superlattice ferromagnetic nanowires, *Journal of Magnetism and Magnetic Materials*, Vol. 316, No. 2, September 2007, E56-E58, 0304-8853

Zhao, S.; Chan, K.; Yelon, A. & Veres, T. (2007). Preparation of open-through anodized aluminium oxide films with a clean method, *Nanotechnology*, Vol. 18, No. 24, June 2007, 245304, 0957-4484

8

Impact of Nanowires on the Properties of Magnetorheological Fluids and Elastomer Composites

R. C. Bell[1], D. T. Zimmerman[1], and N. M. Wereley[2]
[1]*The Pennsylvania State University, Altoona College*
[2]*University of Maryland*
USA

1. Introduction

Magnetorheological (MR) fluids are a type of smart material whose effective viscosity may be rapidly changed (~few ms) in a nearly reversible manner by the application and removal of an externally-applied magnetic field. Conventional MR fluids are composed of micron-scale, ferromagnetic spherical particles (typically 30 to 40 volume percent) suspended in a hydrocarbon, silicone, or aqueous carrier fluid (Klingenberg, 2001). The viscosity and apparent shear strength of these suspensions can be controlled by varying the strength of an applied magnetic field. Without an applied magnetic field (off-state), MR fluids are a viscous liquid/particle suspension with a viscosity in the range of 0.1 – 3 Pa · s. Upon application of a magnetic field (on-state), the particles acquire a magnetic polarization and attract one another forming chain-like structures that join to form columnar structures parallel to the applied field (schematic, Fig. 1). The newly formed columns span the surfaces of the device parallel to the field lines resulting in a material that behaves as a Bingham plastic fluid, with increased viscosity and apparent yield stress under shear. The viscosity and yield stress of the fluid is scalable with the magnitude of the applied magnetic field until magnetic saturation of the particles is reached (Jones & Saha, 1990). At high fields, the fluid is converted to a semi-solid with a five to six orders-of-magnitude change in apparent viscosity (Genç & Phulé, 2002). As the particle loading approaches 40 vol. %, the field-induced yield stress can reach 100 kPa. Exceeding the yield stress of the fluid causes the fibril chains and columnar structures to continuously break and re-form, resulting in a post-yield viscosity.

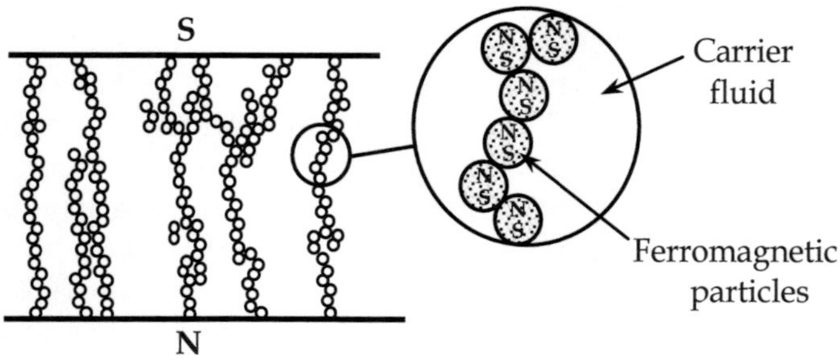

Fig. 1. Schematic representation of the magnetorheological effect with spherical particles.

The rheological properties of MR fluids depend not only on the strength of the applied magnetic field, but also on the loading (vol. % or wt. %), size, composition, magnetic properties, and morphology of the particles, as well as on the concentration and type of additives within the suspension. As the applied magnetic field is increased, the shear strength of the fluid eventually reaches a maximum as the magnetization saturation of the particles is reached. Particle loading is the most important factor affecting the achievable yield strength of MR fluids. While increasing the volume fraction of particles in the suspension results in an increase in the yield strength of the material, it also results in an increase in the off-state viscosity. The shear strength of the suspensions increases with an increase in size of the particles (Genç & Phulé, 2002); however, spherical particles larger than about 10 µm tend to settle rapidly, even with the addition of special additives. Suspensions composed of nanometer-sized particles tend to settle more slowly, if at all, due to Brownian motion. However, the effectiveness of Brownian motion combined with a smaller magnetic moment, also hinders the formation of the chain-like structures in these energized fluids, greatly reducing the yield stress of the material (Rosenfeld et al., 2002; Poddar et al., 2004; Chaudhuri et al., 2005). Given these competing factors, the ideal diameter of particles is in the range of 1 to 10 µm. In this size range, sedimentation still occurs due to the inherent density difference between the particles and the carrier fluid. Upon settling, the particles tend to form tightly packed sediments resulting from remnant magnetization within the particles and are not easily re-dispersed (Phulé & Ginder, 1998). Since the yield stress of MR fluids is directly related to the volume fraction of particles in the suspension, the behavior of the fluid is less predictable during this remixing period. The settling rate can be reduced slightly and less dense sediments formed, by the addition of special additives such as surfactants, nanoparticles, and other non-spherical particles (Poddar et al., 2004; Chaudhuri et al., 2005; Chin et al., 2001; Weiss et al., 1997; Wereley et al., 2006).

There are a growing number of applications that exploit the MR fluid's continuously controllable, field-dependent yield stress including variable dampers, brakes, clutches, and optical polishing machines (Kamath et al., 1999; Zipser et al., 2001; Ramallo et al., 2002; Harland et al., 2001; Kordonski et al., 2006). Conventional MR fluids have had some difficulty finding widespread commercial use, owing to relatively high manufacturing costs, as well as a limitation on the volume loading necessary to maintain their low off-state viscosity. The use of MR fluids in more diverse applications have been limited, due to the settling of the particles in the absence of continual mixing (Chen & Chen, 2003) and particle wear (Carlson, 2003) that tends to reduce the efficacy of the fluid over time and eventually lead to device failure.

Both the yield stress and settling properties of magnetorheological fluids have been shown to depend on particle morphology; however, few studies have explored the particle shape-dependence in either MR fluids (Chin et al., 2001; Nishiyama et al., 2005; Kuzhir et al., 2009; López-López et al., 2009) or electrorheological (ER) fluids (Kanu & Shaw, 1998; Lengálová et al., 2003; Sanchis et al., 2004; Yin & Zhao, 2006). Studies involving electrorheological fluids utilizing elongated titanate (TiO_2) whiskers demonstrated greatly reduced sedimentation. Moreover, with a 10 vol. % loading, these fluids displayed an increase in yield stress from 80 to 1100 Pa when ~80 nm diameter particles were exchanged with TiO_2 whiskers (with 10 nm diameters by several hundreds or thousands nanometers) (Yin & Zhao, 2006). However, it is unclear whether the increase in yield stress was due to the differences in particle morphology or due to the large mass differences between the particles. Studies on MR fluids with up to 7 vol. % 60 μm long, 4.8 μm diameter cobalt microfibers displayed substantial mitigation of sedimentation and an increase in yield stress as compared to suspensions containing 1.34 μm diameter cobalt spheres (López-López et al., 2009). It was suggested that the enhancement of the yield stress was due to inter-fiber solid friction. Yet, the fibers used in these studies had non-uniform dimensions, rough surfaces, and the difference in the size of the particles made a direct comparison between the properties of the two types of suspensions difficult to interpret. All of these studies suffered from broad size distribution, inconsistency in particle shape, surface roughness, and particle fragility, making it difficult to form conclusions or infer the underlying physics. Thus, while MR/ER fluids employing elongated particles demonstrated reduced sedimentation properties, the effect on yield stress remained unclear.

More recent studies utilizing nanowires have shown promising results for not only reducing or preventing settling, but have also increased the apparent yield stress of the materials (Bell et al., 2007; Bell et al., 2008; Ngatu et al., 2008; Zimmerman et al., 2009). Unlike the ferromagnetic oxide particles used in previous studies (Chin et al., 2001) and those formed from spherical particles adjoined in the presence of a magnetic field (López-López et al., 2009), nanowires have well defined structure and controllable length distributions. This facilitates systematic experimental and theoretical studies and will lead to increased understanding and design control. Nanowire-based MR fluids have two distinct advantages over suspensions that contain only spherical particles (Bell et al., 2007; Bell et al., 2008; Ngatu et al., 2008). First, the maximum achievable yield stress can be twice that of conventional fluids (or greater, even at the same metal loading) and nanowire-based fluids also provide more sensitive control over the yield stress at field strengths below magnetic saturation of the suspension. Second, sedimentation is greatly reduced, if not eliminated. However, one drawback to pure nanowire fluids is the limit on particle loading to 10 vol. %, resulting in fluids with a maximum yield stress much lower than most conventional MR fluids. The wires occupy a much larger effective volume compared to spheres, due to the excluded volume concept. Even so, fluids using only nanowires display some very interesting and useful properties. For applications that require the highest yield stress (and thus maximum particle loading of 35-40 vol. %), a "dimorphic" fluid was generated that contains both spherical and nanowire particles. These fluids not only display reduced sedimentation, but some formulations also display an increase in yield stress over conventional fluids.

2. Synthesis of magnetorheological fluids

Two types of ferromagnetic particles were used in this study. To replicate conventional MR fluids, we used spherical iron particles having diameters of 1 - 3 μm or 6 - 10 μm (Alfa Aesar), cobalt particles with nominal diameters of 1.6 μm (Alfa Aesar), and nickel particles with diameters in the range of 1 to 10 μm (Novamet). Nanowire-based fluids were synthesized using wires generated via template-based electrodeposition techniques.

2.1 Nanowire generation and characterization

The nanowires were generated via template-based electrodeposition using commercially available, anodized alumina membranes (Whatman) as templates and various electrolytic solutions depending on nanowire composition. The electrolyte solution for nickel nanowires consisted of 1.1 *M* $NiSO_4$, 0.19 *M* $NiCl_2$, and 0.6 *M* H_3BO_3, with the pH of the solution adjusted to 3.5 using H_2SO_4. A 99.5% nickel foil was used as the working electrode. Iron nanowires were fabricated using an electrolytic solution consisting of 0.9 *M* $FeSO_4$, 0.03 *M* $FeCl_2$, 0.10 *M* NH_4Cl, 0.01 *M* $C_6H_8O_6$, and 0.5 *M* H_3BO_3 at a pH of 3 using H_2SO_4. A 99.99% iron foil was used as the working electrode. Cobalt nanowires were generated from a solution containing 0.2 *M* $CoSO_4$, 0.02 *M* $CoCl_2$, and 0.3 *M* H_3BO_3 adjusted to a pH of 3.5 using H_2SO_4. A 99.97% cobalt foil was used as the working electrode. The working electrodes were suspended approximately 2 cm from the alumina template in the electrolytic solutions. Wires were electrodeposited using a current density ranging from 4.8-5.5 mA cm^{-2} under ambient conditions without agitation. The diameter of the wires was fixed by the diameter of the channels within the membrane and the lengths of the wires were controlled by the current and deposition time. The nanowires were recovered by dissolving the wire-filled templates in a 1 *M* NaOH solution.

Particle dimensions and size-distribution information for both the spherical particles and nanowires were obtained using Hitachi S570 and Hitachi S800 scanning electron microscopes (SEM). Hundreds of images were examined to determine an average and standard deviation for the diameter and length. The size distribution of the wires can be seen in the SEM micrographs in Fig. 2.

Using induction-coupled plasma spectrometry (ICP), the stoichiometric purity of all wire compositions was determined to be greater than 99.4%. In particular, the ICP data indicates that the wires do not contain significant amounts of gallium, indium, copper, aluminum, or other elements that come in contact with the wires during fabrication.

The coercivity, remanence, and magnetic saturation all influence the magnetorheological properties of the fluids and depend on the composition and aspect ratio of the particles. The magnetic properties of the particles and MR fluids were determined at room temperature using a Lakeshore 7400 series vibrating sample magnetometer (VSM). Figure 3 shows hysteresis curves exhibited in iron nanowires aligned parallel to one another within an alumina template, where the easy axis is parallel to the field flux and the hard axis is perpendicular to the flux. This reveals a shape-anisotropy in the wires due to their aspect

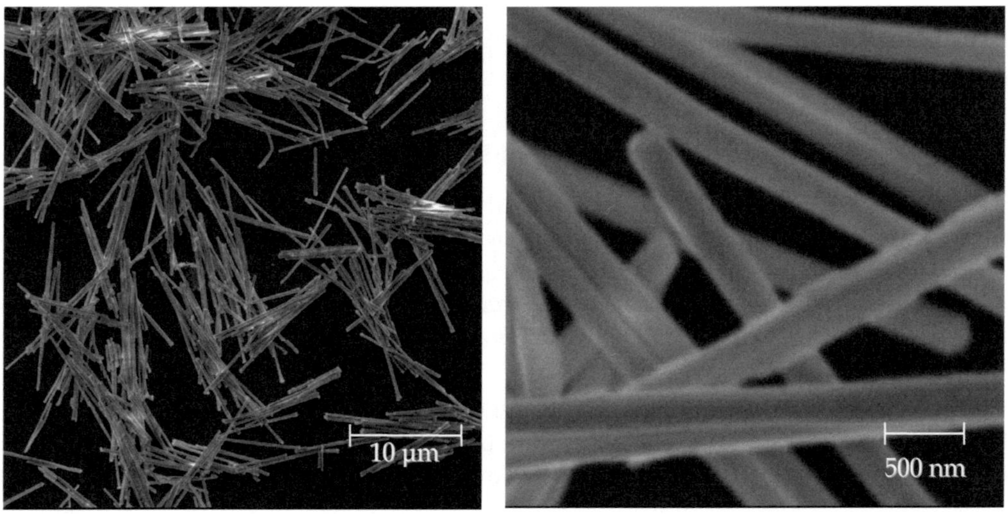

Fig. 2. SEM micrographs of 6.8 ± 1.7 µm long cobalt nanowires with diameters of 296 ± 39 nm.

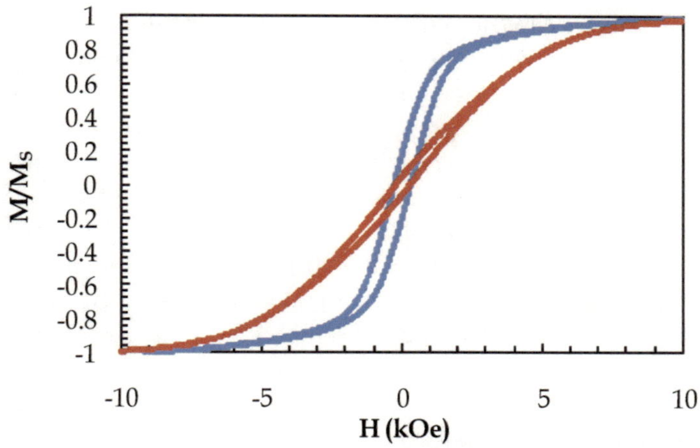

Fig. 3. VSM measurements for 7.89 μm iron nanowires still contained within the alumina template and therefore aligned parallel to one another. Wires aligned parallel to the field are represented by the red curve and wires aligned perpendicular to the field are represented by the blue curve.

ratio. This anisotropy causes the suspended wires within an MR fluid to experience a torque when exposed to the slightly non-uniform magnetic field, resulting in the self-aligning behavior of the nanowires discussed in *section 3.2*. Since the saturation magnetization (M_S) is obtained when all magnetic moments in the material are aligned in the same direction, it is an intrinsic material property which is unrelated to sample morphology. Therefore, we would expect both the spheres and nanowires to have a value that is relatively similar to the bulk material of the same composition. The magnetic saturation of the iron spheres and nanowires is 200.8 emu cm^{-3} and 174.1 emu cm^{-3}, respectively. When both are assumed to have a bulk density, the magnetic saturation for cobalt spheres is 126.3 emu g^{-1} and for the nanowires, it drops to 111.6 emu g^{-1}. This deviation from the bulk value is not surprising due to the oxide layer that is present. It is not unexpected that the nanowires would display a lower value due to increased surface area and the concomitant increase in oxide. As expected, spherical iron particles exhibit a near zero coercivity.

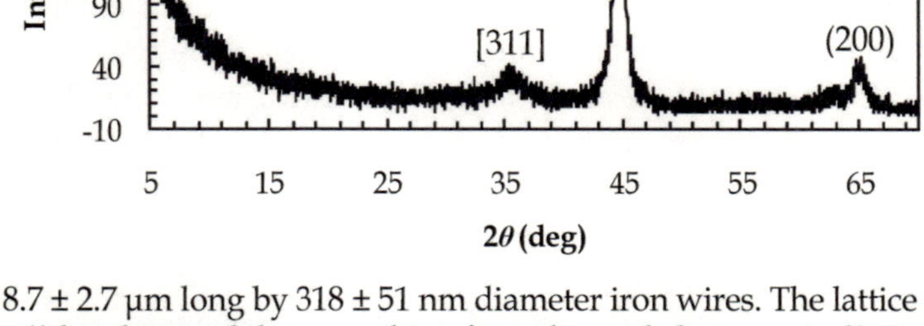

Fig. 4. XRD of 8.7 ± 2.7 μm long by 318 ± 51 nm diameter iron wires. The lattice planes for iron are indicated in ( ) brackets and those resulting from the oxide layer are indicated in [ ] brackets.

To determine the bulk crystal structure and approximate grain size of the crystallites forming the spheres and nanowires, x-ray diffraction (XRD) patterns are acquired using a Scintag X2 automated diffractometer equipped with a Peltier solid-state detector. XRD of the free wires indicates that iron wires grow in a bcc polycrystalline structure and that cobalt wires grow in a hexagonal polycrystalline structure. Using Scherrer's formula, the approximate crystallite size for as-deposited wires was determined to be 23.8 nm, 14.8 nm, and 42.2 nm for the nickel, iron, and cobalt, respectively. The XRD revealed the susceptibility of the iron nanowires to oxidation (some samples display a composition as high as 28% iron oxide) while those of nickel and cobalt displayed only minor peaks (or none at all) for their corresponding oxides. We observed minor metal oxide peaks for the iron spherical particle distributions, but none for those of nickel or cobalt.

2.2 Magnetorheological fluid composition

The MR fluids were prepared by thoroughly mixing the ferromagnetic particles in silicone oil. Lecithin was added to the oil prior to the addition of particles as a surfactant for producing good dispersions with 2 wt. % total particle mass. For the nickel-based MR fluids, the particles were suspended in 0.2 Pa · s silicone oil (Alfa Aesar). Magnetorheological fluids used for the percolation studies (*Section 4*) were prepared using a 0.45 Pa · s viscosity silicone oil. A more viscous fluid was used for these low vol. % suspensions to avoid sedimentation and to prevent the oil from being expelled from the rheometer. All other MR fluids were prepared using a 0.18 Pa · s viscosity silicone oil (GE SF96-200).

2.3 Rheological fluid testing

To determine the flow curves (shear stress vs. shear rate), measurements were carried out on an Anton-Paar Physica MCR300 parallel plate rheometer equipped with a MRD180 magnetic cell. A standard gap of 1 mm was maintained between the plates and a nominal 0.3 mL sample of fluid was placed between them. A Hall probe (F.W. Bell FH301) was placed within the gap to calibrate the electromagnet current of the rheometer in terms of the magnetic flux density within the fluid-containing gap. For the percolation studies, a gap of 0.5 mm was maintained between the plates and a nominal 0.15 mL sample of fluid was used. A magnetic flux density from 0 to 1.0 T was used for the various studies and the temperature of all samples was maintained at 25°C with a closed-cycle cooling system. To avoid sedimentation, the tests were started as soon as the thoroughly-mixed fluid was injected between the disks.

For the preliminary studies of the nickel nanowire-based MR fluids, a custom built rheometer was constructed. This device consisted of a C-shaped electromagnet capable of generating fields up to 0.6 T as measured by a Hall probe between the poles of the magnet. The sample was then placed tightly between the poles. To measure effective viscosity, a thin aluminum plate was submersed 1.3 cm into the fluid with 1 mm of the MR fluid on either side of the plate. With the magnetic field applied, a force transducer (range of 0.02 - 45 N) measured the force required to pull the plate out of the fluid at a rate of 1.06 mm/s. However, due to the poor resolution of this force transducer, the maximum field of 0.6 T was necessary to measure the yield stress within a reasonable uncertainty. Thus, all future studies were conducted on the aforementioned Anton-Paar rheometer.

2.4 Sedimentation testing

The sedimentation velocity of particles within the suspensions was determined by exploiting the magnetic properties of MR fluid particles. As seen in Fig. 5, the inherent density difference between the carrier fluid and magnetic particles results in sedimentation of the particles, leaving a volume of supernatant fluid (the clarified fluid above the sediment mudline). To quantify and compare the sedimentation velocity of conventional sphere-based MR fluids with those containing nanowires, an inductance-based solenoid sensor was constructed to track the mudline location of the settling fluid (Ngatu & Wereley, 2007). The sedimentation velocity is defined as the rate at which the mudline descends due to particle settling, until the particles pack tightly at the bottom of the vial without further sedimentation. The magnetic permeability of the MR fluid is highly dependent on the volume fraction of fluid. Therefore, the inductive sensor solenoid is maintained within this region. The permeability of the MR fluid within the sensor region is related to the sensor inductance L as

$$L = \frac{N^2 A \mu_0}{l} \mu_r \tag{1}$$

where N is the number of turns of the solenoid, A is cross-sectional area of the coil, l is solenoid length, μ_0 is the vacuum permeability, and μ_r is relative permeability of the MR fluid. Thus, the rate of change of the mudline can be determined as it traverses through the sensor by measuring the rate of change of magnetic inductance within the sensor.

Another method to quantify the sedimentation properties of fluids is the percentage of sedimentation, θ, given by

$$\theta = \left(\frac{\text{volume of supernatant fluid}}{\text{total volume of fluid}} \right) \times 100\% \tag{2}$$

The percentage of sedimentation is an indication of how tightly-packed the particles become once settled and is directly related to the ease with which the suspensions can be re-dispersed.

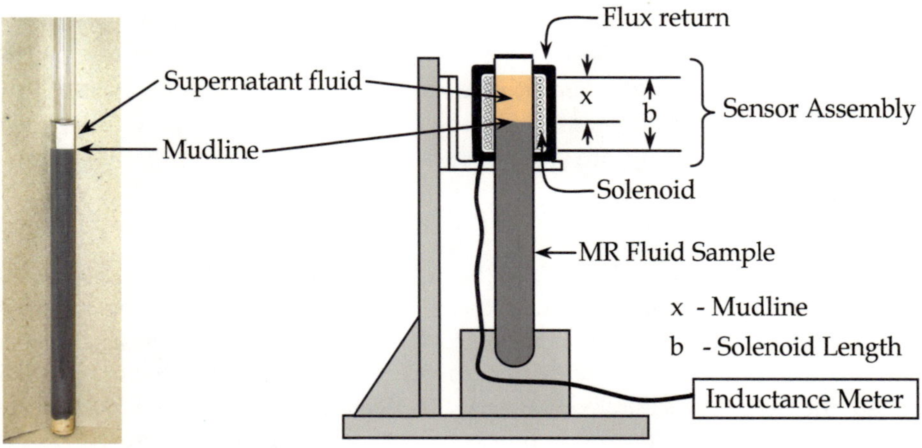

Fig. 5. Sedimentation testing apparatus for determining settling velocity of MR fluids. The inductance measuring instrument is comprised of the sensor assembly, an inductance meter, and a stand to mount the apparatus and sample.

3. Nanowire-based magnetorheological fluids

3.1 Nickel-based MR fluids

To investigate the feasible performance of the new nanowire-based MR fluids, we conducted a preliminary investigation of the rheological properties and settling characteristics of fluids containing nickel nanowires with an average length of 5 - 25 µm and diameters of 300 ± 30 nm (Bell et al., 2007). Using the custom-designed rheometer described in section 2.3, the rheological properties of these fluids were compared to conventional fluids containing nickel-carbonyl spheres (range of 1-10 µm dia.) having the same volume percent of suspended particles. The rheological and settling properties of the MR fluids consisting of 7.6 vol. % nickel nanowires were compared to fluids containing 7.6, 17.5, and 40.0 vol. % nickel spheres. These initial studies indicated that there was no difference in the maximum yield stress of the suspensions. However, the settling properties of the fluids were significantly different. The nanowire-based fluids displayed no settling after several months (discussed further below). However, these studies suffered from several shortcomings. The custom-built rheometer lacked the sensitivity to examine fluids below magnetic saturation and could not measure the rheological properties at high shear rates. In addition, the wires were not well-characterized and an average wire length could not be determined from the limited SEM images acquired for this study. However, the settling results were so promising that these initial studies paved the way for more carefully constructed fluids using instrumentation capable of examining the rheological and sedimentation properties with greater sensitivity.

3.2 Iron-based MR fluids

We then focused our attention on iron nanowire-based MR fluids with two distinct length distributions: 5.4 ± 5.2 µm and 7.6 ± 5.1 µm with diameters of 260 ± 30 nm. These were compared with conventional sphere-based MR fluids (Bell et al., 2008). These studies illustrated the distinct differences between conventional MR fluids employing spherical particles to nanowire-based fluids with the same ferromagnetic material and particle loading. The rheological properties of the fluids were determined using the parallel plate rheometer modified with an electromagnet to allow application of a variable magnetic field across the sample. We plotted steady-state flow curves of shear stress (τ) versus shear rate ($\dot{\gamma}$) as seen in Fig. 6. A Bingham-plastic (BP) constitutive model (Jolly et al., 1999) was used to fit the flow curve to determine the apparent yield stress (τ_y) and the post-yield viscosity, η. The BP model for the viscoplastic flow with yield stress is given by

$$\tau = \tau_y + \eta\dot{\gamma} \qquad (\dot{\gamma} > 0) \tag{3}$$

The values of τ_y and η were determined by fitting with a weighted least-squares-error minimization for each fluid tested at all values of the applied field. By using the measured shear rates as the weighting factor, the model provided a better fit to the high shear rate data. As shown in Fig. 7, the dynamic yield-stress displays a dependence on the applied field for all the fluids studied. At a saturated magnetic flux density, it approaches a maximum value of 4.72 kPa for fluids containing 6 vol. % iron spheres. For fluids containing the 5.4 µm iron nanowires in silicone oil, we observed maximum yield stresses of 0.65, 2.23, and 4.76 kPa for the 2, 4, and 6 vol. % fluids, respectively. The yield stress increased to 8.23 kPa for the 6 vol. % fluid when the average length of the wires increased to 7.6 µm.

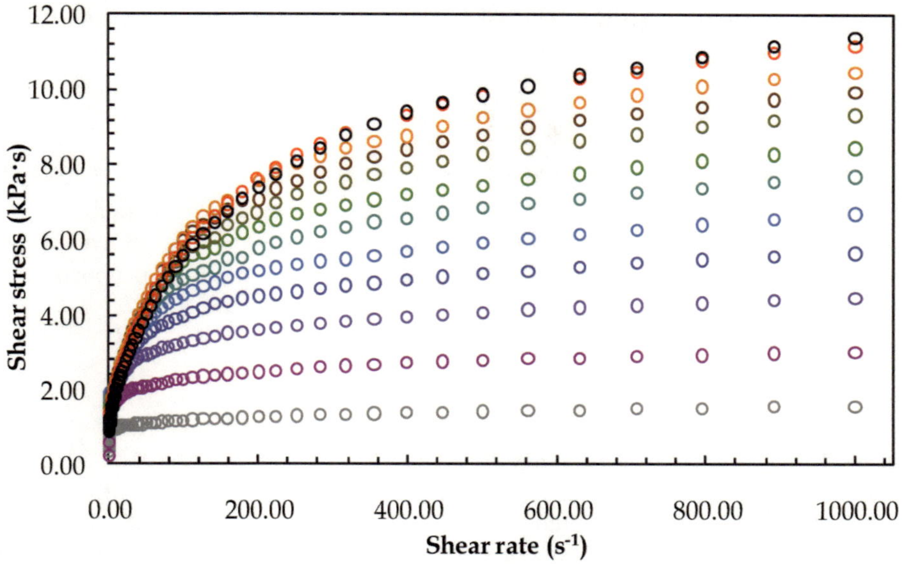

Fig. 6. Experimental flow curves for fluids containing 6 vol. % 7.6 μm Fe nanowires. The applied magnetic field was varied from 0.0 T (bottom) to 0.72 T (top).

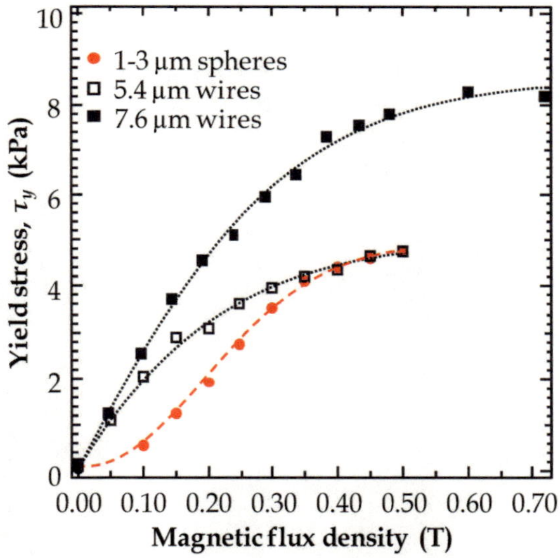

Fig. 7. Yield stress versus magnetic flux density for MR fluids containing 5.4 μm and 7.6 μm iron nanowires and for those fluids with 1 - 3 μm iron spheres. All fluids have a 6 vol. % particle loading. The dotted curves are included simply as a visual aid.

As seen in Fig. 7, the nanowire-based fluids exhibit a more sensitive response displaying a higher yield stress at low field as compared to the sphere-based fluids. This enhancement in the response is likely attributed to magnetic shape-anisotropy of the wires. The nanowires experience a torque within the applied magnetic field forcing them to align with their long axis parallel to field (Qi et al., 2003) and they experience a greater magnetization due to the shape anisotropy (Sun et al., 2005).

Models of the yield stress (τ_y) for sphere-based fluids predict a characteristic response at intermediate fields below saturation as given by

$$\tau_y \propto \phi \mu_0 M_S^{1/2} H^{3/2} \tag{4}$$

where μ_0 is the permeability of free space, ϕ is the volume fraction, M_S is the magnetic saturation of the material, and H is the strength of the applied magnetic field. The sphere-based suspensions used in this study displayed similar behavior. However, for the nanowire-based suspensions, the trend in the yield stress is found to be proportional to the square root of the applied field,

$$\tau_y \propto \phi \mu_0 H^{1/2} \tag{5}$$

How the magnetic saturation and shape anisotropy of the particles quantitatively affects the yield stress of the nanowire-based suspensions is still not well understood. The magnetic saturation of the nanowires is a function of their length and diameter, with longer wires (with fixed diameter) having a higher magnetic saturation (Han et al., 2002). Yet the difference in the magnetic saturation of the particles cannot account for the substantial difference in the yield stresses of the fluids containing the two wire lengths observed in our studies. Therefore, other factors such as magnetostatic coupling between neighboring wires (this tends to create a magnetic easy axis perpendicular to the wire axis), inter-fiber friction, and structural differences of the two types of suspensions in an applied magnetic field must also be considered. Some structural differences between sphere-based fluids and nanowire-based fluids are apparent in the SEM micrographs of Fig. 8.
Knowing the structure of the suspension both in the presence (on-state structure) and absence (off-state structure) of a magnetic field is essential to understanding the properties of these fluids. To generate these images, the wires were suspended in a drop of water and a uniform magnetic field was applied across the droplet. The droplet was then allowed to

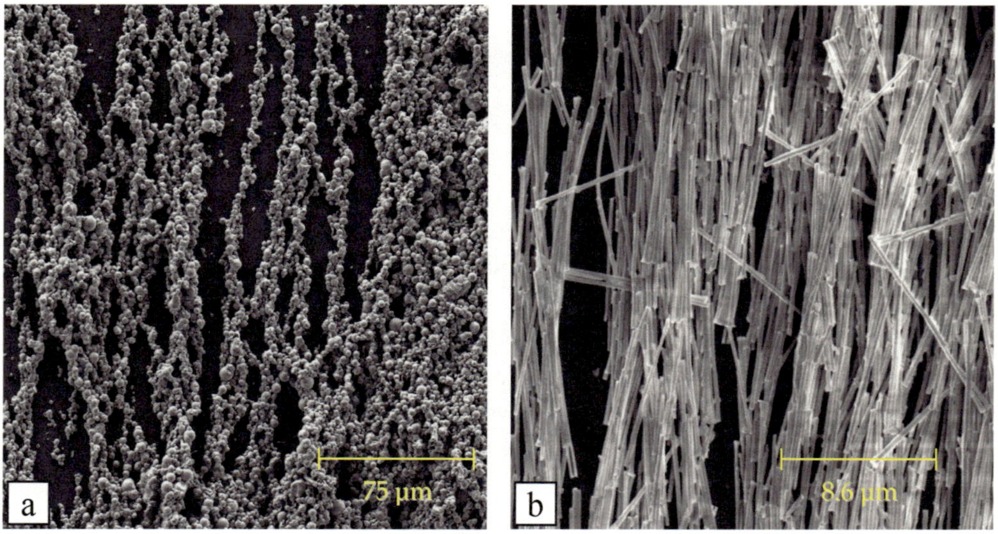

Fig. 8. SEM micrograph showing the columnar structures of (a) spherical 1 - 3 µm diameter iron particles and (b) 10.6 µm long iron nanowires with 260 nm diameters formed in the presence of a magnetic field.

evaporate while the field was maintained. The spheres line up to form interconnected columns as depicted in Fig. 1. In addition to forming columnar structures, the nanowires of the wire-based fluids also tend to interlock with one another due to their elongated shapes.

There are a number of factors that must be considered to understand the increase in the yield stress of the nanowire-based fluids as compared to the sphere-based fluids. As noted previously, for sphere-based fluids, yield stress scales with particle size. If the increase in yield stress depended on the volume of the individual particles alone, we would expect the fluids containing the 1 - 3 μm spheres to exhibit much higher yield stresses than the wire-based fluids. (The volume of a 2 μm diameter sphere is 4.19 μm^3 and the volumes of 5.4 and 7.6 μm long wires with 260 nm diameters are 0.29 and 0.40 μm^3, respectively.) At fields below magnetic saturation, the wires have a smaller demagnetization factor and thus it is easy to understand their more sensitive response at lower magnetic fields. But above saturation, differences in magnetic saturation no longer apply.

Given that the suspensions contain the same mass of iron particles (and therefore, the same volume), the magnetic permeability of the two types of suspensions is expected be very similar. However, the local packing density of the nanowires may be slightly higher than that of the pure spheres (Bell et al., 2008), which would result in a higher local permeability and thus stronger columnar structures within the suspension. The packing density of the two nanowire distributions should be similar and we would expect the yield stress of these two fluids to be similar. However, it is apparent from Fig. 7 that the longer nanowires display a much greater yield stress than the shorter wires at the same loading. This suggests that the packing efficiency of the fluids may only play a minor role in the overall yield stress of the fluid.

The off-state viscosity of the fluids is also an important factor that must be accounted for when developing new types of MR fluids; the lower the off-state viscosity, the larger the range of achievable viscosities. The pure silicone oil used in these studies had a viscosity of 0.41 Pa · s and the off-state viscosity for the 6 vol. % fluids containing spheres was 0.66 Pa · s. The nanowire-based fluids demonstrate a similar off-state viscosity of 0.68 Pa · s for the 6 vol. % fluids, which drops to 0.50 Pa · s for the 2 vol. % fluids. However, as we now discuss, the nanowire-based fluids display a much higher degree of viscoelasticity.

Nanowire-based fluids display no discernable settling at volume fractions greater than ~4 vol. % of ~11 μm long wires after 2 months of sitting undisturbed. The exact vol. % required to effectively stop sedimentation (under vibration free conditions) depends on the length of the wires: shorter wires require higher volume percentages than longer ones. A disadvantage of the nanowire-based suspensions is that the maximum volume fraction of nanowires (< 10 vol. %) is much less than the desired 30 - 40 vol. %, achievable with spherical particles, which is necessary for high yield-stress applications. For these reasons, suspensions that contain strictly nanowires would be more useful in applications where sedimentation would be extremely detrimental or where low yield-stress is sufficient.

On the other hand, fluids employing 6 vol. % spheres displayed 79% settling in just 48 hours. The difference in the sedimentation (and the viscoelastic properties) of the two fluid types is likely a result of several factors, including nanowire entanglement, increased surface area compared to spheres, and hydrostatic and magnetic interactions between nanowires, particularly those due to remnant magnetization. Comparing the total surface area of the particles in fluids with the same volume fraction of particles, fluids containing 5.4 μm long nanowires with 260 nm diameters have more than 5 times the total surface area per volume than fluids with 2 μm diameter spheres (this difference varies little with wire length).

However, entanglement of the nanowires is likely the primary factor responsible for the differences between the sedimentation and viscoelastic properties of the two types of fluids. Entanglement occurs when the volume fraction of wires is high enough so that neighboring wires restrict the end-over-end rotational and translational motion of the wires perpendicular to the long axis (Doi & Edwards, 1978). However, this does not greatly affect the off-state viscosity of the nanowire-based suspensions under shear because the wires tend to line up and slide past one another easily, resulting in a viscosity that is similar to that of the carrier fluid at low particle concentrations (Bercea & Navard, 2000).

4. Percolation phenomena in MR fluids

The ability to predict the effective physical properties of MR-fluid composites from the known properties of the constituents would have a tremendous impact for numerous industrial processes. However, formulating an effective-medium model is complicated when the physical properties of the constituents vary greatly. In such cases, the composite material typically exhibits a phase transition at a critical volume fraction of the component of interest. The prototypical example is that of a metal-insulator composite with electrical conductivity that can change by 16-20 orders of magnitude as a critical volume of the conducting component is reached. Such a phase change in the physical properties of materials is known as a percolation transition. Numerous studies over the past few decades have focused primarily on electronic and thermodynamic transport and to some extent on the elastic and mechanical properties of materials near the critical point.

Various percolation models involving static matrices have been developed using both discrete-lattice networks and continuum networks (Stauffer & Aharony, 1994, and references therein). Each of these models attempts to determine the scaling behavior of the system using different critical exponents (s and t, in the case of conductivity networks) to describe the nature of the transition as the critical volume fraction (ϕ_c) is approached. In so-called elastic percolation networks (EPNs), a two-component mixture consists of bonds with finite elastic modulus together with completely non-rigid bonds. As ϕ_c is approached from above, the general elastic modulus (G) of an EPN is thought to vanish as (Feng et al., 1987; Deptuck et al., 1985; Sahimi, 1996),

$$G \sim (\phi - \phi_c)^f \tag{6}$$

In contrast, a superelastic percolation network (SEPN) consists of a mixture of perfectly rigid bonds and bonds with finite elastic constants. Below the critical point, the elastic moduli diverge as (Bergman, 1986),

$$G \sim (\phi_c - \phi)^{-c} \tag{7}$$

Some studies indicate that for $\phi > \phi_c$, the elastic exponent is equal to the conductivity exponent, $f = t$ (Tokita & Hikichi, 1987; Craciun et al., 1998). Other studies on the rheology of magnetic suspensions exhibiting elastic behavior suggest that f depends on concentration, varying from 1.0 to 2.26 (Kanai et al., 1992). However, most experimental measurements and theoretical models indicate that $f > t$, with typical f-values in the range of 3.5 – 4.0 (Feng et al., 1987; Deptuck et al., 1985; Sahimi, 1996), which are attributed to the higher tensorial order of EPNs. Studies of SEPN behavior in two dimensions indicate that the critical exponent c should scale as the superconductivity exponent s (Bergman, 1986), while numerical simulations have shown c to be slightly smaller than s (Feng, 1987).

To investigate further the rheological behavior of nanowire-based MR fluids, we have systematically varied ϕ in cobalt-nanowire MR fluids and measured the dynamic yield stress for varying applied magnetic field strengths (Zimmerman et al., 2009). In doing so, we observe a percolation transition in the yield stress and find the behavior near the critical point to depend on the magnitude of the applied field.

We varied the applied magnetic field incrementally over the range of 0 – 1.0 T for samples having cobalt-nanowire volume fractions in the range ϕ = 0 – 0.06 (note that we are using volume fraction rather than vol. %). Figure 9 shows a series of τ_y versus ϕ curves that reveal the dependence of the critical volume fraction and the transition onset on the applied field. To characterize the data we employed a normalized, two-parameter fit based on Eq. 8 and 9 using values of the yield stress extrapolated to $\phi = 0$ ($\tau_{low} \sim 5 - 12$ Pa) and far beyond ϕ_c ($\tau_{high} \sim 10^5$ Pa).

$$\tau_y = \tau_{low}\left(\frac{\phi_c - \phi}{\phi_c}\right)^{-c}, \phi < \phi_c \tag{8}$$

$$\tau_y = \tau_{high}\left(\frac{\phi - \phi_c}{1-\phi}\right)^{f}, \phi > \phi_c \tag{9}$$

The data exhibit several trends: c decreases with increasing field, while ϕ_c increases, and f appears to be relatively constant, independent of the applied field. The extracted values of f for all field-strengths range from 1.0 to 1.2. These are intriguingly reminiscent of the

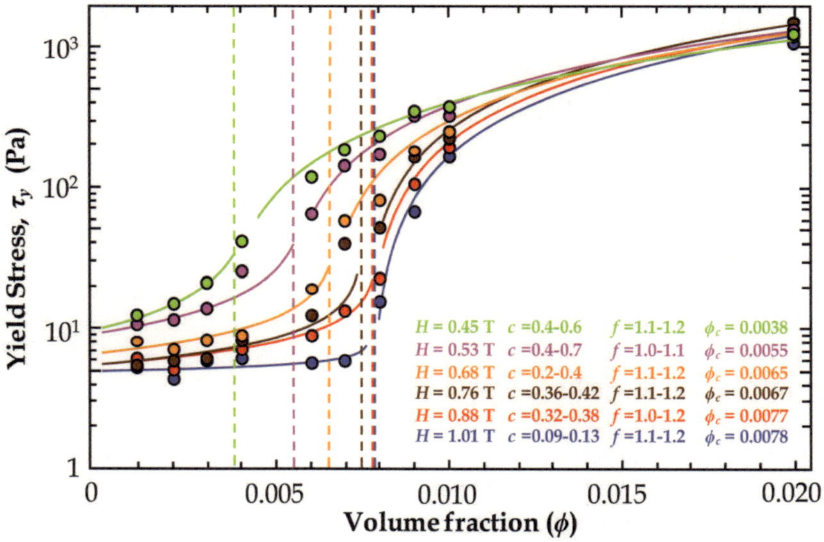

Fig. 9. The apparent yield stress (τ_y) of on-state, cobalt-nanowire MR fluids as a function of nanowire volume fraction (ϕ) for varying applied magnetic fields. The solid curves represent the best fit to expected power-law dependence given in Eq. 8 and 9. Data up to ϕ = 0.06 was used in the analysis, but not shown for the sake of clarity near the critical point.

conductivity exponent in two-dimensional static matrices ($t \sim 1.3$) and the values of the elastic exponent ($f \sim 1.3$) found in two-dimensional, static-matrix numerical studies (Plischke & Joós, 1998; Farago & Kantor, 2000). This suggests that an MR fluid system subjected to a relatively uniform magnetic field behaves two-dimensionally, possibly because the applied field defines a preferred spatial direction for chain formation. The variation in c and ϕ_c are difficult to explain in light of known scaling models. In conductivity networks, the critical volume fraction is expected to decrease as the metal/insulator particle-size ratio decreases (Kusy, 1978), and at the same time, increase with aspect ratio due to the effect of excluded volume (Balberg et al., 1984). Yet, neither of these is a variable in the MR fluid system under study. The critical exponent, c, presumably related to the growth of elastic clusters below the transition (ferromagnetic chains in this case), varies considerably with the magnetic field. The variation in c may be describing the changes in the random reinforcement of the MR fluid by chain growth below the critical volume fraction that does not span the sample. In all likelihood, the fluid-chain morphology of the on-state MR fluids and the dynamic nature of the rheological measurements may also play a significant role in the character of the percolation transition. The importance of this issue cannot be underestimated in light of the numerous applications that exploit the MR effect.

While the values of the critical volume fraction are reasonable for nanowires having a high aspect ratio (~30), the dependence of critical volume fraction on the applied field is unexpected and likely attributed to the dynamic nature of the measurement technique and the inhomogeneities of the rheometer magnetic field. Unlike the present study, involving a dynamic-matrix fluidic system with both short-range contact and long-range magnetic interactions, the vast majority of other studies have considered static systems with short-range interactions or those involving direct particle-particle contacts. In future studies, we will vary the nanowire aspect-ratio, use other ferromagnetic materials (e.g., iron and nickel), and employ quasistatic, rather than dynamic measurements, to help separate these issues from the underlying physical mechanisms.

5. Dimorphic magnetorheological fluids

Dimorphic MR fluids were conceived as a way to realize the benefits of both nanowire-based and conventional MR fluids (Ngatu et al., 2008). In particular, we wanted to synthesize an MR fluid having low sedimentation properties yet having sufficiently high particle-loading to achieve high yield stress. We investigated the rheology and sedimentation stability of these dimorphic fluids, which substitute a given percentage of spheres within a conventional MR fluid for nanowires. For these studies, we used iron spheres of diameter 8 ± 2 μm and iron nanowires with diameters of 230 nm and lengths ranging 7.6 ± 5.1μm suspended in a 0.18 Pa · s silicone oil. Samples were generated for total particle loadings of 50, 60, and 80 wt. % combined with nanowire loading from 0 to 8 wt. % in 2 wt. % increments. This resulted in off-state viscosities for fluids with a total loading of 80 wt. % ranging from 1.3 Pa · s for the pure sphere fluids to 4.4 Pa · s for the 8 wt. % nanowire-substituted dimorphic fluids.

Dimorphic MR fluids with moderate particle loading of 50 and 60 wt. % displayed yield-stress behavior very similar to conventional fluids as seen in Fig. 10. However, we observe slight differences in the yield-stress of dimorphic MR fluids at 80 wt. %. At low field strengths, a slight increase in yield stress with increasing wire substitution was observed, consistent with the enhanced response observed in pure nanowire MR fluids at lower field strengths. At moderate field strengths there were no differences in performance. At high fields approaching the magnetic saturation of the particles, there was an increase in yield-stress of 10% for 2 wt. % substituted dimorphic fluids. Yet, with a further increase in nanowire substitution we observe a ~5 % *decrease* for 8 wt. % substituted dimorphic fluids. This was hypothetically explained by considering the chain structures formed in conventional fluids compared to those formed by dimorphic fluids. At higher loadings, SEM micrographs showed that the wires began to interfere with the columnar structure of the on-state fluids.

Sedimentation stability analysis of dimorphic fluids showed drastic improvements across all ranges of particle loading. Time constraints were a factor in this study since only one sample could be tested at a time and each test required a minimum of one month. Thus, not all samples underwent both rheological and sedimentation tests. For fluids in the moderate particle loading ranges of 50 and 60 wt. %, only one conventional and one dimorphic sample were tested so that the more highly-loaded fluid samples of 80 total wt. %, which are more commonly employed throughout industry, could undergo more extensive testing.

Sedimentation tests of the moderately loaded fluids showed a decreased (order of magnitude lower) sedimentation velocity when comparing 50 wt. % conventional fluids to dimorphic fluids with 6 wt. % substitution, as well as 60 wt. % conventional fluids to 8 wt. % substituted dimorphic fluids. Dimorphic fluids with a total loading of 80 wt. % also showed

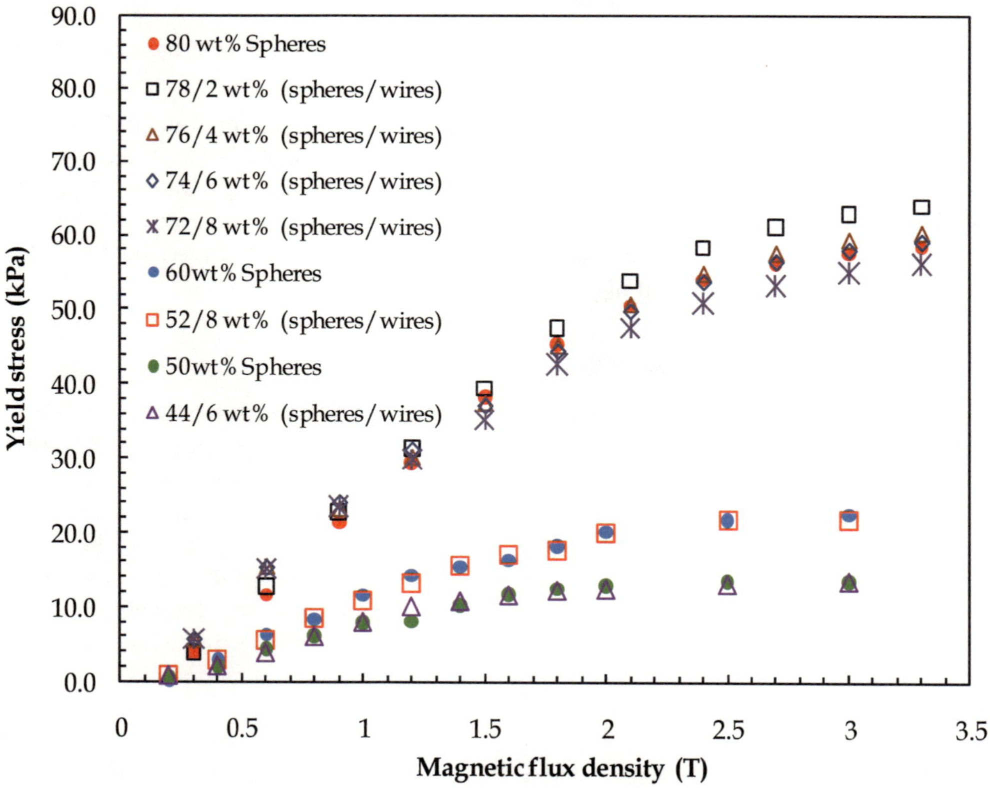

Fig. 10. Yield stress versus magnetic field for conventional sphere-based fluids and for dimorphic fluids of the same total particle loading.

a significant sedimentation rate decrease. The rate of 2 wt. % substituted fluids decrease by a factor of two, while the rate of 4 wt. % substituted dimorphic fluids displayed an order of magnitude decrease. Similarly, the difference in the degree of sedimentation between conventional and dimorphic fluids for the same samples showed a total sedimentation decrease of 50% in 50 wt. % loaded fluids and an 80% decrease in 60 wt. % loaded fluids. The 80 wt. % fluids showed a decrease in total sedimentation of 40% in 2 wt. % substituted fluids and 60% in 4 wt. % substituted fluids. Results of the sedimentation tests are summarized in Table 1.

Composition	Sedimentation Velocity [μm/s]	Sedimentation Percentage (%)
50 wt. % spheres	1.9	72.9
44 wt. % spheres + 6 wt. % wires	0.036	23.4
60 wt% spheres	0.86	66.7
52 wt% spheres + 8 wt% wires	0.017	14.0
80 wt% spheres	0.0254	14.8
78 wt% spheres + 2 wt% wires	0.0122	8.8
76 wt% spheres + 4 wt% wires	0.0021	5.7

Table 1. The sedimentation velocity and percentage sedimentation of iron-based dimorphic fluids containing various wt. % substitutions of spheres by nanowires.

In summary, the benefits of using dimorphic MR fluids over conventional MR fluids are effected by the heterogeneous chain structures formed in dimorphic fluids - the nanowires appear to structurally support the chains that are primarily formed by spheres (Ngatu et al., 2008). However, as wire substitution is increased, there appears to be a threshold reached where the wires begin to interfere with the columnar structures formed by the spheres resulting in a lower yield stress at higher wire substitution.

6. Magnetorheological elastomers

Elastomers (i.e., silicone, polyurethane, natural rubber, etc.) have been commonly used for "passive" shock and vibration attenuation, yet passive damping performance is limited due to the fixed viscoelastic properties of these materials. To overcome these performance limitations, magnetorheological elastomers (MREs) have been recently introduced and are being actively studied. MREs are multifunctional materials as they can also be used as sensors or actuators similar to other controllable, active-damping materials such as piezoelectrics, magnetostrictives, electroactive polymerics, and so on. However, in contrast to other active-damping materials, MREs can operate in a passive mode as well. MREs operate in the active mode by application of a variable applied magnetic field that changes the frequency-dependent elastic modulus.

We examined both the static and dynamic character of nanowire-based MRE composites (Song et al., 2009a; Song et al., 2009b). These MREs were synthesized using a silicone rubber embedded with ferromagnetic particles of iron and cobalt of varying weight fraction (10, 30, and 50 wt. %). To assess the effect of particle morphology, we compared nanowire-based

MRE composites to those made with spherical particles. By means of a modified material testing machine using both static and sinusoidally-oscillating compressive loads, we measured such field-dependent properties as static and dynamic stiffness, elastic modulus, yield stress, and equivalent damping. To investigate magnetic anisotropy effects in nanowire-based MRE composites, samples were cured both in the presence and absence of a magnetic field to yield samples with nanowires both aligned and randomly oriented with respect to the field, respectively.

The static stiffness of both microsphere- and nanowire-based MRE composites increased linearly as the particle weight fraction increased as shown in Fig. 11. Iron nanowire-based MRE composites displayed a greater sensitivity to the weight fraction than that of iron microsphere-based MRE composites. The increase in stiffness of the nanowire-based MREs is primarily due to the increased surface area of nanowires compared to that of spheres. This difference in static stiffness is thus attributed to the increased contact surface area in nanowires combined with the greater compressive strength of the nanowires compared to the microspheres.

The predicted elastic modulus of the nanowire-based MRE composite was determined by the modified Halpin-Tsai equation (Halpin & Kardos, 1976) and compared with the experimental results. It was shown that the theoretical model can adequately predict the elastic properties of the composites (Song et al., 2009b).

The nanowire-based MRE composites exhibited larger specific moduli (defined as the elastic modulus over the density) than the microsphere-based samples, implying that the nanowire-based MRE composites will display enhanced performance over their heavier counterparts (i.e., lower wt. % nanowire MREs can be used in place of the heavier, higher wt. % microsphere composites). Compared to the iron nanowire-based MRE composite, the cobalt nanowire-based MRE composite displayed an even higher specific modulus.

Using the Bingham model, the total force of an MRE composite can be modeled in terms of the elastic and MR effect as follows:

$$F_{total}(t) = k_e x(t) + c_e \dot{x}(t) + F_{MRE}(t) \tag{10}$$

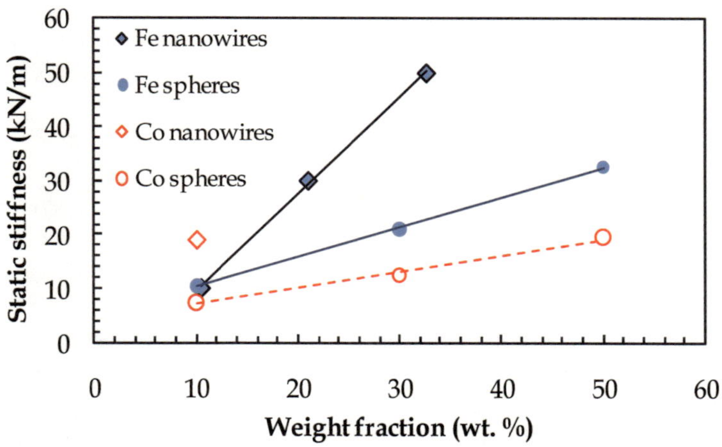

Fig. 11. Measured static stiffness of the nanowire-based MRE composites versus the weight fraction of particles in the absence of magnetic field.

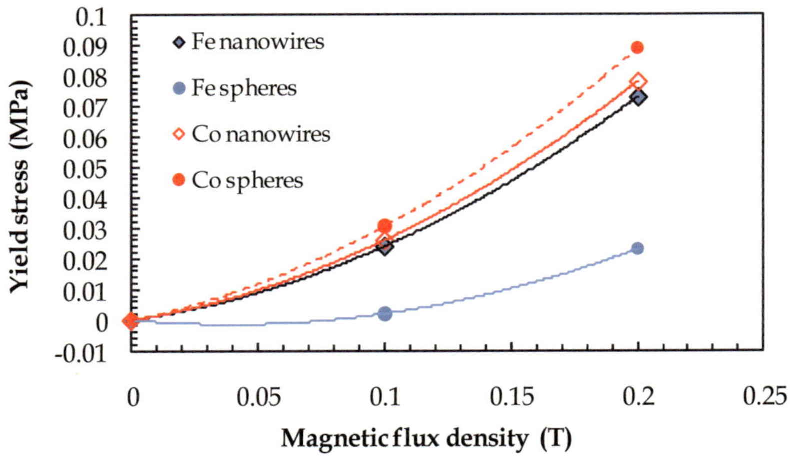

Fig. 12. The yield stress of iron and cobalt nanowire-based MRE composites versus magnetic flux density. The curves are simply a guide for the eye.

with

$$F_{MRE}(t) = \frac{4}{3}\frac{\pi r^3 \tau_y(B)}{\left[x_0 + x(t)\right]}\operatorname{sgn}\left[\dot{x}(t)\right] \quad (11)$$

Here, k_e and c_e are the stiffness and damping coefficients of the MRE composite, respectively. $F_{MRE}(t)$ is the controllable damping force at a given magnetic flux density, r is the radius of MRE composite, and x_0 and $x(t)$ are the initial position and excitation displacement, respectively. In addition, the yield stress can be represented by

$$\tau_y(B) = \alpha B^{\beta} \quad \text{(in MPa)} \quad (12)$$

where a and β are an empirically-determined coefficient and exponent, and B is the magnetic flux density (B) in Tesla. The yield stress of the nanowire-based MRE composites versus the magnetic flux density is shown in Fig. 12. In this figure, the yield stress was calculated from the increment of force (i.e., the difference in yield stress between 0.2 T and zero field) by using Eqs. 10-12. Similar to the yield stress behavior of MR fluids, the yield stress of the MRE composites can be modeled as a power-law function of B.

The dynamic range of the nanowire-based MRE composites was evaluated by measuring field-dependent dynamic stiffness and equivalent damping. The nanowire-based MRE composites had nearly the same dynamic stiffness; however, these composites display greater equivalent damping ranges than the microsphere-based MRE composites (Fig. 13). In addition, the 10 wt. % cobalt nanowire-based MRE composite had an increased damping capacity over that of the 10 wt. % iron nanowire-based sample.

The effect of aligned vs. nonaligned particles in the samples was observed in 10 wt. % cobalt nanowire-based MREs: aligned nanowire-based MRE composite displayed a much greater damping capacity than that of the nonaligned samples (Fig. 14).

The static stiffness of both spherical microparticle- and nanowire-based MRE composites increases monotonically as the particle-weight-fraction is increased. The nanowire-based MRE composites display a greater sensitivity to the weight fraction than that of spherical

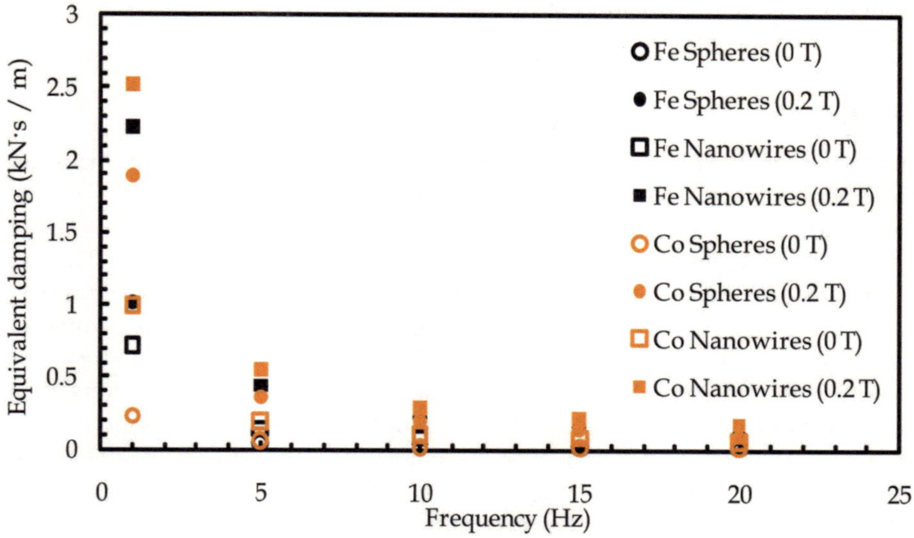

Fig. 13. Equivalent damping for 10 wt% nanowire MRE composites in the frequency domain.

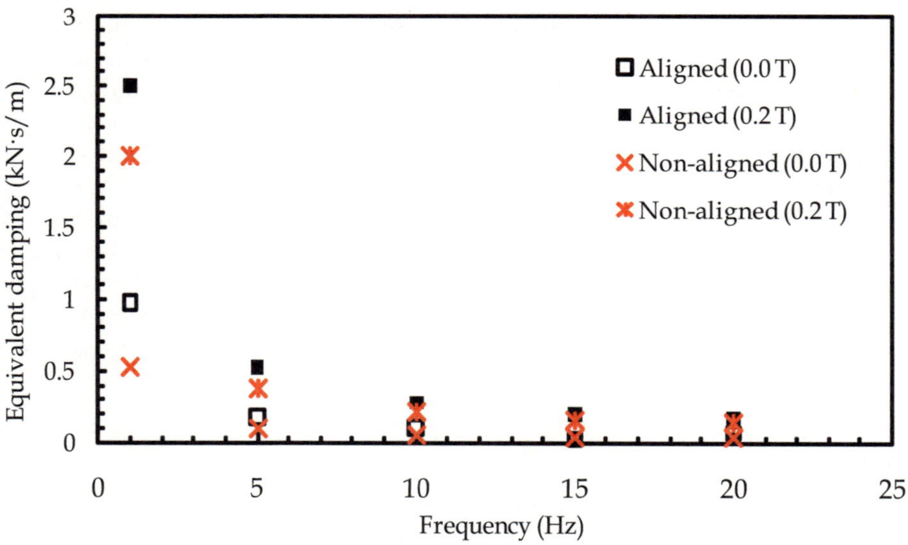

Fig. 14. Equivalent damping of the Co nanowire-based MRE composite: Aligned vs. nonaligned particles.

microparticle-based MRE composites, a result of the increased surface area of the nanowires. The nanowire-based MRE composites exhibited larger specific moduli than the microparticle-based samples, implying that the nanowire-based MRE composites having high specific moduli that can yield enhanced performance over their heavier counterparts. The nanowire-based MRE composites generated greater dynamic stiffness and equivalent damping ranges than the spherical microparticle-based MRE composites. In addition, the 10 wt% Co nanowire-based MRE composite had increased damping capacity over the 10 wt% iron nanowire-based sample.

7. Conclusions

Nanowires are well suited as model particles for the exploration of the shape- and composition-dependent properties of the suspended particles in magnetorheological fluid and elastomer composites. Two major advantages can be seen for the nanowire-based MR suspensions as compared to those that contain strictly spherical particles. The nanowire-based suspensions make for a more efficient MR fluid (higher yield stresses) at low magnetic fields allowing for more sensitive control of the fluid properties. The sedimentation velocity and the percentage of sedimentation are both greatly reduced compared to conventional MR fluids. A disadvantage of the suspensions is that the maximum volume fraction of nanowires is much less than the desired 30-40 vol. % achievable with spherical particles. For these reasons, suspensions that contain strictly nanowires would be more useful in applications where sedimentation would be extremely detrimental and low yield stresses are sufficient. Dimorphic MR fluids exhibit very attractive magnetorheological and sedimentation properties that make them feasible for an expanded set of device applications, since the higher mass loadings of conventional MR fluids can be attained. While the differences in the rheological properties of the fluids containing the spherical and nanowire particles must result from a combination of magnetic and mechanical properties, separating the details of these effects will require more in-depth experimental and theoretical studies. Further work on this new type of MR fluid will consist of examining suspensions with narrow nanowire length distributions and varying the diameters to further understand the dependency of magnetorheological and sedimentation properties on aspect ratio. In the case of nanowire-based MREs, the static stiffness was higher than that of spherical particle-based elastomers in the absence of a magnetic field at the same weight fraction due to the increased surface area of the nanowires as compared to the spherical particles. With nanowire-based MRE samples, the equivalent damping was significantly greater in the presence of a magnetic field. Future studies will include the consideration of aspect ratio dependence, as well as an exploration of the percolation transition within these unique composite materials.

8. Acknowledgements

The authors acknowledge funding support from the National Science Foundation (NSF-CBET-0755696), The Pennsylvania State University, and Altoona College. Additional support provided by a DARPA SBIR Phase 2 Contract No. W31P4Q-06-C-0400 (N. M. Wereley). This publication was supported by the Pennsylvania State University Materials Research Institute Nano Fabrication Network and the National Science Foundation Cooperative Agreement No. 0335765, National Nanotechnology Infrastructure Network, with Cornell University.

9. References

Balberg, I.; Anderson, C. H.; Alexander, S. & Wagner, N. (1984). Excluded volume and its relation to the onset of percolation. *Physical Review B*, Vol. 30, No. 7, (October 1984) pp. 3933–3943, 0163-1829.

Bell, R. C.; Miller, E. D.; Karli, J. O.; Vavreck, A. N. & Zimmerman, D. T. (2007). Influence of particle shape on the properties of magnetorheological fluids. *International Journal of Modern Physics B*, Vol. 21, No. 28-29, (November 2007) pp. 5018–5025, 0217-9792.

Bell, R. C.; Karli, J. O.; Vavreck, A. N.; Zimmerman, D. T.; Ngatu, G. T. & Wereley, N. M. (2008). Magnetorheology of submicron diameter iron nanowires dispersed in silicone oil. *Smart Materials and Structures,* Vol. 17, No.1, (February 2008) pp. 015028-1-6, 0964-1726.

Bercea, M. & Navard, P. (2000). Shear dynamics of aqueous suspensions of cellulose whiskers. *Macromolecules,* Vol. 33, No. 16, (Aug 2000) pp. 6011–6016, 0024-9297.

Bergman, D. J. (1986). Elastic moduli near percolation in a two-dimensional random network of rigid and nonrigid bonds. *Physical Review B,* Vol. 33, No. 3, (February 1986) pp. 2013–2016, 0163-1829.

Carlson, J. D. (2003). Critical Factors for MR fluids in vehicle systems. *International Journal of Vehicle Design,* Vol. 33, No.1-3, (2003) pp. 207–217, 0143-3369.

Chaudhuri, A.; Wereley, N. M.; Kotha, S.; Radhakrishnan, R. & Sudarshan, T. S. (2005). Viscometric characterization of cobalt nanoparticle-based magnetorheological fluids using genetic algorithms. *Journal of Magnetism and Magnetic Materials,* Vol. 293, No. 1, (May, 2005) pp. 206–214, 0304-8853.

Chen, L. S. & Chen, D. Y. (2003). Permalloy inductor based instrument that measures the sedimentation constant of magnetorheological fluids. *Review of Scientific Instruments,* Vol. 74, No.7, (Jul 2003) pp. 3566–3568, 0034-6748.

Chin, B. D.; Park, J. H.; Kwon, M. H. & Park, O. O. (2001). Rheological properties and dispersion stability of magnetorheological (MR) suspensions. *Rheologica Acta,* Vol. 40, No. 3, (May, 2001) pp. 211–219, 0035-4511.

Craciun, F.; Galassi, C. & Roncari, E. (1998). Experimental evidence for similar critical behavior of elastic modulus and electric conductivity in porous ceramic materials. *Europhysics Letters,* Vol. 41, No. 1, (January 1998) pp. 55–60, 0295-5075.

Deptuck, D.; Harrison, J. P. & Zawadski, P. (1985). Measurement of elasticity and conductivity of a three-dimensional percolation system. *Physical Review Letters,* Vol. 54, No. 9, (1985) pp. 913–916, 0031-9007.

Doi, M. & Edwards, S. F. (1978). Dynamics of rod-like macromolecules in concentrated solution 1. *Journal of the Chemical Society-Faraday Transactions II,* Vol. 74, No. 3, (1978) pp. 560–570, 0300-9238.

Farago, O. & Kantor, Y. (2000). Entropic elasticity of two-dimensional self-avoiding percolation systems. *Physical Review Letters,* Vol. 85, No. 12, (September 2000) pp. 2533–2536, 0031-9007.

Feng, S.; Halperin, B. I. & Sen, P. N. (1987). Transport-properties of continuum-systems near the percolation-threshold. *Physical Review B,* Vol. 35, No. 1, (January 1987) pp. 197–214, 0163-1829.

Genç, S. & Phulé, P. P. (2002). Rheological properties of magnetorheological fluids. *Smart Materials and Structures,* Vol. 11, No. 1, (February, 2002) pp. 140–146, 0964-1726.

Halpin, J. C., & Kardos, J. L. (1976). The Halpin-Tsai equations: A review. *Polymer Engineering and Science,* Vol. 16, No. 5, (May 1976) pp. 344–352, 0032-3888.

Han, G. C.; Zong, B. Y. & Wu, Y. H. (2002). Magnetic properties of magnetic nanowire arrays. *IEEE Transactions on Magnetics*, Vol. 38, No. 5, (September 2002) pp. 2562–2564, 0018-9464.

Harland, N. R.; Mace, B. R. & Jones, R. W. (2001). Adaptive-passive control of vibration transmission in beams using electro/magnetorheological fluid filled inserts. *IEEE Transactions on Control Systems Technology*, Vol. 9, No. 2, (March 2001) pp. 209–220, 1063-6536.

Jolly, M. R.; Bender, J. W. & Carlson, J. D. (1999). Properties and applications of commercial magnetorhelogical fluids. *Journal of Intelligent Material Systems and Structures*, Vol. 10, No. 1, (January 1999) pp. 5–13, 1045-389X.

Jones, T. B. & Saha, B. (1990). Nonlinear-interactions of particles in chains. *Journal of Applied Physics*, Vol. 68, No. 2, (July, 1990) pp. 404–410, 0021-8979.

Kamath, G. M.; Wereley, N. M. & Jolly, M. R. (1999). Characterization of magnetorheological helicopter lag dampers. *Journal of the American Helicopter Society*, Vol. 44, No. 3, (July 1999) pp. 234–248, 0002-8711.

Kanai, H.; Navarrette, R. C.; Macosko, C. W. & Scriven, L. E. (1992). Fragile networks and rheology of concentrated suspensions. *Rheologica Acta*, Vol. 31, No. 4, (1992) pp. 333–344, 0035-4511.

Kanu, R. C. & Shaw, M. T. (1998). Enhanced electrorheological fluids using anisotropic particles. *Journal of Rheology*, Vol. 42, No. 3, (May-June, 1998) pp. 657–670, 0148-6055.

Klingenberg, D. J. (2001). Magnetorheology: Applications and challenges. *American Institute of Chemical Engineers (AIChE) Journal,* Vol. 47, No. 2, (February, 2001) pp. 246–249, 0001-1541.

Kordonski, W. I.; Shorey, A. B. & Tricard, M. (2006). Magnetorheological jet (MR Jet (TM)) finishing technology. *Journal of Fluids Engineering – Transactions of the ASME*, Vol. 128, No. 1, (January 2006) pp. 20–26, 0098-2202.

Kusy, R. P. (1977). Influence of particle size ratio on the continuity of aggregates. *Journal of Applied Physics*, Vol. 48, No. 12, (December 1977) pp. 5301–5306, 0021–8979.

Kuzhir, P.; Lopez-Lopez, M. T. & Bossis, G. (2009). Magnetorheology of fiber suspensions. II. Theory. *Journal of Rheology,* Vol. 53, No. 1, (January/February 2009) pp. 127–151, 0148-6055.

López-López, M. T.; Kuzhir, P. & Bossis, G. (2009) Magnetorheology of fiber suspensions. I. Experimental. *Journal of Rheology*, Vol. 53, No. 1, (January/February 2009) pp. 115–126, 0148-6055.

Ngatu, G. T. & Wereley, N. M. (2007). Viscometric and sedimentation characterization of bidisperse MR fluids. *IEEE Transactions on Magnetics*, Vol. 43, No. 6, (June 2007) pp. 2474–2476, 0018-9464.

Ngatu, G. T.; Wereley, N. M.; Karli, J. O. & Bell, R. C. (2008). Dimorphic magnetorheological fluids: exploiting partial substitution of microspheres by nanowires. *Smart Materials and Structures*, Vol. 17, No. 4, (August 2008) pp. 045022-1–8, 0964-1726.

Nishiyama, H.; Katagiri, K.; Hamada, K.; Kikuchi, K.; Hata, K.; Sang-Kyu, P. & Nakano, M. (2005). Evaluations of cluster structure and magneto-rheology of MR suspensions.

International Journal of Modern Physics B, Vol. 19, No. 7-9, (April 2005) pp. 1437–1442, 0217-9792.

Phulé, P. P. & Ginder, J. M. (1998). The materials science of field-responsive fluids. *MRS Bulletin*, Vol. 23, No. 8, (August 1998) pp. 19–21, 0883-7694.

Plischke, M. & Joós, B. (1998). Entropic elasticity of diluted central force networks. *Physical Review letters*, Vol. 80, No. 22, (June 1998) pp. 4907–4910, 0031-9007.

Poddar, P.; Wilson, J. L.; Srikanth, H.; Yoo, J. H.; Wereley, N. M.; Kotha, S.; Barghouty, L. & Radhakrishnan, R. (2004). Nanocomposite magneto-rheological fluids with iniformly dispersed Fe nanoparticles. *Journal of Nanoscience and Nanotechnology*, Vol. 4, No. 1-2, (January-Feburary 2004) pp. 192–196, 1533-4880.

Qi, Y.; Zhang, L. & Wen, W. (2003). Anisotropy properties of magnetic colloidal materials. *Journal of Physics D-Applied Physics*, Vol. 36, No. 1, (January 2003) pp. L10–L14, 0022-3727.

Ramallo, J. C.; Johnson, E. A. & Spencer Jr., B. F (2002). "Smart" base isolation systems. *Journal of Engineering Mechanics*, Vol. 128, No. 10, (October 2002) pp. 1088–1099, 0733-9399.

Rosenfeld, N.; Wereley, N. M.; Radakrishnan, R. & Sudarshan, T. S. (2002). Behavior of magnetorheological fluids utilizing nanopowder iron. *International Journal of Modern Physics B*, Vol. 16, No. 17-18, (July, 2002) pp. 2392–2398, 0217-9792.

Sahimi, M. (1996). Linear and nonlinear, scalar and vector transport processes in heterogenous media: Fractals, percolation, and scaling laws. *Chemical Engineering Journal*, Vol. 64, No. 1, (October 1996) pp. 21–44, 0923-0467.

Sanchis, A.; Sancho, M.; Martínez, G.; Sebastián, J. L. & Muñoz, S. (2004). Interparticle forces in electrorheological fluids: effects of polydispersity and shape. *Colloids and Surfaces A-Physicochemical and Engineering Aspects*, Vol. 249, No. 1-3, (November 2004) pp. 119–122, 0927-7757.

Song, H. J.; Wereley, N. M.; Bell, R. C.; Planinsek, J. L. & Filer II, J. A. (2009a). Field dependent response of magnetorheological elastomers utilizing Fe spherical particles versus Fe nanowires. *Journal of Physics: Conference Series*, Vol. 149, (March 2009) pp. 012097-1–4, 1742-6596.

Song, H. J.; Padalka, O.; Wereley, N. M. & Bell, R. C. (2009b). Impact of nanowire versus spherical microparticles in magnetorheological elastomer composites. *Proceedings of the 50th AIAA/ASME/ASCE/AHS/ASC Structures, Structural Dynamics and Materials Conference*, 9781615671380, Palm Springs, CA, May, 2009, Curran Associates, Inc., Red Hook, New York.

Stauffer, D. & Aharony, A. (1994). *Introduction to Percolation Theory, Revised 2nd Edition*, Taylor and Francis Inc., 0748402535, Philadelphia.

Sun, L.; Hao, Y.; Chien, C. L. & Searson, P. C. (2005). Tuning the properties of magnetic nanowires. *IBM Journal of Research and Development*, Vol. 49, No. 1, (January 2005) pp. 79–102, 0018-8646.

Tokita, M. & Hikichi, K. (1987). Mechanical studies of sol-gel transition: universal behavior of elastic modulus. *Physical Review A*, Vol. 35, No. 10, (May 1987) pp. 4329–4333, 1050-2947.

Weiss, K. D.; Nixon, D. A.; Carlson, J. D. & Margida, A. J. (1997). Thixotropic magnetorheological materials. *US Patent* 5 645 752.

Wereley, N. M.; Chaudhuri, A.; Yoo, J. H.; John, S.; Kotha, S.; Suggs, A.; Radhakrishnan, R.; Love, B. J. & Sudarshan, T. S. (2006). Bidisperse magnetorheological fluids using Fe particles at nanometer and micron scale. *Journal of Intelligent Material Systems and Structures*, Vol. 17, No. 5, (May, 2006) pp. 393–401, 1045-389X.

Yin, J. B. & Zhao, X. P. (2006). Titanate nano-whisker electrorheological fluid with high suspended stability and ER activity. *Nanotechnology*, Vol. 17, No. 1, (January 2006) pp. 192–196, 0957-4484.

Zimmerman, D. T.; Bell, R. C.; Filer II, J. A.; Karli, J. O. & Wereley, N. M. (2009). Elastic percolation transition in nanowire-based magnetorheological fluids. *Applied Physics Letters*, Vol. 95, No. 1, (July 2009) pp. 014102-1–3, 0003-6951.

Zipser, L.; Richter, L.; Lange, U. (2001). Magnetorheologic fluids for actuators. *Sensors and Actuators A*, Vol. 92, No. 1-3, (August 2001) pp. 318–325, 0924-4247.

9

The Applications of Metallic Nanowires for Live Cell Studies

Chiung-Wen Kuo and Peilin Chen
Research Center for Applied Sciences, Academia Sinica
Taiwan

1. Introduction

Recently, there has been increasing research attentions focused on the development of nanomaterials for solving complicated biological problems. Nanoparticles, such as quantum dots or metallic nanoparticles, have been shown to exhibit superior performance to the conventional techniques in biosensing (Nam et al., 2003; Xiao et al., 2003) and biolabeling (Michalet et al., 2005; Medintz et al., 2005). However, the applications of nanoparticles for the studies of living cells are less explored due to the issues of biocompatibility and cytotoxicity (Derfus et al., 2004; Goodman et al., 2004; Chithrani et al., 2006). Noble metals, such as gold, have been used in the biological studies for a long time because of their stability and low toxicity. The use of metallic nanoparticles may offer several advantages in biomedical applications including simple preparation, well-defined size, various available surface modification schemes, and high sensitivity detection. Therefore, there are renewed research efforts in developing metallic nanoparticle based techniques for labeling (Katz & Willner, 2004), drug delivery (Shen et al., 2004; Salem et al., 2003; Sandhu et al., 2002) and gene regulation (Rosi et al., 2006).

A common approach to use nanomaterials for biomedical applications is to chemically modify the surfaces of the nanoparticles such that the nanoparticles can recognize a specific molecule or receptor on the cell surfaces or the nanoparticles can form complexes with drugs or genetic materials to enter the cells. However, in the complicated cellular environments, it often requires individual nanoparticles to possess several functionalities to achieve multiple tasks. The surfaces of nanoparticles may have to be decorated with biomolecules to recognize specific cells or to enhance the uptake efficiency. When the nanoparticles are inside the cells, additional molecules may be needed to help the nanoparticles to escape the endosomes or to reach specific organelles. In addition, the optical properties of nanoparticles may allow monitoring the cellular uptake process and their spatial distribution by optical microscope whereas the magnetic nanoparticles may be used for separation or contrast agent. To engineer the nanoparticles with multiple functionalities, two peptides have been attached to the same gold nanoparticles allowing traversing cell membrane by receptor-mediated endocytosis pathway and endosomal escape (Tkachenko et al., 2003). However, these two peptides were randomly distributed on the surfaces of nanoparticles. It is very difficult to control the spatial distribution of molecules or functionalities on the spherical nanoparticles. This problem can be solved by using non-spherical nanoparticles such as multi-segment nanowires, which could be engineered with different functionalities in a spatial controlled manner through various coupling schemes. It has been shown that two types of molecules have been incorporated onto the gold-nickel nanowires where the gold end was used to bind to a plasmid DNA through electrostatic interaction while the nickel surface was engineered to carry a specific polypeptide for site recognition (Salem et al., 2003). To extend this type of applications, micrometer long metallic nanowires may be useful. Micrometer long multi-segment nanowires have been used as barcodes for biological multiplexing, which could be easily visualized by an optical microscope (Keeting & Natan, 2003). The micrometer long nickel nanowires could be internalized by cells allowing the manipulation of living cells through magnetic field (Tanase et al., 2005). However, it is not known whether the micrometer long metallic nanowires can be internalized by the cells without damaging the cells, which is an important issue for the development of nanowires based living cell probing system. If the nanowires can be internalized by the cells, it is possible to observe the intracellular microenvironment around the individual nanowires through an optical microscope. In this chapter, we will focus our discussion on the fabrication and functionalization of metallic nanowires for probing living cells.

2. Synthesis of metallic nanowires

In general, metallic nanowires can be synthesized by two different approaches: chemical synthesis (bottom up) and template deposition (top down). While the chemical synthesis has been used to produce nanowires with various aspect ratios in great quantity, the compositions of nanowires are very limited due the available precursors and synthesis procedures. On the other hand, template deposition allows not only the fabrication of metallic nanowires with a wide range of compositions but also the synthesis of nanowires with very high aspect ratios. Therefore, the template deposition is widely used in the production of metallic nanowires. To fabricate metallic nanowires, two types of templates are often used: anodic aluminum oxides (AAO) and polycarbonate membranes. The metallic nanowires can be obtained by electro-deposition of the desired materials inside these porous membranes. The back sides of the membranes are coated with a layer of metal, which serves as the working electrode for the electro-deposition. The length of nanowires can be controlled via deposition time and current. The compositions of metallic nanowires can be regulated using various plating solutions resulting in multi-segment nanowires.
For the template deposition, the AAO membrane is the most popular one due to their availability and tunable pore size. AAO templates have been widely used for the fabrication of various one dimensional nanomaterials such as polymers (Jessensky et al., 1998; Xu et al., 2007), semiconducors (Pena et al., 2002; Divliansky et al., 2001) metals (Mitchell et al.,2002) using electrochemical deposition. For example, using sequential electrodeposition, multi-component nanowires have been fabricated (Keating & Natan, 2003). Because of the variations in composition along the length, it has been suggested that different surface chemistry could be used to modify individual segment of the nanowires. These functionalized nanowires have been demonstrated capable of conducting chemical and bio-sensing. In another study, the rotation of the multi-segment nanowires in the present of the hydrogen peroxide was observed due to the catalytic decomposition of hydrogen peroxide on the nickel segment (Fournier-Bidoz et al.,2005). The AAO membrane can be purchased from commercial vendors or prepared in-house by electrochemical etching process. To

prepare AAO templates, thin aluminum sheets are anodized in electrolyte under appropriate anodizing voltage. With different electrolytes such as oxalic acid or sulfuric acid solution and appropriate voltages, it is possible to control the pore size of the AAO membranes. Ordered AAO templates can also be obtained by imprinting an ordered layer of nanostructures on the surfaces and followed by the self-organization process of AAO templates (Masuda & Fukuda, 1995). After the electrodepositon, the AAO templates can be removed by saturated $HgCl_2$ solution.

3. Surface modifications

To utilize the nanoparticles in the biological system, it often requires surface modifications of the nanoparticles to improve their biocompatibility. In addition, the added functionalities through surface modifications would allow the binding of the nanoparticles to specific biomolecules or carry other molecules such as drugs. For metallic nanoparticles, the surface modification scheme normally involves the formation of self assembled monolayer through thiol, amino, cyanide and carboxylic acid group. Charged or hydrophilic groups are often incorporated into the nanoparticle surfaces to provide solubility in water and functionality for interaction with biomolecules. Hydrophilic groups, such as oligo(ethylene glycol)(OEG) and Poly(ethylene glycol) (PEG) are commonly use on the nanoparticle surfaces. Surface adsorption of proteins will result in denaturing of proteins and, in some cases, may limit the interactions of the ligand with the target on the cell surface due to steric hindrance (You et al., 2007). OEG and PEG are known to resist the non-specific interaction with biomolecules, which in turn improves the biocompatibility of nanoparticles.

As for the surface modification of nanorods or nanowires, it is possible to selectively modify the multi-segment nanorods and nanowires. For example, selective surface modification scheme has been employed to modify two component nickel-gold nanowires to achieve bi-functionality (Bauer et al., 2003). In this case, the nanowire were reacted with 11-aminoundecanoic acid and 1,9-nonanedithol. It was shown that the carboxylic acid could bind to the nickel segment while the thiol group was used to modify the gold segment surface. The result of dual functionalization could be observed by the fluorescence microscopy (Bauer et al., 2004). Since multi-segment nanowires could exhibit dual functionalities, they can offer simultaneous detection and separation of multiple species in solution. Mirkin et al. have demonstrated that Au-Ni-Au nanowires could be used for target binding and separation of biomolecules process. To achieve this goal, the gold portions of the nanorods were passivated with 11-mercaptoundecyl-tri (ethylene glycol)(PEG-SH) and fluorescein-tagged poly-His (His > 6) was then bound to the Ni portions of the substrate as evidenced by confocal fluorescence microscopy (Lee et al., 2004; Hurst et al., 2006).

To modify the surfaces of nanowires, a simple approach is to functionalize the surfaces of nanowires with amino, methyl and carboxyl terminal group through corresponding thiols. Additional molecules can be attached to these functional group via different coupling schemes. The surface modification of nanowires could be achieved by mixing the suspended gold or silver nanowires (1×10^8 nanowire/ml) solutions with an ethanol solution of 1 mM of 11-amino-1-undecanethiol, 1 mM of octadecanethiol, or 1 mM of 11-mercaptoundecanoic acid. After 24 hours of incubation, the nanowires were cleaned by the doubly distilled water and the excess thiols were removed by dialysis for 24 hours using a 3.5kD cut off dialysis membrane in the doubly distilled water. For comparison, the serum coated nanowires were also investigated where the serum coated nanowires were obtained by incubating the nanowires in the PBS solution containing serum for 24 hours. All surface modified nanowires can be suspended in aqueous solutions by strong vortexing. However, longer nanowires exhibited higher sedimentation rate. The zeta-potentials of the surface modified nanowires were measured by a Zeta Potential Analyzer at a field of 8-16V/cm.

4. Cytotoxicity of metallic nanowires

There is an increasing trend in using nanomaterails in biomedical applications for the purpose of diagnosis, imaging, and drug delivery. However, it is only until recently that the issues of the toxicity and health effects of the nanoparticles have been widely investigated. In many cases, additional coatings for the nanoparticles are needed to optimize their utility in the cellular studies. For instance, magnetic nanoparticles without polymer coating often suffer from the aggregation in water or tissue fluid, which may limit their applications in isolation and detection. Surface coatings may affect the particle size, physical properties, and the corresponding effects on toxicological properties. For example, it has been reported that the cytotoxicity was not significant for the nickel ferrite nanoparticles with a hydrophilic surface (Yin et al., 2005). However, the cytotoxicity was observed for nickel ferrite nanoparticles modified with a layer of oleic acid. Therefore, it was concluded that the hydrophobic coating was more toxic than the hydrophilic surface. Size is another important factor for cytotoxicity. At higher concentrations, both small and large nanoparticles exhibited similar cytotoxicity. Since the size of the nanoparticles could be related to their surface interaction area and surface energy, it was found larger nanoparticles (about 150 nm) was more toxic than smaller nanoparticles (10 nm). In a separate experiment, it has been shown that the interacellular delivery of anionic iron oxide nanoparticles reduced the ability of PC12 cells to respond to nerve growth factor in a dose dependent manner. The use of anionic magnetic nanoparticles changed cell phenotype and behavior, which could be directly correlated with the level of nanoparticle exposure (Pisanic et al., 2007).

As for the metallic nanowires, the CTAB precursor used in the synthesis process may be the source of the cytotoxicity. (Connor et al., 2005, Takahashi et al., 2006). It has been clearly demonstrated that when the nanowires were coated with phosphatidylcholine molecules, the cytotoxicity was reduced. In another experiment, PEG-modified gold nanorods were used to reduce the nonspecific binding with blood component such as blood proteins and blood cells. For the micrometer long nanowires, we have investigated the cytotoxicity of nanowires with different surface charges for two cell lines, NIH 3T3 fibroblast cells from normal tissue, which is an adherent cell line, the HeLa S3 cells from neoplastic tissue, which can grow in suspension media (Kuo et al., 2007).

To determine the cytotoxicity in the presence of gold nanowires, the cells were first seeded in 96-well plates at a density of $1x10^5$ cell/ml at 37^0C in 5% CO_2 atmosphere. After 24 hours of culture, the wells were refilled with the fresh medium and serial dilutions of nanowires at nanowire concentration ranging from 10^3 to 10^6 nanowire/ml. 90 μl of the nanowires solutions at different concentrations were added to each well. Control experiments were carried out with cells treated with an equivalent volume of serum medium without any nanowires. Cells were then incubated for 24 hours at 37^0C. Cytotoxicity was investigated using an MTT (3-(4,5-dimethylthiazol-2-yl)2,5- diphenyltetrazolium bromide) assay to measure the succinate dehydrogenase mitochondrial activity. PBS solutions containing 10 μl of 5 mg/ml MTT stain were added into each well and incubated for 4 hours. After mixing, 90 μl of MTT solubilization solution was added into each well. The stain was aspirated and the purple color crystal was dissolved with acidic isopropyl alcohol. After 15 minutes, the absorbance in each well was measured at 570 nm in a microplate reader. Background absorbance was measured in PBS solution without the presence of cells and nanowires. All experiments were repeated 3 to 9 times. The cell viability (%) related to the control wells containing the cell culture medium without nanowires was calculated by $[A]_{test}/[A]_{control}$ X 100 where $[A]_{test}$ was the absorbance of the test sample and $[A]_{control}$ was the absorbance of the control sample.

To investigate the cytotoxicity effect of the surface modifications, the aminothiols were used to produce surfaces with positive charge, and the mercapto acids were used to generate negatively charged surfaces whereas the alkanethiols were used as a nonionic surface modification. For comparison, the nanowires coated with serum were also tested. The viability tests for fibroblast 3T3 and HeLa cells were summarized in figure 1(A) and (B). In a typical viability test, 4.5 μm long gold nanowires with various surface modifications were used. The density for both cells was about 10^5 cell/ml, and the density of the nanowires was varied from 10^3 to 10^6 nanowire/ml. After 24 hours incubation of the cells, it was found that the cytotoxicity increased as the density of the nanowires increased and all surfaces modified nanowires except the serum coated nanowires exhibited some degree of toxicity to both cell lines.

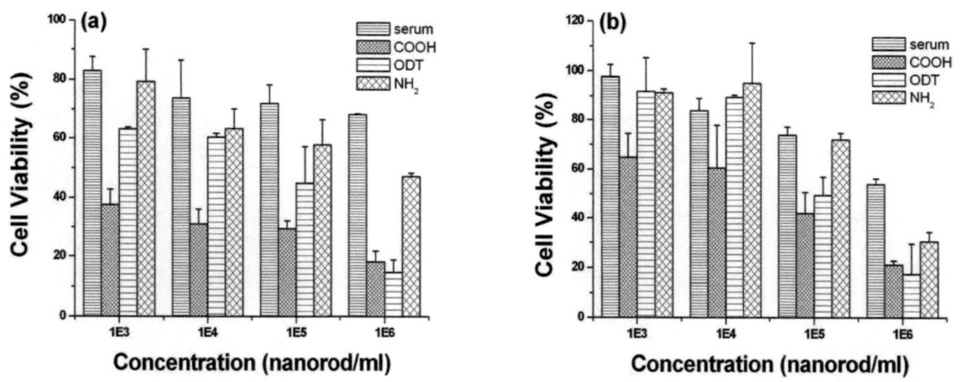

Fig. 1. The cytotoxicity of gold nanowires with various surface modifications for (a) fibroblast cells and (b) HeLa cell.

At a lower density (< 10^4 nanowire/ml), most of the cells were unaffected by the addition of the nanowire solution. However, the carboxyl presenting nanowire surfaces exhibited very strong cytotoxicity even at a very low nanowire density (10^3 nanowire/ml). At a higher nanowire density, most of the cells were injured by the addition of nanowire solution except for the serum coated nanowires. Large aggregation of nanowires was observed at a density of 10^7 nanowire/ml, which made it difficult to evaluate the number of nanowires. At a nanowire density higher than 10^8 nanowire/ml, the bottom of the well was completely covered by the nanowires. The LD_{50} value of the serum coated gold nanowires for 3T3 cells was estimated to be 5 x 10^7 nanowire/ml (~150 μg/ml), which was lower than the value measured for smaller gold nanoparticles (~750 μg/ml) (Salem et al., 2003). The reason for the lower LD_{50} for the gold nanowires could be attributed to larger size nanowires used in this experiment.

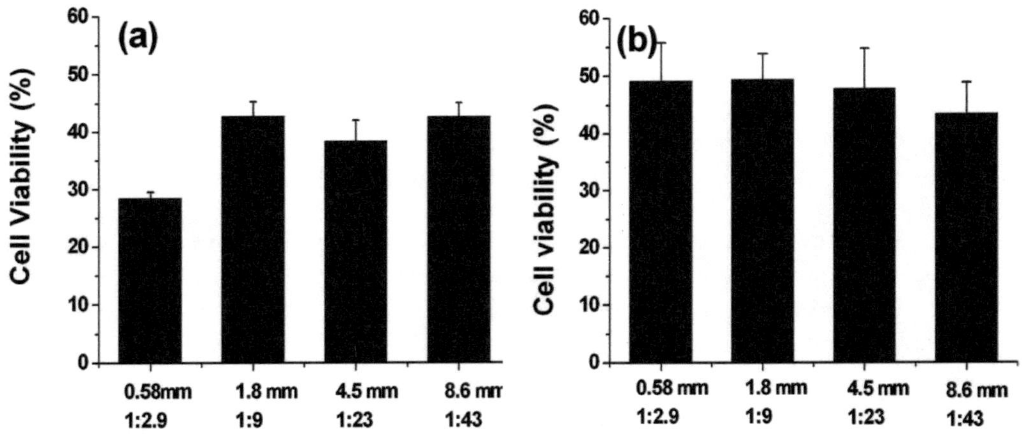

Fig. 2. The cytotoxicity of different sizes of gold nanowires modified with mecapto acid for (a) fibroblast cells (b) HeLa cell.

In a previous study (Chithrani et al., 2006), it was reported that the cellular uptake of gold nanoparticles depended on both the size and the shape of the nanoparticles. With nanoparticle size smaller than 100 nm, it was found that the uptake efficiency of different size of nanoparticles peaked at 50 nm. And it was also concluded that the uptake efficiency decreased as the aspect ratio increased. If the cellular uptake depends on the size of the nanoparticles, the cytotoxicity will also vary with the size of nanoparticles. Therefore, we have investigated the cytotoxicity of different sizes of nanowires with various aspect ratios from 1:2, 1:10, 1:25, and 1:50. In this study, the concentration of mercapto acid modified gold nanowires was 10^5 nanowire/ml for all aspect ratios. The results of cytotoxicity for 3T3 and HeLa cells were depicted in figure 2 (A) and (B). Surprisingly, the cytotoxicity of the micrometer long nanowires exhibits little dependence on their aspect ratios for both cell lines. However, in a separated experiment, the cytotoxicity of 250 nm spherical gold nanoparticles (BBInternational) modified with mercapto acid was tested. The viability for both cell lines was measured to be larger than 80% with nanoparticle concentrations up to 10^7 nanoparticle/ml as shown in figure 3(A) and (B). Considering that the mass concentration of 10^7 nanoparticle/ml 250 nm spherical gold nanoparticles is equivalent to that of 10 μm long nanowires with 200 nm diameter at a concentration of 1.6 x 10^5 nanowire/ml, the nanowires are more toxic than the spherical nanoparticles, which is probably due to the geometry of the nanowires. However, the exact origin of the cytotoxicity of the nanowires requires further investigation.

To further understand the internalization process of the nanowires, we have studied the uptake kinetics of aminothiol modified gold nanowires for both cell lines. The internalized nanowires were obtained by lysing the cells with alcohol and counted by a hemacytometer. The results for the nanowires with various aspect ratios are plotted in figure 4(A) and (B). The uptake kinetics for all sizes of nanowires was very similar in both HeLa and 3T3 cells. Both types of cells exhibited maximum uptake after 8 hours of incubation with the nanowires and decreased after reaching maxima. The maximum internalization efficiency was higher than 50% for both cell lines. The reason for the decrease in the uptake was due to the cell death at high concentration of nanowires as indicated in the viability test (figure 1). Another trend observed in this experiment was that the uptake efficiency for the longer nanowire was lower than the shorter one except the 0.58 μm long gold nanowires, which were too short to determine the number of nanowires accurately. This trend agreed with the previous measurement using smaller nanorods (Bauer et al., 2003). Since nanowires with different aspect ratios exhibited the same cytotoxicity and the uptake efficiency of the nanowires decreased as the aspect ratio increased. We concluded that the internalized nanowires with higher aspect ratio were more toxic to both cell lines. This conclusion agreed with the conclusion that the gold nanowires were more toxic than the spherical (250 nm) gold nanoparticles and also explained why the LD_{50} of the nanowires was lower than the nanorods with smaller aspect ratio.

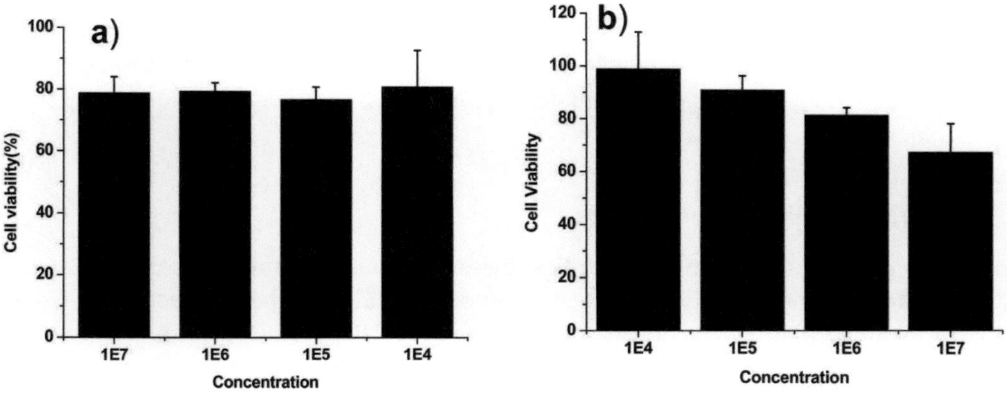

Fig. 3. The cytotoxicity of 250 nm gold nonoparticles at different concentrations for (a) fibroblast cells (b) HeLa cell.

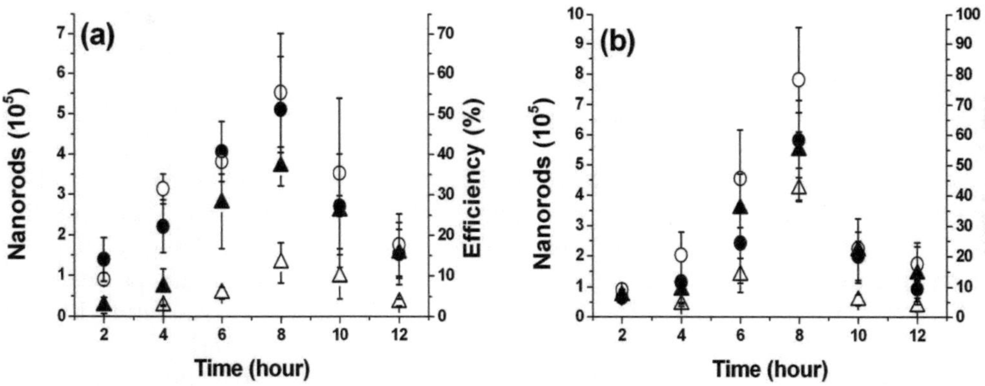

Fig. 4. The uptake of aminothiol modified gold nanowires with four different lengths in (a) fibroblast cell and (b) HeLa cell. 0.5μm (solid circles), 1.8 μm (open circles), 4.8μm (solid triangles), 8.6 μm (open triangles).

The cytotoxicity of silver nanowires was investigated by using 5μm long silver nanowires with four different coatings: positive charge (amino group), negative charge (carboxyl group), hydrophic alkyl group and bovine serum albumin (BSA) coating. After 24 hour incubation, we found that the cytotoxicity of silver nanowires increased as the number of silver nanowires and the cytotoxicity behavior was very similar to gold nanowires for both cell lines as shown in figure 5(a) and (b).

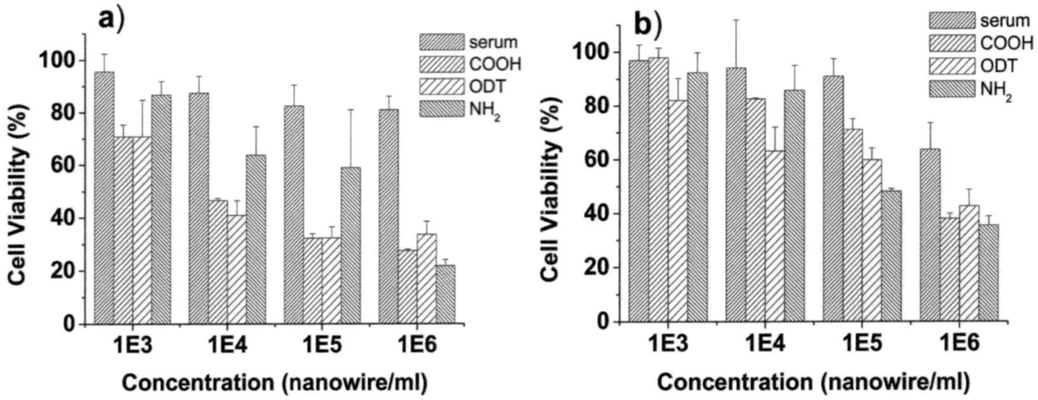

Fig. 5. Cytotoxicity for a) fibroblast and b) HeLa cells in the presence of various surface functionalized silver nanowires.

5. Nanowires as living cell probes

To use nanoparticles for living cell studies, these nanomaterials must remain stable in the intracellular environment and do not disturb normal cellular biochemical activities. Several types of nanomaterials such as magnetic, polymeric, metallic, semiconductor nanoparticles or nanowires have been introduced for the cellular study. For example, the capability of gold nanoparticles to incorporate secondary tags such as peptides to target specific cell types has been investigated (Goodman et al., 2004). In this study, the gold nanoparticles have been functionalized with cationic and anionic side chains. The carboxylated modification on the gold nanoparticles was found to be nontoxic to the cells. In contrast, the cationic side chain bound to the gold nanoparticles exhibited moderately toxicity. It was found that the toxicity of gold nanoparticles was related to their interactions with the cell membranes. In another study, the intracellular delivery of quantum dots (QDs) for live cell labeling and organelle tracking (Defus et al., 2004) were demonstrated. The QDs modified with polyethylene glycol (PEG) were mixed with different transfection reagents and then were delivered to the interior of the HeLa cells. The flow cytometry was used to quantify the amount of QDs delivered to the cells. It was found that QDs often tended to accumulate in vesicles and distributed non-homogeneously in the cytoplasm (Chen & Gerion, 2004).

As regarding to metallic nanoparticles, recent studies have shown that it was possible to use metallic nanoparticles for the targeted nuclear delivery. The nuclear targeting has been achieved by multifunctional gold nanoparticle-peptide complexes (Tkachenko et al., 2003). 20 nm diameter gold nanoparticles were modified with bovine serum albumin (BSA) bound with various cellular targeting peptides. To enter the nucleus of HepG2 cells, these nanoparticles must carry both receptor-mediated endocytosis (REM) and nuclear localization signal (NLS) peptides. Non-sphereical nanorods have also been demonstrated capable of conducting gene delivery. It was shown that the conjugation of DNA plasmid and targeting ligands can be achieved simultaneously in a spatially defined manner. For example, it was demonstrated that gene delivery can be achieved using bifunctional Au/Ni nanorods (Salem et al., 2003). In this approach, the carboxylate terminal group was first attached to the Ni segment. Subsequently, the plasmids were bound to the protonated amines on the surface of nickel segment through electrostatic interactions. The transfferrin was bound to the gold segment of the nanorods through thiolate linkage. These dual-functionalized Au/Ni nanorods were used in a vitro transfection experiment using human embryonic Kindney (HEK 293) mammalian cell line.

We have introduced functionalized gold nanowires as new probes (Kuo et al., 2007). Using various types of functionalization, it was possible to probe the local environment inside the cells by confocal microscopy. To observe the internalization process of nanowires, 5 μm long gold nanowires (200 nm in diameter) coated with serum were added into a glass bottom culture dish, which were placed in a CO_2 incubator on an inverted microscope. The images were recorded by a CCD camera every 10 minutes as shown in figure 6. The internalization process of the gold nanowire to the HeLa cells was recorded by the time lapse microscopy for 4 hours. We found that the micrometer long metallic nanowires can be internalized by cells without damaging the cells, which is important in cellular study.

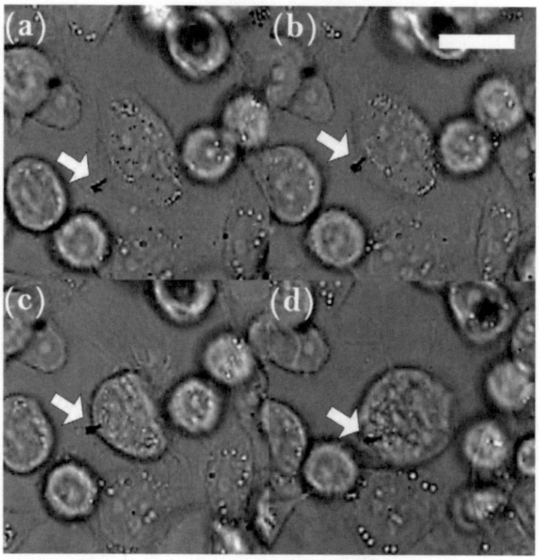

Fig. 6. The phase contrast image of the serum coated gold nanowires internalized by HeLa cell. Bar: 20 μm.

To investigate the capability of delivering DNA molecules into the cells through the nanowires, the 5 μm long gold nanowires were functionalized with aminothiols, which covered the nanowire surfaces with positive charges. The negatively charged plasmid DNA molecules were attached to the nanowires through electrostatic interaction. To demonstrate that the surface functionalized nanowires can carry the plasmids into the cytoplasm, a green fluorescence protein expressing plasmid (pAcGFP1-Actin, BD) was coated on the gold nanowires. For visualization purposes, the plasmid DNA on the nanowire surfaces were further labeled with YOYO-1, which emitted a strong green fluorescence when bound to a double strain DNA, and the cells were stained with Image iT LIVE Plasma Membrane and Nuclear Labeling Kit as shown in figure 7. From the confocal image, it can be clearly seen that the 5μm long functionalizd gold nanowires coated with plasmid were located inside the cell. Knowing that the gold nanowires can delivery plasmid into cells, it is important to investigate the functionality of plasmids on the gold nanowires. Shown in figure 8 is the combined DIC and confocal image of the fibroblast cell expressing GFP using gold nanowires as the gene carrier. This experiment confirmed that the micrometer long aminothiol modified gold nanowires can not only protect plasmid DNA molecules from degradation but also release plasmid NDA molecules inside the cells.

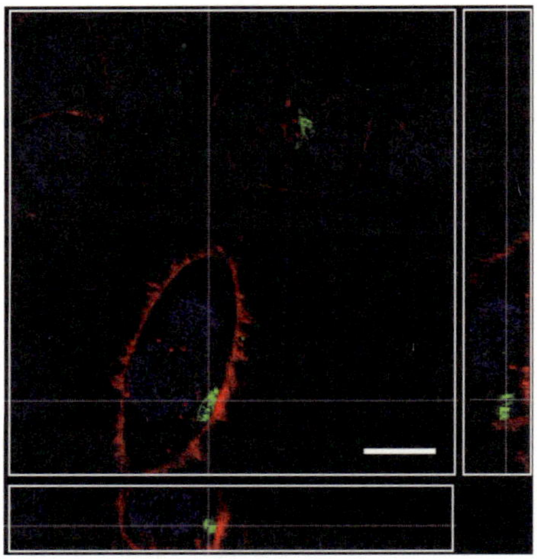

Fig. 7. Stacked confocal image of nanowires in HeLa cell

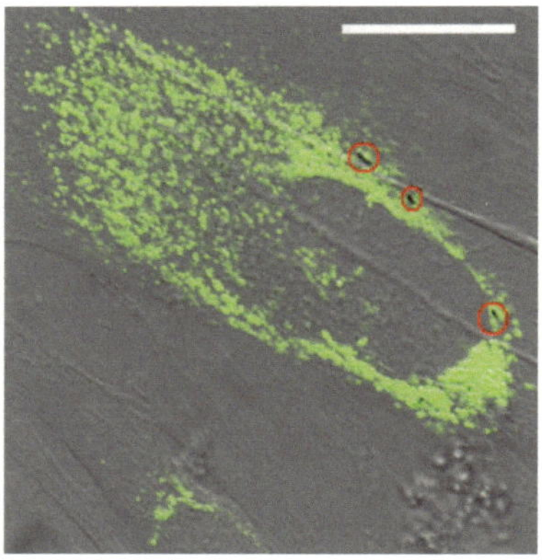

Fig. 8. The combined DIC and fluorescence image of plasmid coated nanowires and fibroblast cell expressing green fluorescence proteins. Nanowires are indicated by the red circles. The bar is 20 μm.

To further demonstrate that it was possible to monitor the local environment of the nanowires, the LysoSensor Yellow/Blue, which exhibited pH dependent dual emission spectra for intracellular environment study, was modified to the surface of nanowires. The LysoSensor coated gold nanowires inside the 3T3 cells were measured by confocal microscope as shown in figure 9. In the acidic environment, the emission of the LysoSensor dyes is in the range of 500 to 600 nm, which shifts to 410 to 500 nm region in the less acidic environment. In this experiment, the LysoSensor Yellow/Blue dyes was incubated with the 2 μm long amino-modified gold nanowires. LysoSensor coated nanowires can be used to probe to the acidic organelles such as endosomes or lysosomes. It was found that the emission from LysoSensor coated nanowires was in the green region indicating the environment around nanowires was less acidic. By monitoring both the DIC image of the nanowires and confocal image of the probe molecules as a function of time, the evolution of the local environment around the nanowires can be explored. In general, this approach can be extended to other types of probe molecules or biological assays.

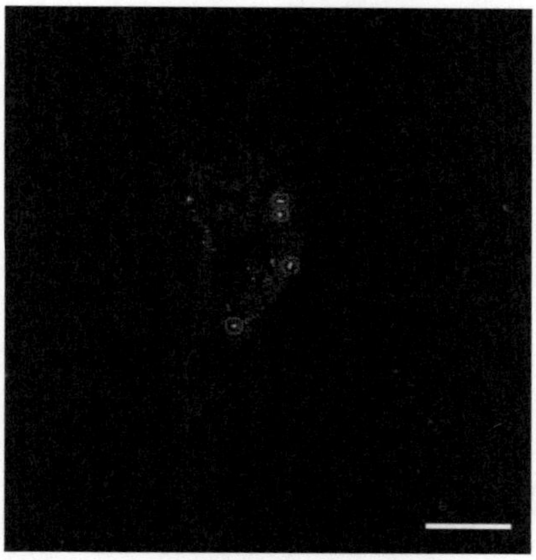

Fig. 9. The confocal image of the gold nanowires coated with LysoSensor inside HeLa cell. Bar: 20 μm.

6. Nanowire for gene delivery

Gene therapy is one of the promising strategies in the treatment of various diseases such as AIDS, cancer. The challenge of *in vivo* gene therapy is to develop safe and efficient gene delivery system. For the gene therapy to be used in the clinical applications, it requires the development of efficient DNA delivery vehicles that can be synthesized both easily and in large quantities. It has been demonstrated that the viral vectors can provide the most efficient gene transfer. Viral vectors were originally developed as an alternative to the transfection of the naked DNA in the molecular genetic experiments. So far, several types of viral vectors have been developed for the gene transfection including retroviruses, lentiviruses, adenoviruses, and adeno-associated viruses. However, the immuno-response induced by the viral vectors has raised a lot of concerns of using viral vector as the gene delivery system.

To reduce the immuno response, non-viral vectors, such as calcium phosphate, cationic lipids, cationic polymer, and dendrimers, have been recently developed. These non-viral vectors have been shown to exhibit lower transfection efficiency than the viral vectors. However, they are believed to be safer than viral vectors. Recent research efforts for the development of non-viral vectors have been focused on the improvement of their transfection efficiency. Non-viral vector approach usually involved the use of plasmid DNA or oligonucleotides, which are considered to be safer than the viral rectors. With recent advances in nanotechnology, researchers are now routinely synthesizing various nanomaterials with desire properties. Therefore, nanomaterials may also be used as the carriers for gene delivery. In recent reports, polymer nanoparticles or nanowires have been introduced as the vehicles for the drug delivery to enhance the cellular uptake efficiency. For this purpose, nanomaterials have been developed to incorporate with DNA molecules such that they can directly facilitate cellular interactions. The nanomaterial based drug delivery systems are being developed to overcome the barrier of size, stability, charge density, and biodistrution, which may lead to novel therapeutic strategies for the treatment of a variety of inherited or acquired diseases.

Since the metallic nanowires are capable of carring DNA into cells, they may be used as the vectors for gene delivery. To measure the optimal condition biding efficiency of the plasmid to the nanowires, we have examined 5 different concentrations of 5 μm long aminothiol modified gold nanowires (5×10^3, 2.5×10^4, 5×10^4, 2.5×10^5, 5×10^5 / ml), which were mixed with 50 ng of the DsRed-Monomer-golgi plasmid DNA for 24 hours (Kuo et al., 2008). The complexes with nanowires were then mixed with XYBR dyes and loaded in an agarose gel. The result is shown in figure 10 where the plasmid bands in land 5 and 6 were missing that at these concentrations all the plasmids were bound to the nanowires. Therefore, we estimated that the binding efficacy for the plasmid DNA molecules on the 5 μm long aminothiol modified nanowires were about 1 pg/nanowire. The zeta potential of the plasmid bonded gold nanowires was measured to be -9.3±0.5mV, which clearly indicated that DNA bound to gold nanowires to form complex. The DNA gold nanowire complex could protect the DNA molecules from the attack of the DNA nucleases.

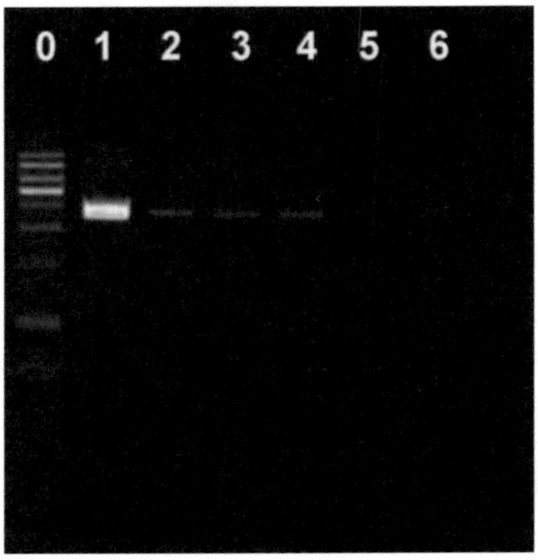

Fig. 10. Agarose gel electrophoresis of the aminothiol modified gold nanowires and plasmid. Land 0 is the DNA marker. Land 1 is the 50 ng plasmid DNA without nanowires. Land 2 to 6 are the mixture of 50 ng plasmid DNA with increasing numbers of the gold nanowires (5×10^3, 2.5×10^4, 5×10^4, 2.5×10^5 ,5×10^5).

To test the transfection efficiency of the aminothiol modified gold nanowires, 5 μm long nanowires with concentration of 10^5 nanowire/ml was mixed with the 50 ng of DsRed-Monomer-golgi vector and incubated over night. Two cell lines, NIH 3T3 and HeLa S3, were used in this experiment. The control experiments were conducted with dendrimer based commercial transfection agent (PolyFect) and calcium phosphate. The transfection efficiency for the naked plasmids was also measured. The viability of the cells incubated with the transfection reagents was checked separately by staining the cell with trypan blue 24 hours after the transfections. The transfection efficiency was calculated by measuring the ratio of the number of cell exhibited red fluorescence to the total number of the cells on the surfaces and normalized to the cell viability. Shown in figure 11 are the combined DIC and fluorescence images of 3T3 cells 24 hours after transfection using four different transfection reagents. The viability of the 3T3 cells 24 hours after transfection was measured to be 90%, 65%, 65% and 71% for gold nanowires, PolyFect, calcium phosphate and naked DNA, respectively. It is clear that aminothiol modified gold nanowires exhibited the highest transfection efficiency with very low toxicity for 3T3 cells. Previously, we have shown that more than 30% of the gold nanowires could be internalized by the 3T3 cells within 8 hours, which may explain the high transfection efficiency of the micrometer long gold nanowires.

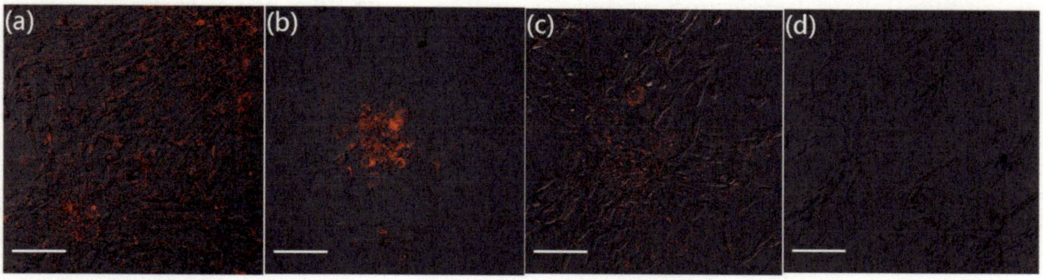

Fig. 11. The combined DIC and confocal images of 3T3 cells 24 hours after using (a) gold nanowires (b) PolyFect (c) calcium phosphate (d) naked DNA as the transfection reagents. Bar: 100 μm

Fig. 12. The combined DIC and confocal images of HeLa cells 6 days after using (a) gold nanowires (b) PolyFect (c) calcium phosphate (d) naked DNA as the transfection reagents. Bar: 100 μm

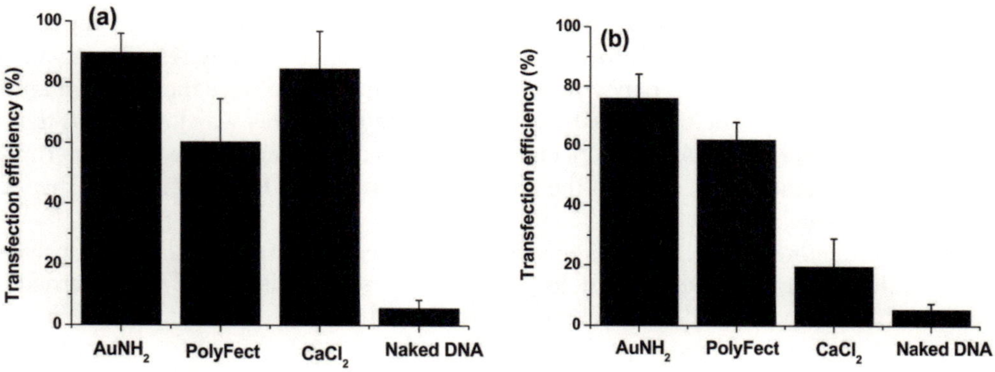

Fig. 13. The transfection efficiency measured for different transfection reagents for (a) 3T3 cells (b) HeLa cells.

As for the HeLa cells, the condition changed. Very little amount of cells were transfected 24 hours after incubation with all transfection reagents despite of the fact that more DNA molecules were used. After a few days, the cells started to exhibited red fluorescence. Shown in figure 12 are the combined DIC and fluorescence images of the transfected HeLa cells six days after transfection. The viability of the HeLa cells 24 hours after incubation with the transfection reagents was measured to be 95%, 87%, 86% and 90% for gold nanowires, PolyFect, calcium phosphate and naked DNA, respectively. HeLa cells seemed to resist to the addition of the transfection reagents. Therefore, lower cytotoxicity as well as lower transfection efficiency was measured for the HeLa cell lines. The transfection efficiencies of four different reagents for both cell lines are summarized in figure 13. In both cases, the gold nanowires exhibited the highest transfection efficiency while very little cytotoxicity to both cells lines was measured. To achieve higher transfection efficiency for other transfection reagents, DNA loading in the μg region was needed (Akita et al., 2004). Therefore, we concluded that the micrometer long nanowires could effectively deliver plasmid DNA into both 3T3 and HeLa cells.

7. Conclusions

Metallic nanowires with various aspect ratios have been synthesized using AAO template deposition. Four different surface modification schemes including serum, alkanethiols, mercapto acids and aminoalkylthiols were used to functionalize the surfaces of the metallic nanowires. It was found that all surface modifications expect aminothiols produced negatively charged gold nanowires. Therefore, the amino presenting gold nanowires were used bind the negatively charged plasmid DNA molecules.

The cytotoxicity of the micrometer long metallic nanowires with various aspect ratios and surface modifications have been investigated. It was found that the serum coated nanowires exhibited the least cytotoxicity with a LD_{50} value around 150 μg/ml, which was less than those measured for the smaller gold nanoparticles. All other surface functionalized nanowires possessed some degree of toxicity, which depended on the surface charge. Among them, the mercapto acid modified nanowires were the most toxic nanowires. For the same type of surface modification, HeLa cell, which can grow in suspension, were found to be more resistant to the addition of the nanowire solution. As for the nanowires with different aspect ratio, the cytotoxicity experiments indicated that nanowires with different aspect ratios exhibited the same degree of toxicity. However, the uptake efficiency for the shorter nanowires was measured to be higher than the longer nanowires. Therefore, we concluded that the internalized nanowires with higher aspect ratio were more toxic than the shorter one, which explained that the LD_{50} value for the nanowires was lower than that of the low aspect ratio nanorods. This conclusion also agreed with the cytotoxicity experiment for the spherical nanoparticles where the nanowires were found to be more toxic than the spherical nanoparticles.

We have also demonstrated that the 5 μm long nanowires with surface modifications were capable of delivering plasmid DNA molecules, which has been visualized and recorded on an optical microscope. In addition, LysoSensor was attached to the nanowires revealing the local environment of the nanowires. By monitoring the color change of the Lysosensor on the nanowires, it was found the nanowires stayed in the less acidic environment indicting that nanowires never escaped from the endosome/lysosome complexes.

The transfection efficiency of the surface functionalized nanowires has also been studied. It was found that the transfection efficiency of the aminothiol modified gold nanowires was the highest among the tested transfection reagents while almost no cytotoxicity was observed for gold nanowires under our experimental condition. It was also shown that it was possible to trace the nanowires inside the cells with sub-micrometer resolution. Therefore, we concluded that the micrometer long nanowires could be used for probing living cells with several advantages including easy fabrication and surface modification process, high transfection efficiency with very low cytotoxicity and readily observable by a microscope.

8. References

Akita, H; Ito, R.; Khalil, I. A.; Futaki, S.; Harashima, H. (2004) *Molecular Therapy* 9, 443.
Bauer, L. A.; Reich, D. H.; Meyer, G. J. (2003) *Langmuir* 19, 7043.
Bauer, L. A.; Birenbaum, N. S.; Meyer, G. J. (2004) *Journal of Materials Chemistry* 14, 517.
Chen, F. Q.; Gerion, D. (2004) *Nano Letters* 4, 1827.
Chithrani, B.D.; Ghazani, A.A.; Chan, W.C.W. (2006) *Nano Letters*, 6, 662.
Connor, E. E.; Mwamuka, J.; Gole, A.; Murphy, C. J.; Wyatt, M. D.; (2005) *Small* 1, 325.
Divliansky, I.B.; Shishido, A.; Khoo, I.C.; Mayer, T.S.; Pena, D.; Nishimura, S.; Keating, C.D.; Mallouk, T.E. (2001) *Applied Physics Letters* 79, 3392.
Derfus, A.M.; Chan, W.C.W.; Bhatia, S.N. (2004) *Nano Letters*. 4, 11.
Derfus, A.M.; Chan, W.C.W.; Bhatia, S.N. (2004) *Advanced Materials* 16, 961.
Fournier-Bidoz, S.; Arsenault, A. C.; Manners, I.; Ozin, G. A. (2005) *Chemical Communications*, 441.
Goodman, C.M.; McCusker, C.D.; Yilmaz, T.; Rotello, V.M. (2004) *Bioconjugate Chemistry*. 15, 897-900.
Hurst, S. J.; Payne, E. K.; Qin, L. D.; Mirkin, C. A. (2006) *Angewandte Chemie-International Edition* 45, 2672
Jessensky, O.; Muller, F.; Gosele, U. (1998) *Applied Physics Letters* 72, 1173.
Keating, C.D.and Natan, M.J. (2003) *Advanced Materials* 15 451-454.
Kuo, C.W.; Lai, J.-J.; Wei K. H.;Chen, P. (2007) *Advanced Functional Materials* 17, 3707.
Kuo, C.W. ; Lai, J.-J.; Wei K. H.;Chen, P. (2008) *Nanotechnology*, 19, 025103.
Lee, K. B.; Park, S.; Mirkin, C. A. (2004) *Angewandte Chemie-International Edition* 43, 3048.
Mitchell, D. T.; Lee, S. B.; Trofin, L.; Li, N.; Nevanen, T. K.; Söderlund, H.; Martin, C. R. (2002) *Journal of the American Chemical Society* 124, 11864 .
Masuda, H.; Fukuda, K. (1995) *Science*, 268, 1446
Nam, J.-W.; Thaxton, C.S.; Mirkin, C.A. (2003) *Science*, 301, 1884.
Michalet, X.; Pinaud, F.F.; Bentolila, L.A.; Tsay, J.M.; Doose, S.; Li, J.J.; Sundaresan, G.; Wu, A.M.; Gambhir, S.S.; Weiss, S. (2005) *Science*, 307, 538.
Medintz, I.L.; Uyeda, H.T.; Goldman, E.R.; Mattoussi, H. (2005) *Nature Materials* 4, 435.
Pena, D.; Mbindyo, J.; Carado, A.; Mallouk, T.; Keating, C.D.; Razavi, B.; Mayer, T. (2002) *Journal of Physical Chemistry B* 106, 7458 .
Pisanic, T. R.; Blackwell, J. D.; Shubayev, V. I.; Finones, R. R.; Jin, S. (2007) *Biomaterials* 28, 2572 .
Rosi, N.L.; Giljohann, D.A.; Thaxton, C.S.; Lytton-Jean, A.K.R.; Han, M.S.; Mirkin, C.A. (2006) *Science*, 312, 1027.
Salem, A.K.; Searson, P.C.; Leong, K.W. (2003) *Nature Materials* 2, 668-671.
Sandhu, K.K.; McIntosh, C.M.; Joseph, J.M.; Smith, S.W.; Rotello, V.M.; (2002) *Bioconjugate Chemistry*. 2002, 13, 3-6.
Shen, H.; Tan. J.; Saltzman, W.M. (2004) *Nature Materials*. 3, 569-574.
Tanase, M.; Felton, E.J.; Gray, D.S.; Hultgren, A.; Chen, C.S.; Reich, D.H. (2005) *Lab on a Chip* 5 598-605.

Takahashi, H.; Niidome, Y.; Niidome,T.; Kaneko, K.; Kawasaki, H.; Yamada, S. (2006) *Langmuir* 22, 2.

Tkachenko, A.G..; Xie, H.; Coleman, D.; Glomm, W.; Ryan, J.; Anderson, M. F.; Franzen, S.; Feldheim, D. L.(2003) *Journal of the American Chemical Society* 125, 4700 .

Xiao, Y.; Patolsky, F.; Katz, E.; Hainfeld, J.F.; Willner, I (2003) *Science,* 299, 1877-1881.

Xu,H.; Hong, R.; Wang, X.; Arvizo, R.; You, C.; Samanta, B.; Patra, D.; Tuominen, M. T.; Rotello, V. M. (2007) *Advanced Materials* 19, 1383.

Yin,H.; Too, H. P.; Chow, G. M. (2005) *Biomaterials* 26, 5818.

You,C.C.; Chompoosor, A. ; Rotello, V. M. (2007) *Nano Today* 2, 34.